11
Springer Series on Fluorescence

Methods and Applications

Series Editor: O.S. Wolfbeis

For further volumes:
http://www.springer.com/series/4243

Springer Series on Fluorescence

Series Editor: O.S. Wolfbeis

Recently Published and Forthcoming Volumes

Fluorescent Proteins II
Application of Fluorescent Protein Technology
Volume Editor: G. Jung
Vol. 12, 2012

Fluorescent Proteins I
From Understanding to Design
Volume Editor: G. Jung
Vol. 11, 2012

Advanced Fluorescence Reporters in Chemistry and Biology III
Applications in Sensing and Imaging
Volume Editor: A.P. Demchenko
Vol. 10, 2011

Advanced Fluorescence Reporters in Chemistry and Biology II
Molecular Constructions, Polymers and Nanoparticles
Volume Editor: A.P. Demchenko
Vol. 9, 2010

Advanced Fluorescence Reporters in Chemistry and Biology I
Fundamentals and Molecular Design
Volume Editor: A.P. Demchenko
Vol. 8, 2010

Lanthanide Luminescence
Photophysical, Analytical and Biological Aspects
Volume Editors: P. Hänninen and H. Härmä
Vol. 7, 2011

Standardization and Quality Assurance in Fluorescence Measurements II
Bioanalytical and Biomedical Applications
Volume Editor: Resch-Genger, U.
Vol. 6, 2008

Standardization and Quality Assurance in Fluorescence Measurements I
Techniques
Volume Editor: U. Resch-Genger
Vol. 5, 2008

Fluorescence of Supermolecules, Polymeres, and Nanosystems
Volume Editor: M.N. Berberan-Santos
Vol. 4, 2007

Fluorescence Spectroscopy in Biology
Volume Editor: M. Hof
Vol. 3, 2004

Fluorescence Spectroscopy, Imaging and Probes
Volume Editor: R. Kraayenhof
Vol. 2, 2002

New Trends in Fluorescence Spectroscopy
Volume Editor: B. Valeur
Vol. 1, 2001

Series Editor

Prof. Dr. Otto S. Wolfbeis

Institute of Analytical Chemistry
Chemo- and Biosensors
University of Regensburg
93040 Regensburg
Germany
otto.wolfbeis@chemie.uni-regensburg.de

Aims and Scope

Fluorescence spectroscopy, fluorescence imaging and fluorescent probes are indispensible tools in numerous fields of modern medicine and science, including molecular biology, biophysics, biochemistry, clinical diagnosis and analytical and environmental chemistry. Applications stretch from spectroscopy and sensor technology to microscopy and imaging, to single molecule detection, to the development of novel fluorescent probes, and to proteomics and genomics. The *Springer Series on Fluorescence* aims at publishing state-of-the-art articles that can serve as invaluable tools for both practitioners and researchers being active in this highly interdisciplinary field. The carefully edited collection of papers in each volume will give continuous inspiration for new research and will point to exciting new trends.

Preface

A plethora of reviews, popular science books, and scientific textbooks have been written on the significance of fluorescent proteins in the life sciences. More than 30,000 references can be found in bibliographic databases which refer to at least one among the members of this protein family (see Fig. 1). Most of these narrate on how fluorescent proteins may be used to label gene products, how they may be visualized in cellular compartments by fluorescence microscopy, or how they may be expressed in individual cells, thus provoking novel findings in ontogenesis. In most of the experiments described, fluorescent proteins are being exploited as miniaturized light bulbs, the length scale is that of microns, and the time scale is that of seconds or longer. There is no doubt that fluorescent protein technology has revolutionized life sciences in that proteins have become universal and standard tools in molecular biology laboratories.

A minor fraction of roughly 5% of all publications deals with the *nanoscopic* properties of fluorescent proteins (FPs) acting as light bulbs. Early achievements include the crystallographic analysis of their molecular structure [1, 2], the discovery of excited-state proton transfer in the naturally occurring FP [3, 4], and the erratic light emission of individual members of FPs [5, 6]. Especially the last experiments, along with low temperature studies [7, 8], have revealed that FPs exhibit a tremendous heterogeneity in terms of structure and dynamics.

It is therefore not astonishing that FPs have had a large impact on other areas of biophysical research, e.g., in studies on protein folding [9–11]. However, the irregular emission of light by FPs also has impacted experiments in the life sciences: most operators of fluorescent protein technology, whom I was talking to, were concerned about weird experimental features like rapid initial fading in time-lapse microscopy, sometimes with sudden fluorescence recovery, or changing FRET-ratios upon continuous illumination. Such annoying findings can be traced back to the wealth of light-driven processes in the proteins, and I am quite sure that more surprises of that kind have been experienced by others. It should be emphasized here that such "strange" photodynamics have initiated seminal studies on protein diffusion and high-resolution microscopy [12–14].

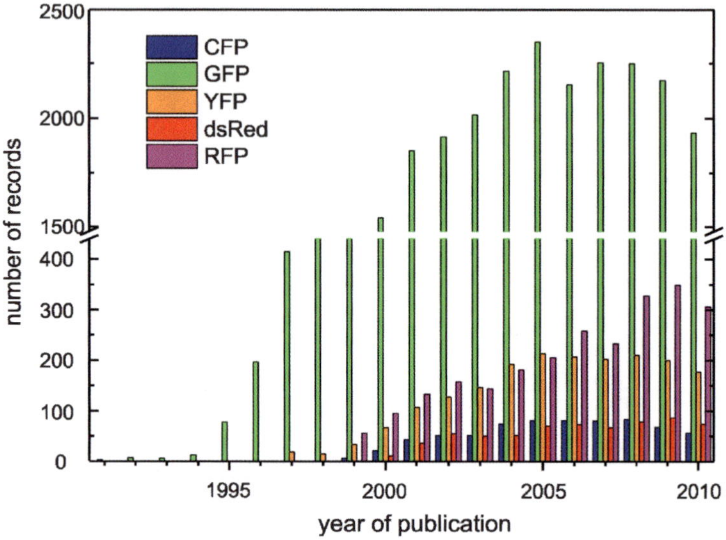

Fig. 1 Number of references related to fluorescent proteins (database: web-of-science). The number of articles dealing with Green Fluorescent Protein has reached saturation now at a level of typically 2,000 articles per year for almost a decade. Those on the Red Fluorescent Proteins are still increasing

Volumes 11 and 12 of the *Springer Series on Fluorescence* deal with various aspects of fluorescent proteins. The first volume (*Fluorescent Proteins I*) is devoted to the molecular, i.e., mainly optical, properties of fluorescent proteins. In the first part, the primary processes leading to fluorescence are discussed: excitation, relaxation, and other processes in the excited state and in emission. Fluorescence proteins are treated as "ordinary" fluorophores, and one article is highlighting our opportunities to circumvent the synthetic limitations given by nature. The second part focuses on the mechanisms that make the difference to conventional fluorophores: isomerization, protonation, as well as reversible and irreversible photochemical reactions. The knowledge on how these processes are affected by the surrounding of the FP allows for tailoring it with respect to spectacular applications, applications that are not conceivable with "ordinary" fluorophores.

In the second volume (*Fluorescent Proteins II*), the key aspect is on applications. Its first part is giving an overview on how many unconventional photophysical properties latently exist in naturally occurring and how double-resonance experiments enable the information to be extracted from microscopy data in an unprecedented way. More on high-resolution microscopy will be found in forthcoming volumes of this series. Quantitation, a central objective of analysis, is the comprehensive caption of the articles in the next part. We may state, justifiably, that researchers have reliable tools at hand to quantify some of the most abundant ions after more than a decade of development. Other physiological parameters of overwhelming importance like the transmembrane potential still need to experience this development. The last part reports on three examples of utmost biological relevance and how ultrasensitivity in bioanalysis, i.e., single-molecule technology, is merged with FP technology. This combination has resulted in an understanding of processes on a molecular level and in detection limits that were not even thought of some 15 years ago.

A preface is also always the occasion to deeply acknowledge the support by others. First of all, I have to thank my family who tolerated my commitment to this experience. I also would like to express my thanks to my coworkers, to my colleagues, and to the representatives of Saarland University for their understanding. In times of growing competitiveness in many academic areas, it is not self-evident to dedicate a substantial amount of time to such a book project. For the same reason, I especially appreciate the immense work of all authors of these two volumes who are all passionate, but busy scientists and who (more or less) voluntarily spared no pains to complete their manuscripts in a wonderful and highly professional way. By now, it also may be appropriate to apologize for my e-mail bombardments!

Saarbrücken, Germany Gregor Jung

References

1. Ormö M et al (1996) Science 273:1392–1395
2. Yang F et al (1996) Nat Biotechnol 14:1246–1251
3. Chattoraj M et al (1996) Proc Natl Acad Sci USA 93:8362–8267
4. Lossau H et al (1996) Chem Phys 213:1–16
5. Dickson R et al (1997) Nature 388:355–358
6. Pierce D et al (1997) Nature 388:338
7. Creemers T et al (1999) Nat Struct Biol 6:557–560
8. Seebacher C et al (1999) J Phys Chem B 103:7728–7732
9. Craggs T (2009) Chem Soc Rev 38:2865–2875
10. Hsu S et al (2009) Chem Soc Rev 38:2951–2965
11. Mickler M et al (2007) Proc Natl Acad Sci USA 104:20268–20273
12. Yokoe E, Meyer T (1996) Nat Biotech 14:1252–1256
13. Patterson G, Lippincott-Schwartz J (2002) Science 297:1873–1877
14. Betzig E et al (2006) Science 313:1642–1645

Contents

Part I
Basics and Manipulation of Light-Matter Interaction in Fluorescent Proteins

One-Photon and Two-Photon Excitation of Fluorescent Proteins

R. Nifosì and V. Tozzini

Abstract Fluorescent proteins (FPs) offer a wide palette of colors for imaging applications. One purpose of this chapter is to review the variety of FP spectral properties, with a focus on their structural basis. Fluorescence in FPs originates from the autocatalytically formed chromophore. Several studies exist on synthetic chromophore analogs in gas phase and in solution. Together with the X-ray structures of many FPs, these studies help to understand how excitation and emission energies are tuned by chromophore structure, protonation state, and interactions with the surrounding environment, either solvent molecules or amino acids residues. The increasing use of FPs in two-photon microscopy also prompted detailed investigations of their two-photon excitation properties. The comparison with one-photon excitation reveals nontrivial features, which are relevant both for their implications in understanding multiphoton properties of fluorophores and for application purposes.

Keywords Fluorescent proteins · Chromophore structures · Computational studies · Isolated chromophores · Multiphoton spectroscopy · Structure-property relationship · Spectral tuning

Contents

R. Nifosì (✉) and V. Tozzini
NEST CNR-NANO, Piazza San Silvestro 12, 56126 Pisa, Italy
e-mail: r.nifosi@sns.it

G. Jung (ed.), *Fluorescent Proteins I*, Springer Ser Fluoresc (2012) 11: 3–40,
DOI 10.1007/4243_2011_26,
Published online: 10 May 2011

1 Introduction

By the term fluorescent proteins (FPs), it is customary to indicate all fluorescent homologues of the original *Aequorea victoria* green fluorescent protein (GFP or *av*GFP). An accessory protein of the bioluminescence system of jellyfish *A. victoria*, *av*GFP was discovered as early as the 1960s [1]. Thirty years later, with the cloning of the gene [2] and the demonstration that its expression in other organisms generates fluorescence [3, 4], interest in GFP began to rise dramatically. Since then, it has triggered a revolution in bioimaging by fluorescence microscopy [5]. Soon, many other fluorescent and nonfluorescent GFP homologues were discovered in a variety of sea organisms, such as reef corals and sea anemones [6]. Further discoveries and mutagenesis engineering have produced a profusion of FPs with optical properties spanning most of the visible spectrum and beyond.

Fluorescence in FPs stems from the presence of a chromophore moiety, formed within the conserved β-barrel fold via a mechanism entailing autocatalytic backbone cyclization at an internal tripeptide sequence. Distinct postcyclization processing leads to different chromophore structures. The multiplicity of optical properties of FPs is surely one of the factors that contribute to their usefulness. It primarily arises from the different chemical structures of the chromophore. A finer tuning originates from the noncovalent interactions of the chromophores with the surrounding molecular matrix.

This chapter focuses on the mechanisms behind this spectral tuning, covering both experimental and theoretical/computational work. The reader is first presented with the more familiar case of *av*GFP. The chromophore structures of other FPs are described in Sect. 3. The following section surveys various studies on synthetic analogs of chromophores of FPs in gas phase and in solution. Section 5 provides a detailed description of spectral modifications due to interactions between chromophore and surrounding protein matrix. Finally, the last section covers two-photon properties of FPs. Several other reviews on FP optical properties are available, some treating more exhaustively the variety of GFP-like fluoro and chromoproteins, and other more focused on application purposes. For recent surveys, see [7–9].

2 *av*GFP: Structure and Optical Properties

The first to be cloned [2] and functionally expressed in other organisms [3, 4], *av*GFP has actually been replaced in most applications by its mutants and homologues. Nonetheless, being one of the best characterized in terms of optical and photophysical properties, it is a suitable starting point to introduce the concepts recurring in this chapter.

2.1 Structure and Chromophore Formation

The 238 amino acid (27 kDa) long sequence of *av*GFP folds in a compact cylindrical form called β-barrel, its lateral wall being an 11-stranded β-sheet (Fig. 1) [10, 11]. Several X-ray structural studies support the notion that also all other FPs share the β-barrel fold. They can differ in quaternary structure, though most natural FPs are tightly bound tetramers and some are dimers, a feature that initially hampered their applications. However, mutagenesis studies were able in most cases to produce viable monomeric variants of the parent proteins.

The β-barrel is capped on both ends by short α-helical sections and traversed by an α-helix segment. This segment contains the chromophore, a 4-(*p*-hydroxybenzylidene) imidazolinone, originating from the posttranslational cyclization of three consecutive amino acids at position 65–67, such as Ser, Tyr, and Gly. As demonstrated by the fact that expression of *av*GFP gene in other organisms leads to fluorescence, the posttranslational synthesis of the chromophore does not require any jellyfish-specific enzyme [3, 4]. It requires, however, exogenous oxygen, in the absence of which GFP does not develop fluorescence [4, 12].

Chromophore formation (Fig. 2) proceeds within the native fold (i.e., no chromophore is formed under denaturing conditions). It first entails backbone cyclization at the Ser65-Tyr66-Gly67 tripeptide through nucleophilic attack of the amide group of Gly67 onto the carbonyl group of Ser65, promoted by Arg96, a conserved residue in all natural FPs (see [13] and references therein). Molecular oxygen is required for the subsequent oxidation reaction. One water molecule is also abstracted from the structure, with most experimental studies supporting the cyclization–oxidation–dehydration sequence of events [13, 14].

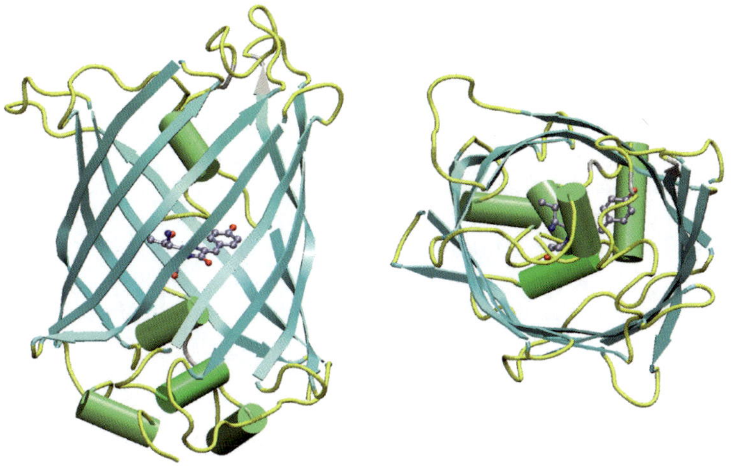

Fig. 1 Tertiary structure of *av*GFP. The usual cartoon representation is used, where α-helices are *cylinders* and β-sheets *arrows*. The chromophore is shown in a *ball-and-sticks* representation with the standard coloration for atom elements (i.e., *gray* for carbon, *blue* for nitrogen, and *red* for oxygen)

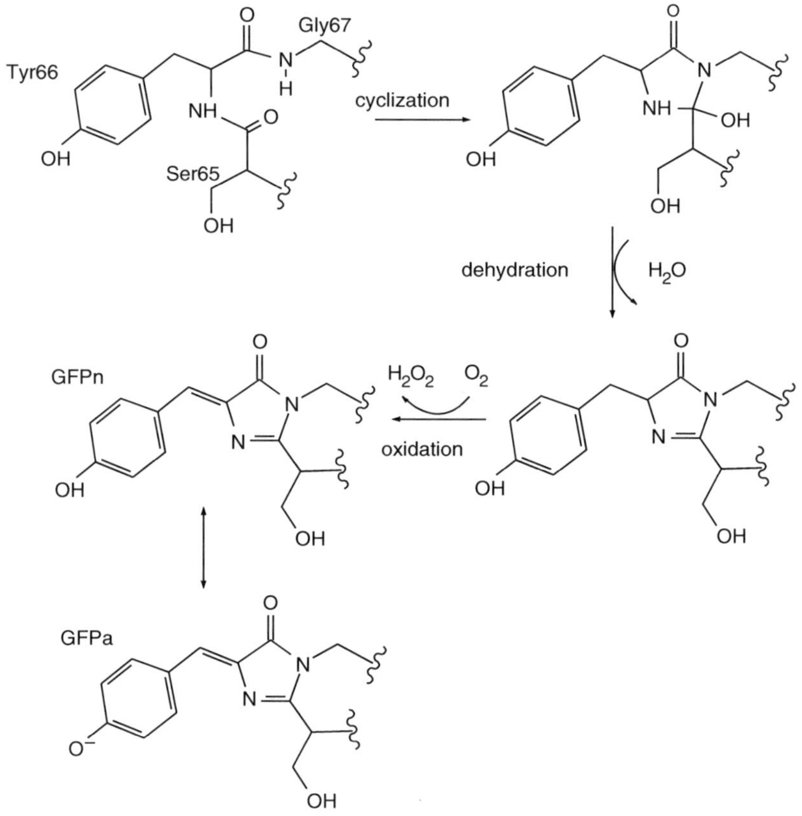

Fig. 2 Proposed mechanisms of biosynthesis of the GFP chromophore. In the last step of the reaction, the two relevant protonation states are shown

The mature chromophore consists of two rings, the phenol ring, coming from the side chain of Tyr66, and the five-member heterocyclic ring (imidazolinone), resulting from the cyclization of the backbone. The imidazolinone core is a common feature of all known FP chromophores. The alternating single and double bonds in the bridge region extend the electron delocalization from the phenolate to the carbonyl of the imidazolinone. Efficient visible-light absorption is ultimately determined by this π-conjugated system, i.e., the single–double bond alternation with connected atomic p-orbitals.

2.2 *Absorption, Excitation, and Emission*

The room temperature absorption, excitation, and emission spectra of *av*GFP are shown in Fig. 3. Apart from the 278-nm band, which is common to proteins containing aromatic amino acids, *av*GFP exhibits a major absorption band at

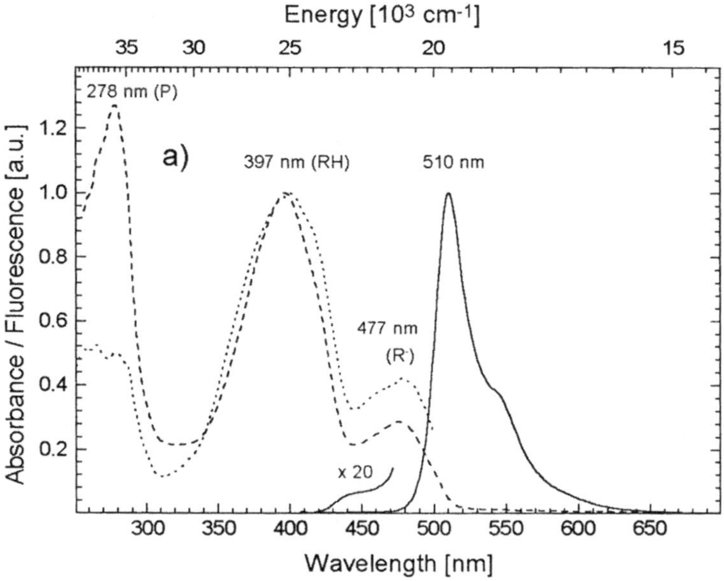

Fig. 3 Room temperature absorption (*dashed line*), fluorescence excitation (*dotted line*) and emission spectra (*solid line*) of *av*GFP, at room temperature and pH 8.0 (adapted from Kummer et al. [15])

397 nm and a minor band at 477 nm, due to absorption of the chromophore. Their relative height depends on proton concentration: increasing pH above 11, the minor lower-energy band increases at the expense of the higher-energy band. Vice versa, at pH below 4 the minor energy band is completely depleted [16]. The relative population in Fig. 3 is rather constant between pH 6 and 11. This behavior arises from the ground-state equilibrium between two states of the chromophore, differing in protonation of the phenolic group from Tyr66. The phenolic oxygen of the chromophore is protonated in the state absorbing at 397 nm (RH in the following) and deprotonated in that absorbing at 477 nm (state R^-). It is commonly accepted that the two other possible protonation sites in the chromophore, i.e., the nitrogen and the carbonyl oxygen of the imidazolinone, are deprotonated in both the absorbing states, thereby giving an overall neutral chromophore in state RH (GFPn in Fig. 2) and anionic in state R^- (GFPa) [17].

Excitation of state RH leads to a fluorescence spectrum peaking at 510 nm (Fig. 3), with a rather high quantum yield of 0.79. State R^- yields a similar fluorescence spectrum, slightly blueshifted and peaking at 503 nm (not shown). In both cases, fluorescence comes from emission of the singlet excited state of the anionic chromophore. Excitation of the neutral chromophore results in ultrafast (4 ps) excited state proton transfer (ESPT) and subsequent emission of the anionic form [18]. The ESPT acceptor has been identified in (deprotonated) Glu222 [18, 19].

Although the anionic chromophore is the emitting species in both states, the configuration of the surrounding residues is different and the decay time of the excited state (few ns) is too short to allow equilibration, resulting in the observed slight shift of emission peak (510–503 nm).

3 Chromophores of FPs

Chromophore structures of FPs and their formation mechanisms are summarized in Fig. 4. The range of excitation and emission wavelengths that they give rise to in different representative FPs can be read in Table 1.

In all natural FPs discovered so far, the prechromophore tripeptide has a X_1-Tyr_2-Gly_3 sequence, where X_1 can be almost any amino acid. Chromophore formation is inhibited upon mutation of the Gly_3 [13], suggesting that the peculiar conformational flexibility of Glycine is necessary at that location. By contrast, substitution of Tyr_2 with an aromatic amino acid (Phe, His, or Trp) preserves the fluorescence. The resulting artificial mutants contain chromophores with the phenol ring replaced by the corresponding aromatic ring, and have blueshifted excitation and fluorescence wavelengths with respect to the parent protein. Blue FPs such as EBFP [118], Azurite [22], and EBFP2 [21] all derive from Y66H *av*GFP (BFP chromophore in Fig. 4) with additional mutations to improve folding efficiency, brightness, and photostability. Y66W *av*GFP mutants (CFP in Fig. 4) such as Cerulean [23] and ECFP [11] emit cyan light.

The other determinant of chromophore structure comes from modifications around the X_1 α-carbon. With respect to the *av*GFP chromophore formation mechanism, an additional step occurs in DsRed [43] and other red fluorescent proteins (RFPs), namely the oxidation of the C–N main-chain bond of X_1, which leaves an acylimine substituent at the corresponding position of the imidazolinone ring. The sequence of events during RFP chromophore self-processing entails first the oxidation leading to the acylimine substituent followed by dehydrogenation of the bridging carbon [14]. The acylimine substituent enlarges the extension of the π-conjugated system, and correspondingly lowers the excitation energy, resulting in red/orange fluorescence. Analogously to the case of the GFP chromophore, substituting the Tyr_2 in this chromophore with Phe or Trp (RFP Y67F and RFP Y67W in Fig. 4) results in blueshifted spectral properties, with fluorescence in the blue (mBlueberry2 [21]) and orange (mHoneydew [32]) domain, respectively.

In eqFP611 and Rtms5, the RFP chromophore is in the *trans* (or E) isomer instead of the *cis* (or Z) form normally present in other FPs [39, 44]. It adopts a coplanar conformation in eqFP611, whereas in Rtms5 it is highly non-coplanar. This feature is linked with the very different quantum yields of the two proteins, presumably because noncoplanar conformations favor non-radiative de-excitation pathways [44].

Further reactions can take place around the acylimine moiety, such as side chain cyclization by nucleophilic addition of the Thr (in mOrange), Cys (mKO), or Lys (zFP538) side chain, the latter followed by backbone cleavage [45–47]. Backbone

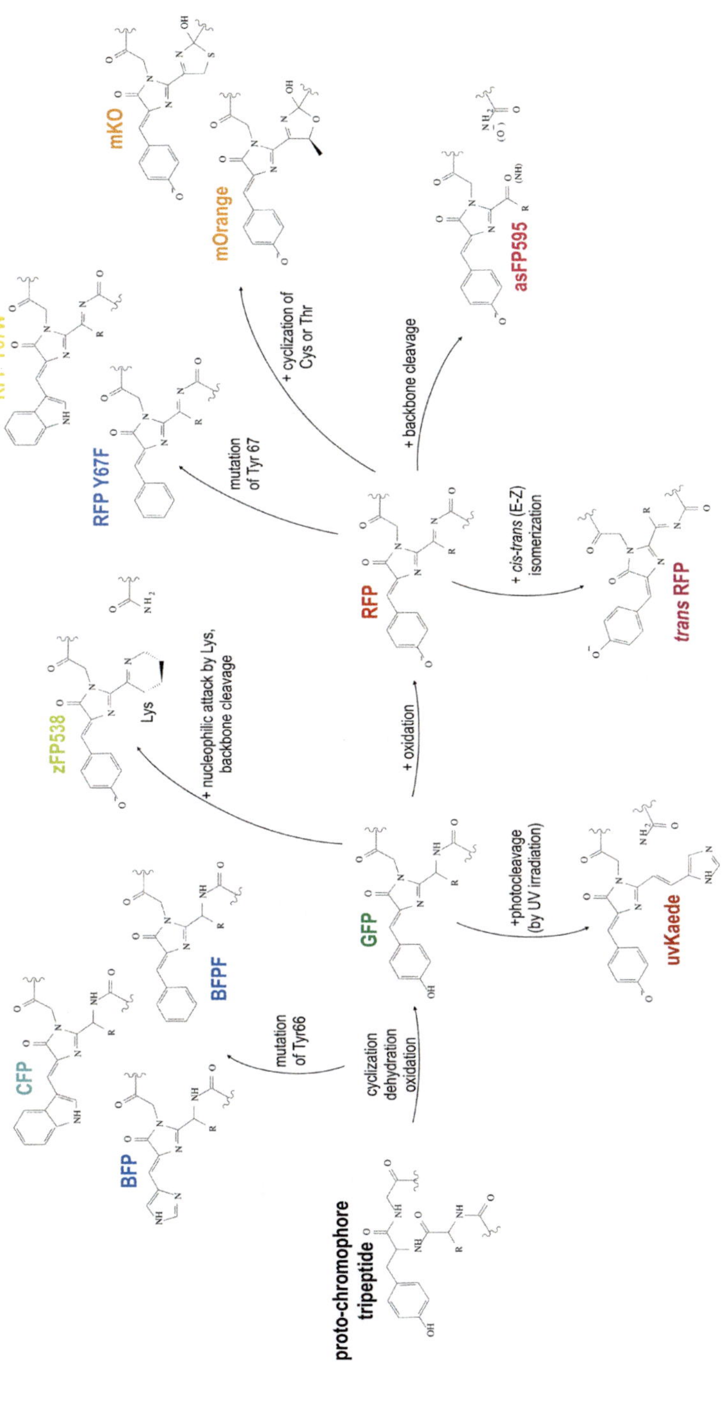

Fig. 4 Chromophore structures and reactions leading to mature chromophores in fluorescent proteins

Table 1 Optical properties of representative fluorescent proteins

Name[a]	Original protein	Chromophore[b]	Exc/Abs peak (nm)	Em peak (nm)	Ext. coeff (10^3 M^{-1} cm^{-1})	QY[c]	References
GFP Y66F	*av*GFP	BFPF	360	442	-	(low)	[11]
EBFP	*av*GFP	BFP	377	446	30	0.15	[20]
EBFP2	*av*GFP	BFP	383	448	32	0.56	[21]
Azurite	*av*GFP	BFP	384	450	26.2	0.55	[22]
mKalama1	*av*GFP	GFPn	385	456	36	0.45	[21]
mBlueberry2	DsRed	RFP Y67F	402	467	51	0.48	[21]
Cerulean	*av*GFP	CFP	433	475	43	0.62	[23]
ECFP	*av*GFP	CFP	434	476	32.5	0.4	[115]
mTFP1-Y67W	amFP486	CFP	424/440[d]	461/482[d]	13	0.02	[24]
cFP484 (tetr)	-	GFPa	456	484	35.3	0.48	[6]
amFP486 (tetr)	-	GFPa	458	486	40	0.24	[6]
mTFP1	amFP486	GFPa	462	492	64	0.85	[24]
EGFP	*av*GFP	GFPa	484	507	53	0.6	[20]
***av*GFP R^- state[e]**	-	GFPa	477	508	9.5	0.72	[116]
***av*GFP RHstate**	-	GFPn	397	510 (ESPT)[f]	25	0.78	[116]
T-Sapphire	*av*GFP	GFPn	399	511 (ESPT)	44	0.60	[25]
Dronpa	22G	GFPa	503	518	95	0.85	[26]
EYFP	*av*GFP	GFPa	514	527	84	0.61	[27]
Venus	*av*GFP	GFPa	515	528	92.2	0.57	[28]
YPet	*av*GFP	GFPa	517	530	104	0.77	[29]
PhiYFP (dim)	-	GFPa	525	537	115	0.60	[30]
zFP538 (dim)	-	zFP538	528	538	-	0.42	[6]
mKO	KO	mKO	548	561	51.6	0.60	[31]
mHoneydew	DsRed	RFP Y67W	487/504[d]	537/562[d]	17	0.12	[32]
mOrange	DsRed	mOrange	548	562	71	0.69	[32]
Dendra pa[g]	DendFP	uvKaede	557	575	20	0.72	[33]
EosFP pa (tetr)	-	uvKaede	571	581	41	0.55	[34]
Kaede pa (tetr)	-	uvKaede	572	582	60.4	0.33	[35]
DsRed (tetr)	-	RFP	558	583	57	0.79	[6]
asFP595 pa (tetr)	-	asFP595	568	595	56.2	$<10^{-3}$	[36]
mStrawberry	DsRed	RFP	574	596	90	0.29	[32]
KFP1 pa (tetr)	AsFP595	asFP595	580	600	59	0.07	[117]
mRFP1	DsRed	RFP	584	607	44	0.25	[37]
mCherry	DsRed	RFP	587	610	72	0.22	[32]
eqFP611 (tetr)	-	*trans* RFP	559	611	78	0.45	[38]
Rtms5 (tetr)	-	*trans* RFP	592	620	80	$<10^{-3}$	[39]
mPlum	DsRed	RFP	590	649	41	0.10	[40]
mNeptune	eqFP578	RFP	600	650	67	0.20	[41]
cjBlue	-	RFP	610	-	66.7	-	[42]

[a]The oligomerization state, i.e., dimer (dim) or tetramer (tetr), is reported for nonmonomeric proteins
[b]Chromophore name refers to Fig. 4
[c]*QY* = Quantum yield
[d]Two excitation/fluorescent peaks are present
[e]RH state and R^- state for *av*GFP refer to the two bands assigned to neutral (GFPn) and anionic (GFPa) protonation state of the chromophore
[f]ESPT in the fourth column indicates that emission from the anionic species takes place after excited state proton transfer
[g]*pa* = after photoactivation

cleavage also occurs in asFP595 [48, 49].[1] AsFP595 is normally nonfluorescent and contains a *trans* (E) chromophore. Its fluorescent form is accessed either temporarily or permanently by photoactivation (hence the alternative name kindling fluorescent protein, or KFP), and contains a *cis* (Z) chromophore.

A different mechanism of redshifting occurs in Kaede and related proteins (KikGR, EosFP, Dendra), where UV excitation photoconverts the originally green-emitting proteins (with a GFP-like chromophore) into RFPs by cleaving the backbone between main-chain N and Cα of the Histidine at X_1. The cleavage is followed by double-bond formation between Cα and Cβ giving rise to the uvKaede chromophore structure [34, 35].

As in *av*GFP, chromophores with the phenol group exist as either neutral (protonated phenol) or anionic (deprotonated phenol). The absorption of the neutral form is always significantly blueshifted with respect to the anionic form. Changing the buffer pH generally results in a fluorescence titration curve that reveals the changing equilibrium between these two forms. Several FPs at physiological pH contain only the anionic chromophore. *av*GFP and some of its mutants are exceptions to this general behavior, because they can present both states, existing in a persistent equilibrium over a broad pH range.

It is worth mentioning that according to some computational investigations, in some proteins, such as asFP595, the zwitterionic form (with the imidazolinone nitrogen protonated) might be the emitting one [51]. In addition, the zwitterionic and other protonation state such as the cationic form (protonated phenol and imidazolinone) may be involved in FP photophysics [52]. The cation, however, exists at too acidic conditions ($pH < 2$) from studies of chromophore analogs in solution [17, 53], in which the protein is denatured. Hence, it is not considered relevant to FPs optical response at physiological conditions.

4 Isolated Chromophores

Synthetic analogs of chromophores of FPs were the starting point for a variety of experimental and computational studies aiming to examine all the factors influencing optical properties of FPs. Isolated chromophore models in solution are poorly, if at all, fluorescent at room temperature. However, they become fluorescent when the temperature is lowered to 77 K [54]. Presumably the conformational freedom at higher temperature enables non-radiative pathways of de-excitation.

The established experimental model for the GFP chromophore is a *p*-hydroxybenzylidene-2,3-dimethlimidazolone or *p*-HBDI (Table 2). Such a model encompasses the relevant π-conjugated system. It lacks, however, the side chain of X_1 (the first residue of the tripeptide). That *p*-HBDI is a fairly accurate model for the

[1] AsFP595 chromophore structure has been subject to debate, regarding the possible presence of an imino (NH) group in place of the keto group at position 2 of the imidazolinone (see discussion in [50]).

Table 2 Synthetic model chromophores and their absorption peaks in gas phase and in solution

Name	Chromophore model	Gas-phase abs peak in nm		Solution abs peak in nm (extinction coefficient $mM^{-1}\ cm^{-1}$, when available)		
		Anionic	Cationic	Neutral	Anionic	Cationic
BFPF[a]	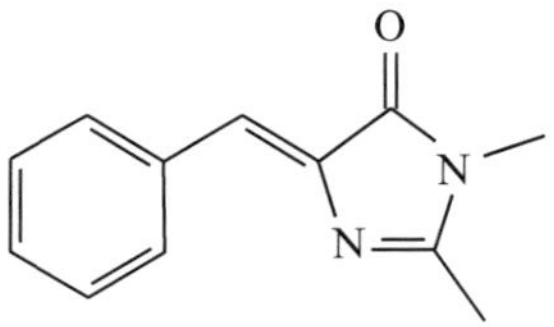	–	–	348 water 348 metOH 353 CH_3CN 355 dioxane 357 toluene	–	–
CFP[a,b]	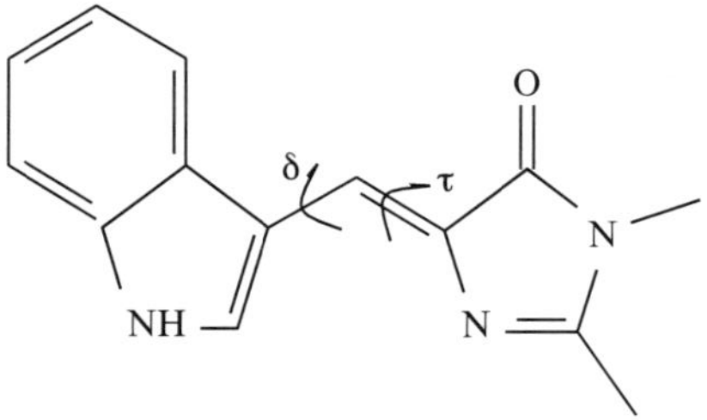	–	–	405 water 402 metOH 388 CH_3CN 390 dioxane 389 toluene	–	–
p-HBDI[c] (GFP model chrom.) *p*Ka = 8.1[d]	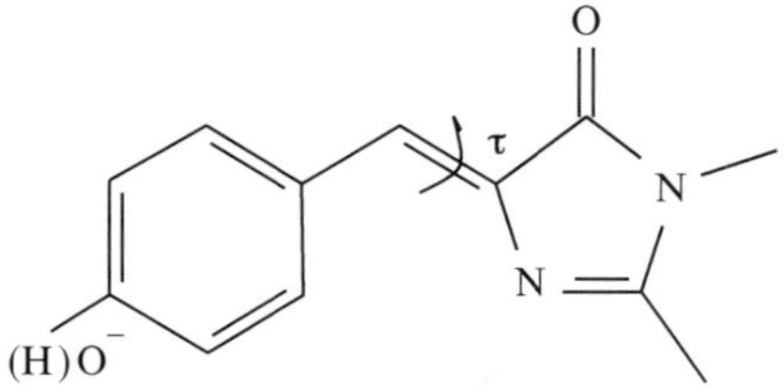	479	406	372 water 367 metOH 362 dioxane 368 DMF 370 DMSO 372 pyridine	425 water 428 metOH 433 dioxane 462 DMF 471 DMSO 482 pyridine	387 water 398 metOH 397 dioxane 378 DMF 383 DMSO

Model chrom. of zFP538[e]	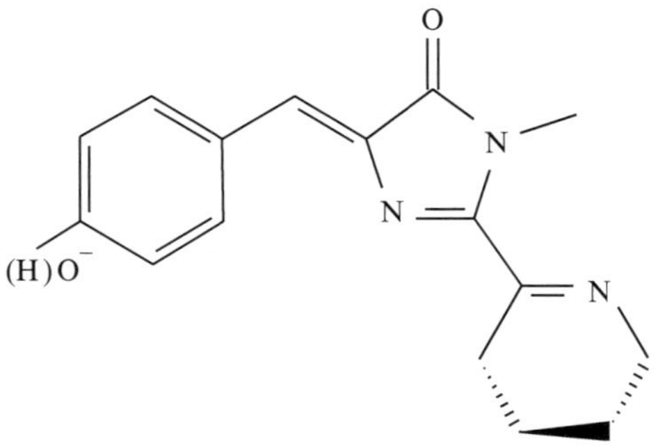	–	–	417 water	530 water	–
p-HBMPI[f] (RFP1, RFP model chrom.)	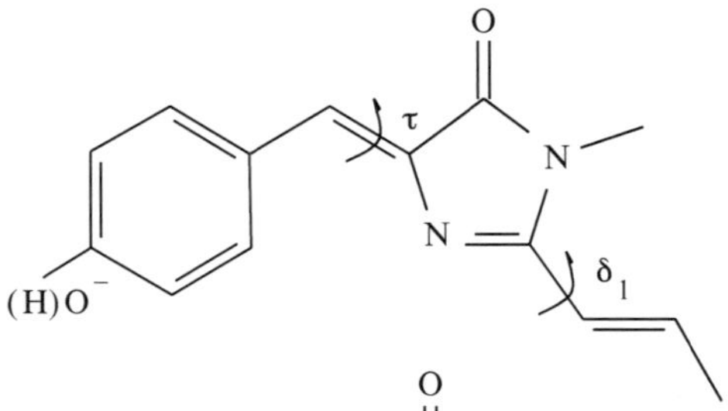	521	441	401 water 397 CH_3CN	460 water 492 CH_3CN	416 water 419 CH_3CN
p-HBMPDI[g] (RFP2, RFP model chrom.)	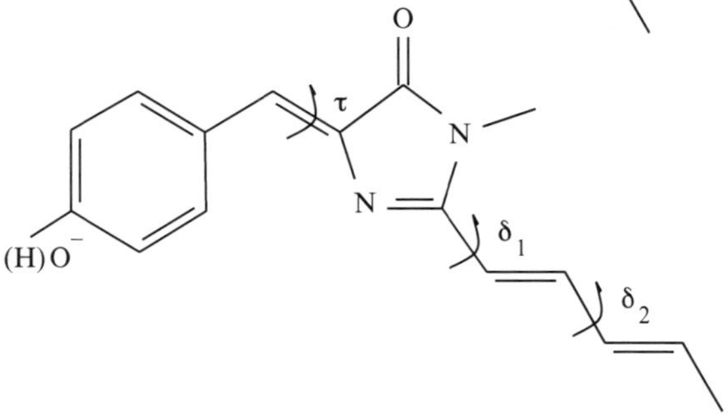	549	448	421 water 419 CH_3CN	482 water 515 CH_3CN	443 water 448 CH_3CN

(*continued*)

Table 2 (continued)

Name	Chromophore model	Gas-phase abs peak in nm		Solution abs peak in nm (extinction coefficient mM^{-1} cm^{-1}, when available)		
		Anionic	Cationic	Neutral	Anionic	Cationic
Model of uvKaede chrom. (red form, after photocleavage)[h] *p*Ka = 7.7	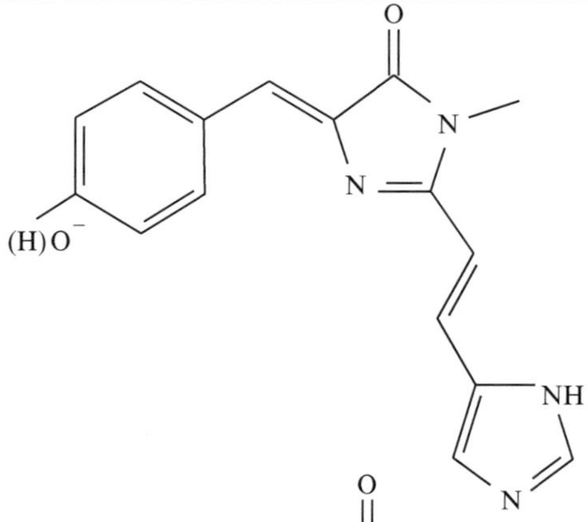	–	–	430 (41) water 443 (40) etOH 444 (41) prOH 446 (40) DMF 448 (37) DMSO	490 (53) water 502 (79) etOH 514 (80) prOH 557 (77) DMF 565 (80) DMSO	–
AHBMI (Model of asFP595 chrom.)[i] *p*Ka = 7.1	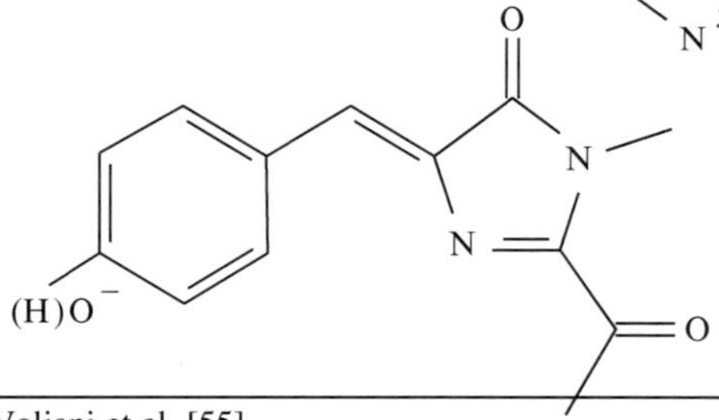	–	–	418 (35) water 425 (40) etOH 428 (39) 2-prOH 422 (38) DMF	520 (47) water 542 (72) etOH 552 (73) 2-prOH 572 (87) DMF	–

[a]Solution values from Voliani et al. [55]
[b]Dihedrals describing different conformational isomers are indicated
[c]Gas-phase values from Nielsen et al. [56]; solution values from Dong et al. [57]
[d]The *p*Ka for the anionic–neutral transition is indicated
[e]Solution values from Yampolsky et al. [58]
[f]*p*-4-Hydroxybenzylidene-1-methyl-2-propenyl-imidazolinone. Gas-phase values from Boy et al. [59]; solution values from He et al. [84]
[g]*p*-4-Hydroxybenzylidene-1-methyl-2-penta-1,3-dien-1-yl-imidazolinone. Gas-phase values from [59]; solution values from He et al. [84]
[h]Solution values from Yampolsky et al. [60]
[i](2-Acetyl-4-(*p*-hydroxybenzylidene)-1-methyl-5-imidazolone. Solution values from Yampolsky et al. [50]

isolated chromophore is implied by the similarity between its (solution) absorption spectrum and that of the denatured GFP [61], where the chromophore is presumed to be completely exposed to the solvent. Varying the solution pH, anionic, neutral, and cationic forms of *p*-HBDI are obtained, with *p*Ka ~ 8 for the anionic to neutral reaction and *p*Ka ~ 2 for the neutral to cationic reaction [17, 57].

Analogous blue fluorescent protein (BFP, with the Y66H mutation), BFPF (BFP, with Y66F), and cyan fluorescent protein (CFP, with Y66W) chromophore models were synthesized [55]. Attempts to synthesize models for the RFP chromophore revealed that the compound with the acylimine group is not stable in solution, due to the susceptibility of acylimines to nucleophilic attack. Hence, currently available models for the RFP chromophore, namely *p*-HBMPI and *p*-HBMPDI, contain an olefinic substituent in place of the acylimine group [59].

4.1 Model Chromophores in Gas Phase and in Solution

4.1.1 Gas Phase

Gas-phase absorption spectra of anionic *p*-HBDI (*p*-HBDI$^-$) were measured in photo destruction spectroscopy experiments [56]. Charged molecules trapped in an electrostatic ion storage ring are irradiated with laser light, and the yield of neutral fragments, produced by photochemical processes upon photon absorption, is measured as a function of the excitation wavelength. Using this technique, Andersen and coworkers reported an absorption peak of 479 nm for *p*-HBDI$^-$ [56]. Such a value is very close to the absorption of the R$^-$ state (anionic chromophore) in *av*GFP. The authors conclude that, in this protein, the arrangement of the side chains and water molecules around the chromophore produces a situation more similar to the gas phase than to solution, where the absorption is blueshifted (see below).

A new set of experiments by Forbes and Jockusch [62] on *p*-HBDI$^-$ were able to expand the spectral window of detection, and to distinguish two photo degradation channels contributing to the total spectrum. One, closely matching the spectrum previously measured by Andersen and coworkers, corresponds to the loss of one of the two methyl groups. The other appears at higher energies (peak at 410 nm) and is attributed to electron detachment. No fluorescence is detected upon excitation [62].

Similar experiments on the anionic *p*-HBMPI and *p*-HBMPDI (RFP model chromophores) yielded absorption peaks at 521 and 549 nm, respectively [59, 63]. The latter value is closer to the absorption maximum of DsRed, at 558 nm [6]. The spectra also display a well-resolved progression of vibronic peaks with vibrational spacing of 10 nm (382 cm^{-1}) and 16 nm (518 cm^{-1}) for *p*-HBMPI and *p*-HBMPDI, respectively. The width of the spectra is hence due to both vibronic and inhomogeneous broadening effects.

Cationic forms *p*-HBDI [64], *p*-HBMPI, and *p*-HBMPDI [63] were also investigated, resulting in peaks at 406, 441, and 448 nm.

The electrostatic storage clearly requires charged molecules. Thereby, the neutral protonation state, despite relevant to GFP (and other FPs) optical properties, was initially neglected. Recently, a model for the neutral chromophore was devised by adding a methyleneammonioum cation to *p*-HBDI, charging the molecule at a site sufficiently detached from the conjugated system. The spectrum of this model compound peaks at 399 nm [65], close to absorption of the neutral band of *av*GFP. How well this model describes the neutral chromophore was discussed in [66].

4.1.2 Solution

With respect to gas-phase experiments, absorption of *p*-HBDI$^-$ in water is blueshifted and exhibits a rather broad band peaking at 425 nm [17, 57] (see Table 2). In nonpolar solvents, the absorption maximum is at 440 $\pm$ 5 nm. Increased solvent polarity yields redshifted values. The longest wavelength peak, 482 nm, was recorded in pyridine. The wavelength shift in going from water to pyridine is rather sizable (57 nm) and corresponds to an energy shift of 2,755 cm^{-1} (0.34 eV). Using the multivariant Kamel–Taft fit on the data in these various solvents, one can extrapolate a value of 437 nm (2.84 eV) for the gas-phase, i.e., a solute with zero acidic, basic, and polar parameters [57].

The significant solvatochromic shift of the anionic state points to a marked sensitivity to the surrounding environment. Such sensitivity emerges also when the chromophore is embedded in the protein matrix: in different FPs, the absorption corresponding to the anionic chromophore shows a similar, if not larger, range of variation (see Sect. 5.1). Absorption of neutral and cationic *p*-HBDI peaks at 368 and 393 nm respectively, and displays a much narrower solvatochromism (variation of ~20 nm, around 1,000 cm^{-1}).

Besides the main band, reported in Table 2, the absorption spectra of both neutral and anionic *p*-HBDI reveal features at shorter wavelength (around 300 and 250 nm for *p*-HBDI and at 320 nm for *p*-HBDI) [55, 57]. Similar blueshifted bands are also measured in AHBMI (model chromophore of asFP and KFP) [50] and in model chromophores of BFPF, BFP, [55], and Kaede [60]. Features around 330 nm are also detectable in red FPs [67]. In *av*GFP, by contrast, these excitations merge with the absorption of other aromatic residues in the protein. Excitation to these states is characterized in some cases by strong two-photon cross-section (see Sect. 6).

As stated above, solvated model chromophores are generally poorly fluorescent at room temperature [54, 68]. For instance, the chromophore fluorescence quantum yield in DMF (dimethylformamide) is 0.00005 [60]. More appreciable fluorescence is instead displayed by AHBMI and by uvKaede chromophore models in DMF solution, with fluorescence quantum yields of 0.0021 and 0.005, respectively [50, 60].

4.2 Chromophores of FPs: Computational Studies

Quite a number of theoretical/computational studies were dedicated to the optical spectra of model chromophores of FPs in the gas phase [66, 69–76]. Attempts to include solvent effects were performed using implicit solvent methods [71] or by considering solvent–solute electronic interactions using liquid-structure simulations [72, 77]. Despite being relatively small molecules (though not so small for the most sophisticated quantum chemistry methods), there are subtle issues to be considered when trying to reproduce FPs chromophore optical spectra [66]. What is usually reported in computational analyses is the vertical excitation energy. This value is compared with the peak of the experimental spectrum either in the solution or in the gas phase. Such comparison neglects vibronic and other broadening effects, which influence the shape and the location of the absorption band. As long as photon absorption is well described by the vertical excitation picture, the comparison between the mentioned values is justified. It should be kept in mind, however, that a complete description requires taking into account vibronic and inhomogeneous broadening effects.

Starting from the most studied *p*-HBDI$^-$, excitation energies in the gas phase were evaluated with a broad range of methods summarized in Table 3. All studies agree that the absorption band is due to excitation from the ground to the first bright excited state, with a prevalently HOMO → LUMO character, where the HOMO and the LUMO are π and π* orbital, respectively. The excitation involves a limited (0.1e) charge displacement from the phenolate ring to the imidazolinone, and to the bridging carbon [70] (see Fig. 5). As discussed in [74], this π–π* excitation is embedded in the photodetachment continuum. The interaction with the ionization continuum might affect the broadening of the gas-phase spectrum and the photodynamics of the chromophore.

Table 3 Theoretical predictions for GFP chromophore excitation energy, in eV (nm)

Method	*p*-HBDI$^-$ eV (nm)	*p*-HBI$^{-\,a}$ eV (nm)
Expt. (gas)	2.59 (479)[b]	
Expt. (sol. ext.)	2.84 (437)[c]	–
MRMP2	2.52 (491)[d]	–
MBGFT	2.67 (464)[e]	–
CASPT2	2.67 (464)[f]	2.63 (471)[g]
CASPT2 (IPEA)	2.92 (425)[h]	2.96 (419)[h]
TD-DFT/SAOP	2.93 (423)[i]	2.99 (415)[i]
DMC	3.04 (408)[j]	3.06 (405)[j]
TD-DFT/B3LYP	3.05 (407)[k]	–

[a]*p*-HBI stands for 4-(*p*-hydroxybenzylidene)imidazolinone, with two hydrogen atoms replacing the methyl groups of *p*-HBDI
[b]Photodestruction experiments [56]
[c]Extrapolation from Kamel–Taft fit [57]
[d]MRMP2 based on sa-CASSCF(14/12)/cc-pVDZ, DFT (PBE0/cc-pVDZ) geometry [74]
[e]Many-body Green's function theory, DFT(PBE) geometry [75]
[f]CASPT2/CASSCF(12/11)/6-31 G(d), CASSCF geometry [74]
[g]CASPT2/CASSCF(12/11)/6-31 G(d), CASSCF geometry [70]
[h]CASPT2/CASSCF(14/14)/6-31 G(d) with IPEA Hamiltonian, DFT/BLYP geometry [66]
[i]TD-DFT/SAOP/ETpVQZ, DFT/BLYP geometry [66]
[j]Diffusion Monte Carlo/CASSCF(14/14), DFT/BLYP geometry [66]
[k]TD-DFT/B3LYP/6-31++G(D), DFT/B3LYP geometry [78]

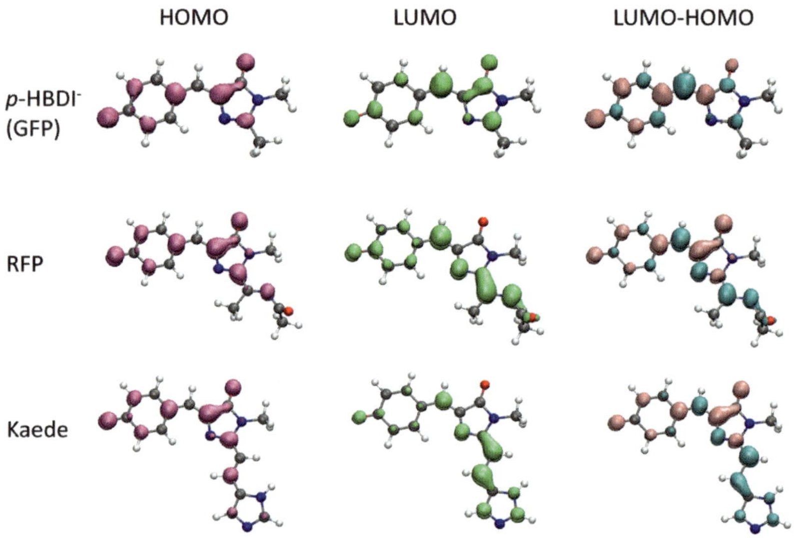

Fig. 5 Isodensity surfaces of HOMO, LUMO, and their difference for three chromophore structures, from Nifosì et al. [71]. In the third column, *pink* (*cyan*) regions indicate charge depletion (increase) upon HOMO → LUMO excitation

The picture emerging from the comparison between gas-phase experimental and theoretical results on *p*-HBDI$^-$ would highlight the good performance of highly correlated methods (CASPT2, many-body Green function, see Table 3), time-dependent density functional theory (TD-DFT) methods predicting too blueshifted values. A recent investigation by Filippi and coworkers showed how using a more accurate zero-order Hamiltonian for CASPT2 (with the IPEA shift) leads to a significantly blueshifted value [66]. For in-depth discussion of these discrepancies, the interest reader is referred to [66, 74].

The case of RFP model chromophores (i.e., *p*-HBMPI$^-$, *p*-HBMPDI$^-$, and the RFP chromophore itself) shows similar trends, with TD-DFT predicting blueshifted values with respect to CASPT2 (Table 4). In contrast to the case of GFP model chromophores, the experimental photo destruction value lies midway between the values predicted by the two classes of methods. A better agreement between TD-DFT and experiment is found for the larger *p*-HBMPDI model. Likewise, for the RFP chromophore (not stable in solution) CASPT2 and TD-DFT predictions are in rather good agreement. CASPT2 prediction for different isomers yields blueshifted values for the *trans* chromophore (180 rotation around τ, defined in Table 2) of around 0.10–0.15 eV, rather uniformly for the various models used [72, 79].

TD-DFT low computational cost makes it an affordable method for a systematic and homogeneous investigation of the large set of chromophore structures [71]. Using this technique, the trends among the various FPs are reproduced. As already mentioned, chromophore structure is the principal determinant of the excitation energy. In term of the computed gas-phase vertical excitation energies, the most

Table 4 Excitation energies in eV (nm) of anionic RFP model chromophores

Method	p-HBMPI$^-$ ex. energy in eV (nm)	p-HBMPDI$^-$ ex. energy in eV (nm)	RFP chrom. ex. energy in eV (nm)
Expt. (gas)[a]	2.40 (521)[a]	2.26 (549)[a]	–
CASPT2	2.26 (549)[b]	–	2.20 (564)[c]
TD-DFT B3LYP	2.68 (463)[d]	2.39 (519)[d]	2.36 (525)[e]
DFT MRCI	–	–	2.01 (616)[f]

[a]Photodestruction experiments [59, 63]
[b]CASPT2/CASSCF(12/11)/6-31 G(d), B3LYP geometry [72]
[c]CASPT2/CASSCF(12/11)/6-31 G(d), CASSCF geometry, connections saturated with hydrogen atoms [79]
[d]TD-DFT B3LYP/631+G*, B3LYP geometry [71]
[e]TD-DFT B3LYP/631+G*, B3LYP geometry, connections saturated with methyl groups [71]
[f]DFT MRCI, B3LYP geometry [80]

blueshifted is the BFP chromophore (340 nm), whereas the most redshifted is the uvKaede chromophore at 577 nm. Substantial variations do occur among different proteins containing the same chromophore (see Sect. 5).

The systematic study allowed the authors to correlate the variation in spectral properties with the extension of the π-conjugated system. The HOMO → LUMO, π → π* character of the excitation is confirmed also for non-GFP chromophore structures (HOMO, LUMO isodensities, and their difference are reported in Fig. 5). Chromophores with a larger charge displacement, measured as the variation of dipole moment between the HOMO and the LUMO, generally display a lower excitation energy (as apparent from Fig. 5, uvKaede involves a larger amount of charge displacement than p-HBDI$^-$ and RFP). This observation reframes quantitatively the idea that larger π-conjugated systems are associated with lower excitation energies.

By minimizing the excited state structure at a planar conformation, it is possible to calculate the emission energy. Such a kind of analysis was performed for p-HBI and for the RFP chromophore using CASSCF/CASPT2, yielding values of 2.45 eV (507 nm) [70] and of 1.76 eV (703 nm) [79], respectively. The first value is in good agreement with the low temperature (77 K) fluorescence peak of a GFP chromophore model[2] in ethanol (490 nm) [54]. Exploration of the excited state energy surface reveals the presence of twisted intermediates and nearby conical-intersections seams [79, 81] that are relevant to the photophysics of the chromophore, for example explaining their very low emission quantum yield when not embedded in the protein matrix.

[2]The early study by Niwa et al. used ethyl 4-(4-hydroxyphenyl)methylidene-2-methyl-5-oxo-1-imidazolacetate, a model chromophore with the methyl group in p-HBDI substituted with CO2C2H5.

5 FPs Spectra: Spectral Tuning by the Protein Environment

The sheer number of FPs discovered and engineered so far eludes an attempt of an exhaustive review. Rather, this section will highlight the cases of two main families of FPs, i.e., those containing a GFP-like or an RFP-like chromophore, focusing on the structural mechanisms operating the spectral shifts. It is useful, before starting to analyze the single cases, to bear in mind how the interaction between chromophore and environment can influence light absorption and emission. The environment can act on the chromophore structure by deforming some bond lengths mainly through hydrogen bonds (H-bonds), and/or by distorting the planarity of the chromophore. In addition, it can differentially stabilize the ground and excited states by electrostatic interactions. Recent experimental [82] and theoretical [83] studies were devoted to understanding the tuning of spectral properties by the protein matrix electrostatic field.

With regard to bond deformation, early Raman spectroscopy studies on the GFP chromophore in solution and on some *av*GFP mutants revealed a tight correlation between the shift in absorption peak and the frequency of a specific Raman-active band [17]. This correlation was rationalized on the basis of the resonance structures of the chromophore (depicted in Fig. 6) [84, 85], as arising from the selective stabilization of one of the two resonance structures by the protein environment. Variations observed in the Raman spectrum reflect a changing stabilization between a benzenoid-like resonance structure (I in Fig. 6) and a quinonoid-like resonance structure (II). For example, the ubiquitous H-bond between the imidazolinone carbonyl and conserved Arg at 96 (in *av*GFP numeration) increases the contribution from the quininoid form (II), thus lowering the excitation energy. H-bonds to the phenolate side, by contrast, tend to decrease the contribution of the quininoid form in the anionic chromophore. Resonance structures are less accessible to the neutral form of the chromophore, because the protonated phenol enforces a benzenoid-like structure (the optical properties of the neutral chromophore are accordingly blue-shifted with respect to the anion).

The same effects can be described in a complementary picture, recalling that excitation involves charge displacement from the phenolate to the imidazolinone and to the bridging carbon. H-bonds to the imidazolinone will preferentially stabilize the excited state, and thus lower the excitation energy. Vice versa, H-bonds to the phenolate will increase the excitation energy. Other electrostatic interactions can be involved, particularly in the proximity of the bridging carbon, another site of charge redistribution. The interplay among these factors can be such that the final outcome becomes difficult to predict. Nonetheless, this general scheme is useful to decipher the mechanisms of spectral tuning.

In the RFP chromophore, a third resonance structure is possible (Fig. 6) [79], thanks to the presence of the conjugated acylimine. H-bonds to this additional site will tend to lower the excitation energy, stabilizing the excited state charge displacement over the region.

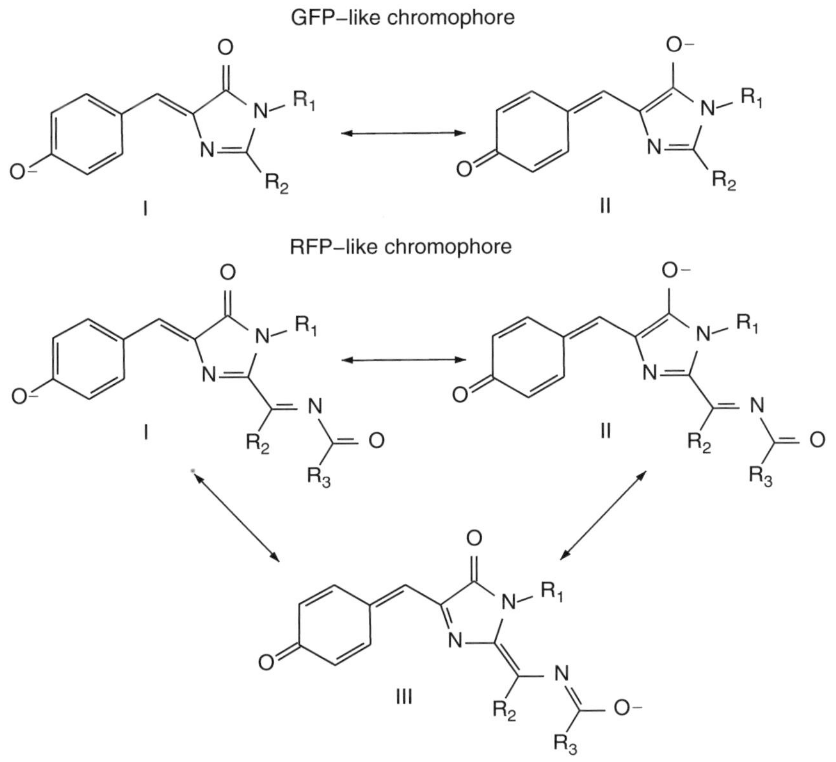

Fig. 6 Resonance forms of the anionic states of GFP chromophore (*top*) and DsRed chromophore (*bottom*)

5.1 *Proteins with GFP Chromophore*

The electronic excitation mostly involves the chromophore region supporting the electronic π-conjugated system. In the GFP-like chromophores, this encompasses the phenolic ring and the imidazolinone, while the side chain of the first position in the tripeptide is not included. Hence, it is justified to consider proteins with a different residue at this position as having the "same" chromophore, as long as it does not undergo additional maturation reactions (like, for instance, the acylimine formation in RFP chromophore).

Starting from the anionic protonation state, Table 5 reports absorption and emission peaks of the anionic form in different representative proteins all containing a GFP-like chromophore (GFPa), together with the nature of the residues surrounding the chromophore at various positions. The proteins can be grouped according to their absorption/emission features in cyan/teal, green, and yellow. In order to specify equivalent positions in the sequences of these homologues, we will use the *av* prefix, referring to the corresponding position in *av*GFP after sequence

Table 5 Proteins containing the GFP-like anionic chromophore

Protein name	Environment residues (*av*GFP #)										Ex./ab. max (nm)[a]	Em. max (nm)	References
	av 62	*av* 65	*av* 68	*av* 69	*av* 94	*av* 96	*av*148	*av*150	*av*167	*av*203			
Cyan/teal													
dsFP483	P	Q	N	K	W	R	S	E	H	T	443ab/437ex	483	[6, 86]
cFP484	N	Q	N	R	W	R	S	E	H	H	456ab	484	[6]
amFP486	T	K	N	R	Y	R	S	E	A	H	454ab/450ex	486	[87]
mTFP0.7 (from cFP484)	N	A	N	R	W	R	S	E	H	H	453ab	488	[88]
mTFP1 (from cFP484)	T	A	N	R	W	R	S	E	H	H	462ex	492	[89]
*av*GFP R96M	T	S	V	Q	Q	M	H	V	I	T	466ab	490	[90]
Green													
mTFP1 H163M (from cFP484)	T	A	N	R	W	R	S	E	M	H	487ex	503	[91]
*av*GFP S65A	T	A	V	Q	Q	R	H	V	I	T	476ex	507	[92]
RSGFP4 (from *av*GFP)	T	G	V	L	Q	R	H	V	I	T	490ex	505	[93]
*av*GFP	T	S	V	Q	Q	R	H	V	I	T	475ex	504	[5]
EGFP (from *av*GFP)	T	T	V	Q	Q	R	H	V	I	T	488ex	507	[5]
dis3GFP	T	Q	N	R	W	R	S	E	M	H	503ex	512	[94]
cmFP512	M	Q	F	R	W	R	A	E	K	H	502ex	511	[95]
Dronpa	T	C	N	R	W	R	S	E	M	H	503ex	518	[26]
CpYGFP	H	G	F	Y	N	R	P	D	R	R	509ex	517	[96]
Yellow													
*av*GFP T203F	T	S	V	Q	Q	R	H	V	I	F	506ex	522	[97]
CpYGFP H52F	F	G	F	Y	N	R	P	D	R	R	514ex	522	[96]
EYFP (from *av*GFP)	T	G	L	Q	Q	R	H	V	I	Y	514ex	527	[27]
E^2GFP (from *av*GFP)	T	T	V	Q	Q	R	H	V	V	Y	515ex	525	[98]
phiYFP	T	T	A	Q	Q	R	H	L	I	Y	525ex	537	[30]

[a] Few-nanometer discrepancies can be expected between absorption and fluorescence excitation spectra, when there are inhomogeneities in the structure

alignment. When no such prefix is present, the number refers to the normal position within the sequence.

5.1.1 Cyan/Teal Group

The cyan/teal group contains FPs isolated from reef corals, such as *Discosoma striata* (dsFP483), *Anemonia majano* (amFP486), and *Clavularia* (cFP484), as well as two cFP484 mutants, mTFP0.7 and mTFP1. These cyan-emitting proteins are naturally tightly bound tetramers; hence, *av*GFP mutants bearing a Trp at *av*66 are more popular as cyan-emitting FPs, their cyan fluorescence being due to a different chromophore structure (CFP in Fig. 4). Recently, however, the mTFP mutants were engineered to avoid oligomerization and exist as monomers [89]. Understanding the origin of the blueshifted spectral properties of this group has proven rather challenging. The broad absorption/excitation band hints at structural heterogeneity of either the chromophore or the surroundings [86]. By contrast, the narrower emission spectrum signals a unique emitting state.

AmFP486, cFP484, and the mTFPs (but not dsFP483) feature a quadrupole network of salt bridges beneath the chromophore, involving one Arg (at *av*69), one His (at *av*203), and two Glu (at *av*150 and *av*222). The His, being hydrogen bonded on both sides to the Glu, is presumably cationic [87]. In amFP486, substitution of this His with Thr resulted in redshifted excitation and emission (from 454 to 470 nm in absorption and from 486 to 515 nm in emission), accompanied by a decline in fluorescence quantum yield and heterogeneity of chromophore structure [87]. In mTFP1, a major role is played by His163 (*av*165). Mutation of His163 into Met abolishes its H-bond with the chromophore phenolate and redshifts the excitation/emission maximum (from 462/492 to 487/503 nm) [89, 91]. Introducing the further T73A mutation additionally shifts the excitation/emission to 498/515 nm, possibly by perturbing the quadrupole arrangement supporting the cationic His197 [91]. The quadrupole arrangement is, however, neither sufficient nor necessary for cyan emission. It is present in Dronpa, a *green*-emitting FP, and it is missing in cyan dsFP483, in which His at *av*203 is replaced by a Thr. However, in dsFP483, a different charged residue, a lysine (K70), is in close contact with the chromophore-bridging carbon and points toward the phenolate [86], suggesting that such proximity of a charged residue is indeed necessary.

Inspection of the various chromophore-environment configurations in the relevant X-ray structures (reported in Fig. 7) highlights another important – and possibly necessary – feature common to these cyan-emitting proteins, namely the presence and arrangement of three H-bond donors facing the chromophore phenolate. Indeed, a water molecule and Ser at position *av*148 are present in all examined structures and at H-bonding distance to the phenolate. The third H-bond is provided either by His at *av*167 (in dsFP483, cFP484, and in the mTFPs), or by a water molecule in amFP486, occupying a cavity set empty by the small bulk of Ala at *av*167. Neither Dronpa nor cmFP512 (both belonging to the green group), though featuring the quadrupole arrangement, has all three H-bond donors: His *av*167 is

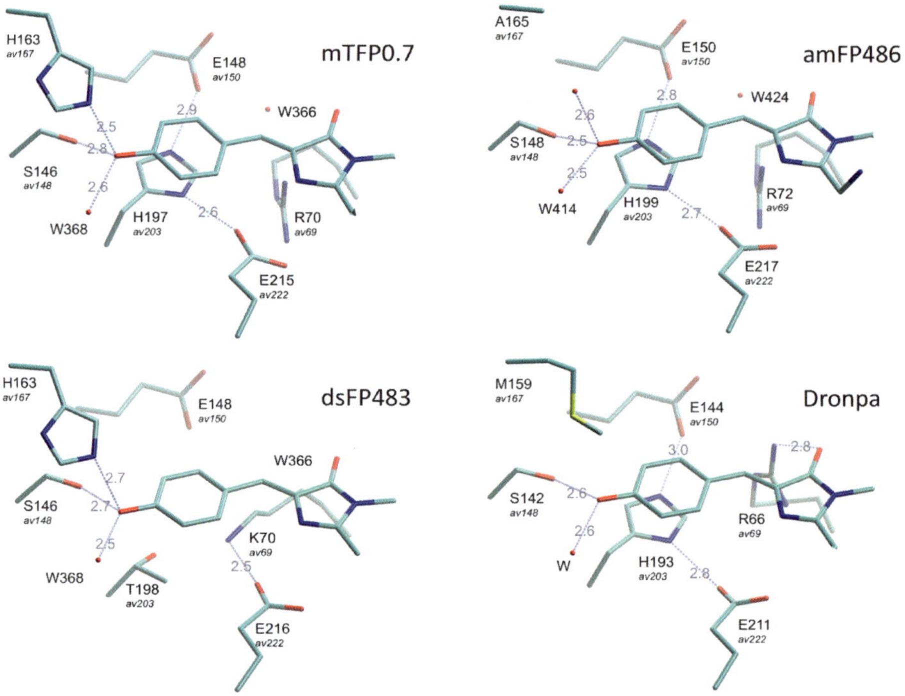

Fig. 7 Relevant residues and water molecules in the chromophore environment of three *cyan* FPs dsFP483, amFP486, mTFP0.7 and the *green* FP Dronpa. The coordinates were taken from the Protein Data Bank, with codes 3CGL [86], 2A46 [87], 2OTB [88], and 2IOV [99]. The Arginine residue at position *av*96 is omitted for clarity. The *av* prefix in the residue number refers to the corresponding position in *av*GFP after sequence alignment

replaced by Met in Dronpa (see Fig. 7) and Ser *av*148 by Ala in cmFP512 (this time with no emptied cavity).[3]

The case of the R96M *av*GFP mutant will be discussed below.

5.1.2 Green Group

In considering the cyan FPs as blueshifted, it is implicitly assumed that the *av*GFP chromophore in the gas phase emits green light. Although no emission is detected in the gas phase [62], the gas-phase absorption would suggest that it is indeed so.

[3]Somewhat upsetting this picture, Wachter and coworkers pointed out that a very similar arrangement of three H-bond donors would also be present in the immature form of the DsRed protein, which, however, emits green, not cyan, light [86]. However, no isolated structure of DsRed with immature green chromophore is available, the X-ray structures being a mixture of mature and immature structures.

However, since in *av*GFP and its green variants, the chromophore is still involved in relevant interactions with surrounding residues and water molecules, a more accurate picture is that of a cancellation of the various contributions. Computational studies [85] point at a counteraction between the redshifting H-bond with Arg at *av*96 (a conserved residue in FPs, being a requisite for chromophore formation) and the blue shifting of H-bonds to the phenolate. Slow-maturing *av*GFP variants with mutated Arg96 do exist, and also shed light on the relevance of Arg96 on the optical properties. Getzoff and coworkers [90] structurally solved the mature R96M mutant, where the H-bond with the imidazolinone carbonyl group is abolished, and replaced with a, somewhat weaker, H-bond with a water molecule (donor–acceptor distance is 2.9 Å in PDB structure 2AWK). In line with theoretical predictions [85], the R96M mutant displays blueshifted optical properties (Table 5). Its application as a cyan-emitting protein is, however, severely hampered by the extremely slow chromophore maturation (months).

5.1.3 Yellow Group

The common feature of proteins in the yellow group is the Tyr or Phe at *av*203 (see Fig. 8). The T203Y mutation was introduced in *av*GFP mutants with the rationale that an aromatic amino acid at that position would be π-stacked to the chromophore phenolate, thereby lowering the excitation energy by increasing the polarizability of the environment around the chromophore [27]. The mutation indeed resulted in redshifted variants (see Table 5), and X-ray studies confirmed the π stacking configuration [27]. Notably, the same structural stratagem for redshifting is presumably at play in a natural FP named phiYFP [30], also featuring a Tyr at *av*203.

One additional important feature arose from T203Y mutation, i.e., a rather enhanced extinction coefficient (see Table 1). Similar spectral shifts are obtained with a phenylalanine replacement in the *av*GFP T203F mutant [97]. Actually, the absorption/excitation red shift in these Thr203 mutants is only partially accounted for by the π-stacking. Indeed, also T203V and T203I mutants show a redshifted absorption with respect to *av*GFP (from 477 nm of *av*GFP to 502 of the T203V and T203I mutants [97]), associated with the removed Thr203 H-bond to the phenolate. Redshifted *fluorescence* is instead almost totally ascribed to the π-stacked Tyr or Phe.

The case of CpYGFP is worth describing. CpYGFP features a Histidine at *av*63, which is stacked *above* the chromophore (i.e., on the opposite side with respect to Y203 in the orientation of Fig. 8) and presumably neutral, giving rise to rather redshifted excitation/emission [96]. Replacement of this His with Thr (the residue present in the equivalent position in *av*GFP) blueshifts the excitation/emission to 495/511 nm. Instead, replacement with Phe additionally redshifts to 514/522 nm, as a consequence of a larger amount of π-stacking.

Both EYFP (enhanced YFP) and CpYGFP X-ray structures show a peculiar H-bond between Glu at *av*222 and the imidazolinone of the chromophore. It would be tempting to assign a role in redshifting to this H-bond. Although X-ray structures of other yellow-emitting *av*GFP variants such as E^2GFP [19] (PDB Code 2H9W)

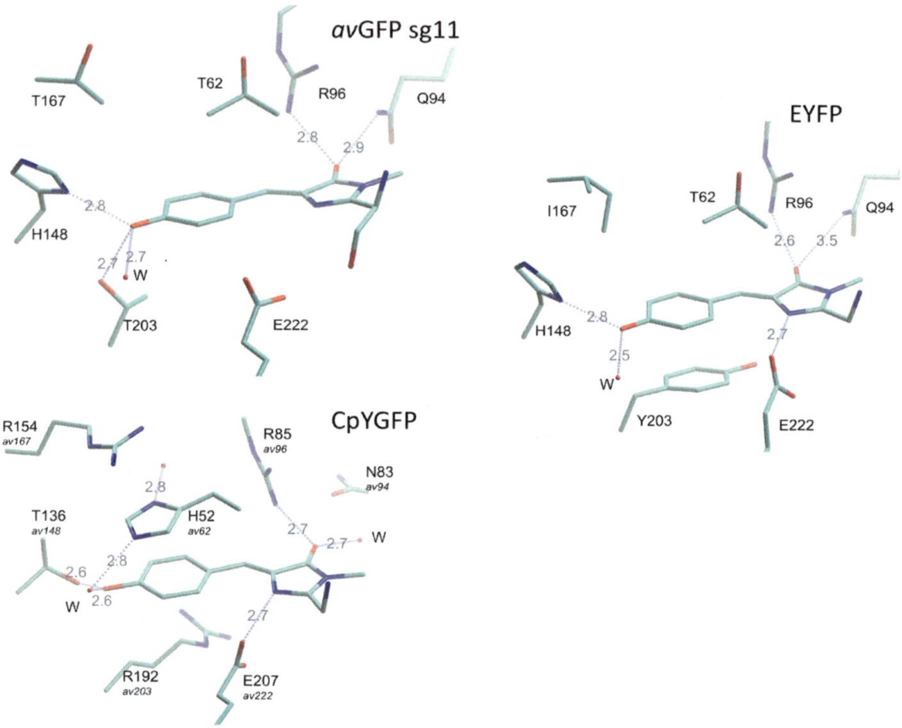

Fig. 8 Relevant residues and water molecules in the chromophore environment of *av*GFP sg11 (F64L/I167T/K238N mutant, displaying very similar optical properties to the original *av*GFP), CpYGFP, and EYFP. The coordinates were taken from the Protein Data Bank, with codes 2WUR [100], 2DD7 [96], and 1YFP [27]. The *av* prefix in the residue number refers to the corresponding position in *av*GFP after sequence alignment

display no such H-bond, its involvement in redshifting was assessed in X-ray studies on DsRed mutants (see below) [45].

5.1.4 Neutral Chromophore (GFPn)

While spectral tuning by the protein matrix on the anionic *av*GFP chromophore gives rise to a rather broad range of optical properties, the neutral species seems less tunable, in line with the findings on isolated chromophores in solution [57] and with the suppressed contribution of the quinonoid-like resonant structure. The neutral absorption peak in various *av*GFP mutants (including the yellow ones) is 400 $\pm$ 3 nm [19, 97],[4] 385 nm in mKalama1 (an *av*GFP mutant [89]), and 388 nm in Dronpa

[4]In *av*GFP mutants, such as EYFP, featuring the anionic band only, the neutral band emerges by decreasing the pH below the protein *p*Ka.

[101]. A particularly redshifted value (415 nm) is achieved in the S65T/H148D *av*GFP variant, where the D148 carboxyl oxygen forms a very tight H-bond (distance 2.4 Å) with the chromophore hydroxyl group [102].

The emission in variants with a neutral chromophore state is generally achieved via ESPT; so the emitting state is the anionic chromophore [18]. The fluorescence of mKalama1, peaking at 456 nm, is, however, due to intrinsic emission of the neutral chromophore [89].

5.2 *Proteins with RFP Chromophore*

Like the anionic *av*GFP chromophore, also the spectral characteristics of the anionic RFP chromophore are strongly tunable by the protein matrix (Table 6). Particular care needs to be put in selecting the right variants for the comparison, because in some DsRed-derived FPs, the chromophore undergoes covalent modifications that alter its structure. Another *caveat* concerns the presence of a *trans* (E) chromophore instead of the more common *cis* (Z) conformation, like in eqFP611

Table 6 Proteins containing the RFP-like anionic chromophore

Name	Environment residues (*DsRed #*)								Ex. max (nm)	Em. max (nm)	References
	*dr*16	*dr*44	*dr*66	*dr*70	*dr*83	*dr*146	*dr*163	*dr*197			
Red											
DsRed	V	V	Q	K	K	S	K	S	558	583	[6]
DsRed K83M	V	V	Q	K	M	S	K	S	564	602	[103]
mStrawberry (from DsRed)	V	A	T	K	L	S	M	I	574	596	[32]
mCherry (from DsRed)	V	A	M	K	L	S	Q	I	587	610	[32]
DsRed	V	V	Q	K	K	S	K	S	558	583	[6]
DsRed K83M	V	V	Q	K	M	S	K	S	564	602	[103]
Far red											
mKate (from eqFP578)	L	M	M	K	F	S	M	R	585	635	[104]
mGrape3 (from DsRed)	T	A	M	K	L	S	M	Y	608	646	[41]
mPlum (from DsRed)	E	A	M	K	L	S	M	I	590	649	[40]
mNeptune (from eqFP578)	L	G	M	K	F	S	M	R	600	650	[41]

(Table 1). When the X-ray structure is not available, the assignment of the chromophore structure may be doubtful. The chromophore of mOrange and mBanana contains a structural modification at the acylimine tail; so the blueshifted optical properties of these FPs cannot be wholly attributed to chromophore–protein noncovalent interactions. As such, the original DsRed is among the most blueshifted FPs with the DsRed chromophore structure (Table 6).

Mutation K83M was found to redshift the optical properties by dislocating Lys70, which in wild-type DsRed is in close contact with the chromophore-bridging carbon (at 3.6 Å distance) [45, 103]. A similar mechanism is at play in mCherry and mStrawberry, both featuring the analogous K83L mutation. Further redshifting in these two mutants is provided by the H-bond between protonated E215 and the nitrogen in the chromophore imidazolinone (see Fig. 9). It is indeed observed that at basic pH conditions (pKa ~10), at which this H-bond is presumably removed due to deprotonation of E215, the absorption peak of mCherry and mStrawberry blueshifts from 584 to 566 nm and from 574 to 548 nm, respectively. A third source of redshifting is achieved in mCherry/mStrawberry by the additional K163Q/K163M mutations eliminating one H-bond to the chromophore phenolate.

Like in yellow mutants of *av*GFP, π-orbital stacking of the chromophore with a Tyr residue results in a redshift. For instance, the protein mGrape3 with the I197Y mutation (corresponding to *av*203) is remarkably redshifted (608 ex and 646 em).

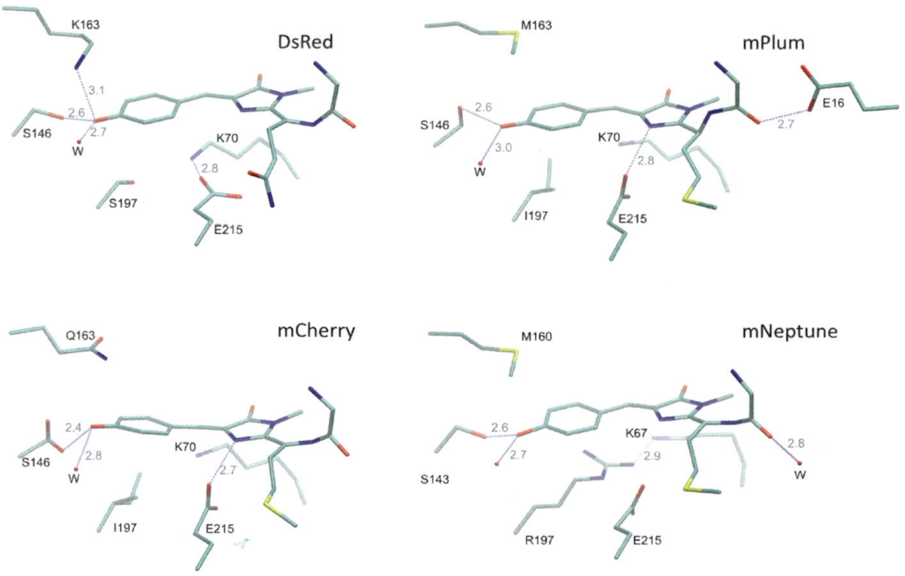

Fig. 9 Relevant residues and water molecules in the chromophore environment of red (DsRed and mCherry) and far-red (mPlum and mNeptune) FPs. The coordinates were taken from the Protein Data Bank, with codes 1ZQO [105], 2H5Q [45], 2QLG [106], and 3IP2 [41]. The DsRed structure contains both mature and immature chromophore conformations. Only the mature chromophore is shown here. The Arginine residue at position 95 is omitted for clarity

It is, however, characterized by low quantum yield (0.03) and complex photophysics possibly connected with *cis–trans* isomerization of the chromophore [41].

As the π-conjugated system in the RFP chromophore reaches down to the acylimine group, variations in the surrounding environment can also modify the optical properties. For example, mPlum, one very redshifted DsRed variant, features (besides the redshifting K163M mutation and the E215-imidazolinone H-bond) a H-bond between (presumably neutral) Glu16 and the acylimine oxygen. This H-bond is absent in DsRed and most other mutants, where position 16 is occupied by hydrophobic amino acids. It is found that mutation of Glu16 strongly affects the emission peak while leaving the excitation almost unchanged. For example, replacing Glu16 with Leu blueshifts the emission peak from 649 to 630 nm, while the excitation peak is only slightly modified from 590 to 588 nm [106].

Likewise, in mNeptune [41], a far-red FP derived from eqFP578 of *Entacmaea quadricolor* (not to be confused with eqFP611) [107], the acylimine oxygen is H-bonded to a water molecule. The slightly blueshifted optical properties of mKate (another eqFP578 mutant) can be attributed to the absence of this H-bond, caused by Met41 displacing the hydrogen-bonded water molecule. As is the case with many FPs, the red shift of mNeptune is, however, the result of multiple entangled contributions. As a consequence, it may be difficult to predict the outcome of a mutation. Quite surprisingly, for example, introducing the Tyr at 197 in mNeptune (R197Y mutant), at the presumably π-stacked position with the phenolate, blueshifts, rather than redshifts, the optical response [41].

Chromophore deviation from coplanarity is yet another determinant of optical properties. In general, distorted chromophore structures are associated with poor quantum yield of fluorescence. For instance, mCherry is characterized by a particularly distorted chromophore and a rather low fluorescence quantum yield (0.22) [7]. Chromophore deviation from coplanarity can be quantified by the twist around τ and ϕ (see Table 2), or also including the tilt angle defined as the deviation of the phenol oxygen from the perfectly planar chromophore [9].

6 Two-Photon Excitation

Despite their relevance in multiphoton fluorescence microscopy applications, two-photon spectroscopic features of FPs have been somewhat neglected in the past years, appearing in few experimental works centered on BFP, CFP, EGFP, and DsRed [108–110]. More recently, parallel but independent experimental [67, 111, 112] and theoretical [113] investigations extended such analyses to several other FPs.

Figures 10 and 11 report the two-photon excitation spectra of, respectively, blue/cyan/green and orange/red FPs. One-photon spectra, suitably scaled in the abscissae, are also reported for comparison. In all investigated proteins, the long-wavelength band (i.e., the main excitation band) is active also at two-photon excitation, implying that the first singlet excitation is characterized by both one and two-photon oscillator strength. This statement is verified for all chromophores examined

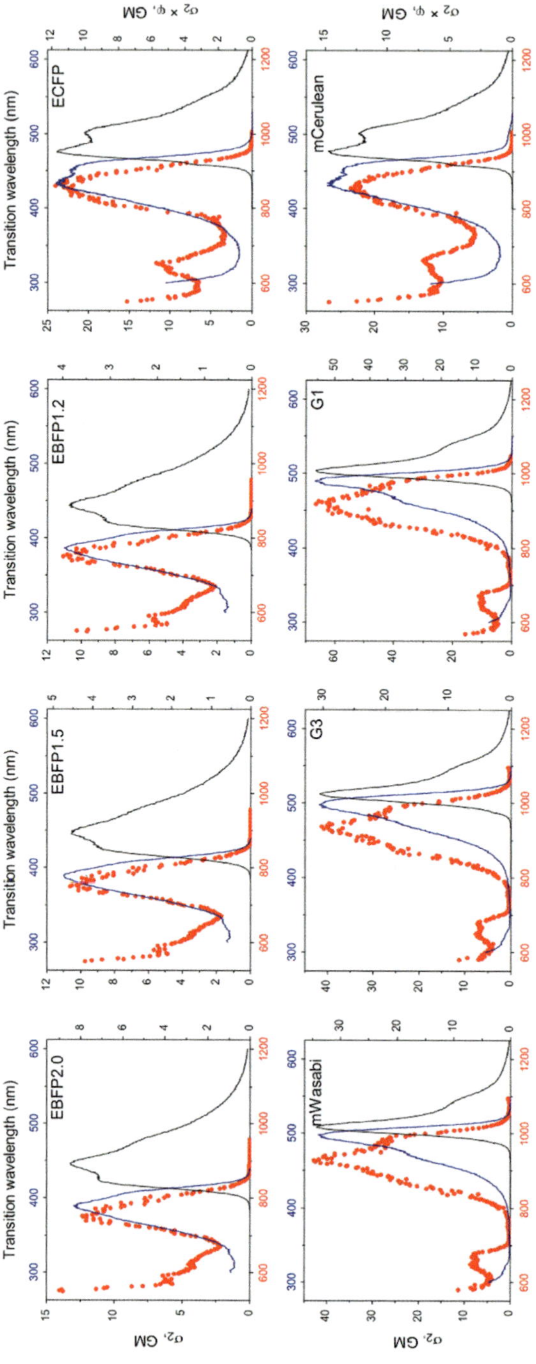

Fig. 10 (continued)

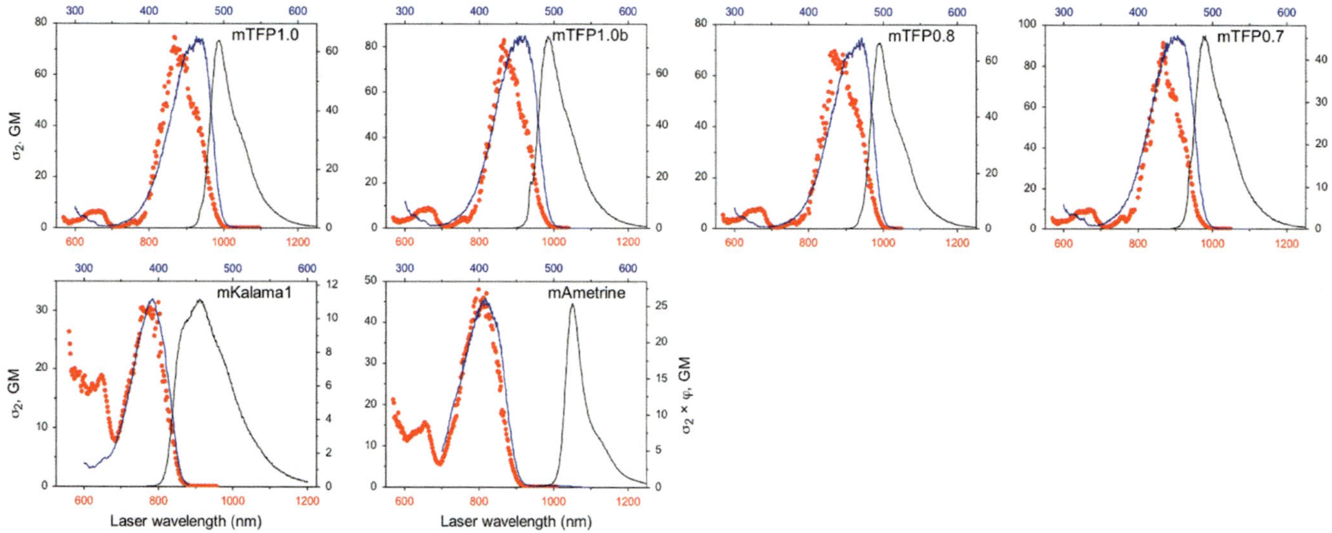

Fig. 10 (continued)Two-photon excitation spectra (*red circles*), one-photon fluorescence excitation spectra (*blue line*), and fluorescence emission spectra (*black line*) of the FPs studied. The left-ordinate axis represents the two-photon cross-section (σ_2) values in GM (Goeppert–Mayer), and the right-ordinate axis represents the two-photon brightness, $\sigma_2 \times \varphi$ (φ being the fluorescence quantum yield), in GM. The bottom x-axis represents the laser wavelength used for excitation and the top x-axis represents the transition wavelength. The excitation and fluorescence emission intensities are shown in arbitrary units. From Tillo et al. [111]

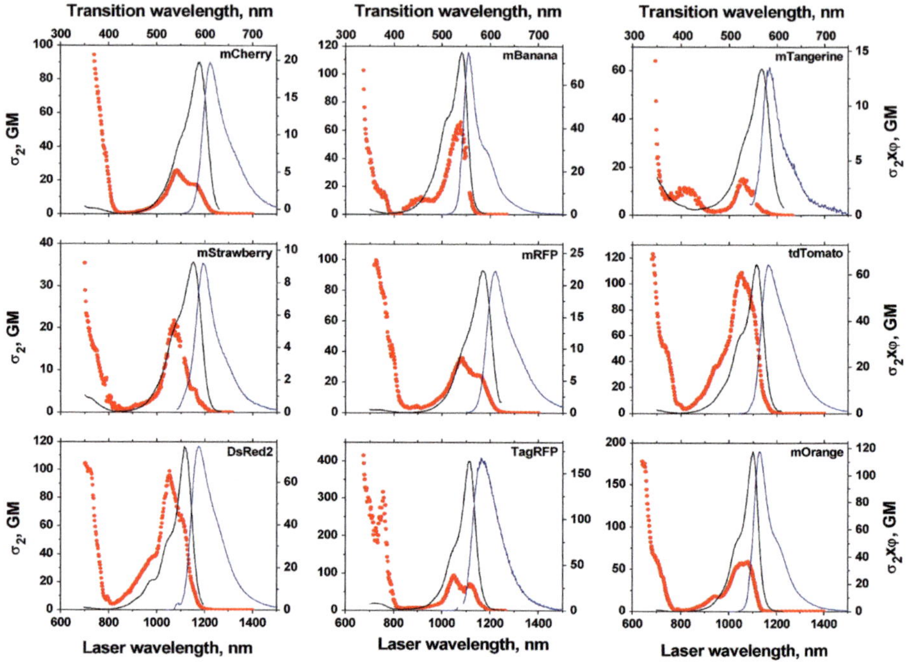

Fig. 11 Two-photon excitation spectra (*red circles*) shown along with fluorescence emission (*blue line*) and one-photon fluorescence excitation (*black line*) spectra. The left-ordinate axis represents the two-photon cross section (σ_2) values in GM (Goeppert–Mayer), and the right-ordinate axis represents the two-photon brightness, $\sigma_2 \times \varphi$ (φ being the fluorescence quantum yield), in GM. The *bottom* *x*-axis represents the laser wavelength used for excitation and the *top* *x*-axis represents the (one-photon) transition wavelength. The excitation and fluorescence emission intensities are shown in arbitrary units. From Drobizhev et al. [67]

and for both neutral and anionic species (for instance, mKalama1 and mAmetrine contain the chromophore in the neutral state [21, 24]).

Interestingly, the profile of the two-photon band is generally blueshifted with respect to the one-photon band and, in certain instances, rather different. Such difference is presumably due to a different vibronic coupling between one- and two-photon excitation. For example, interpreting the shoulder about 50 nm to the blue of the main peak (in mWasabi, G1, G3 of Fig. 10 and in all orange/red FPs in Fig. 11) as a vibronic peak, it is clear that in two-photon excitation such a vibronic mode is enhanced. It is worth mentioning that other authors explain the blueshifted two-photon spectrum as due to the excitation to a different electronic state [114] lying very close in energy to the first excited state. No such state is, however, found in quantum chemical calculations (see Sect. 4.2).

Another common feature of the spectra is the presence of two-photon excitation bands in the high-energy (short-wavelength) side of the spectrum, where the one-photon spectrum is either featureless or shows only limited one-photon excitation. These are located below 700 nm in green/cyan/blue FPs (Fig. 10) and start already

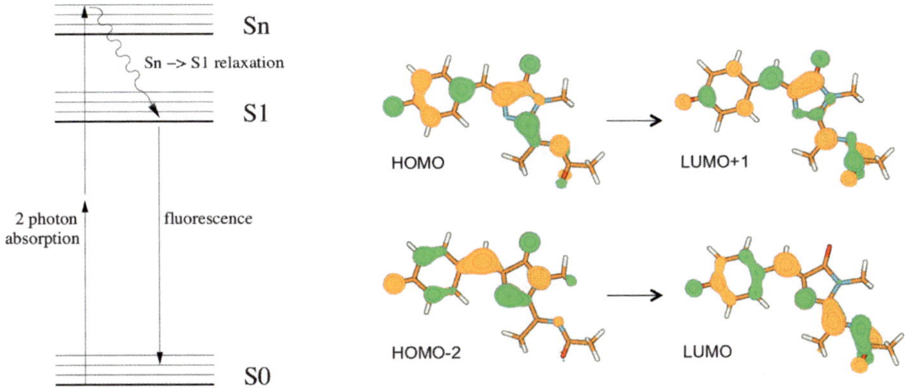

Fig. 12 *Left*: Jablonski diagram for fluorescence following two-photon excitation to the higher S_n state. *Right*: Isodensity surfaces of the Kohn–Sham molecular orbitals relevant to the high-energy two-photon excitation in the RFP chromophore (HOMO-2 is the second, in energy, occupied molecular orbital before the HOMO, and LUMO+1 is the first unoccupied molecular orbital after the LUMO). From Nifosì and Luo [113]

at 800 nm in red/orange FPs (Fig. 11). Such high-energy bands were revealed first in DsRed [108, 110] and were instead absent in cyan and green FPs, because of the limited spectral window examined (750–1,000 nm). Subsequently, thanks to a broader detection window, they were revealed in blue, green, and cyan FPs, and reported to give rise to bright and steady fluorescence, with the same spectral characteristics of the low-energy excitation band.

Theoretical analysis reveals that such bands arise from excitation to higher excited states of the chromophores (Fig. 12). Despite having low one-photon oscillator strength, these excitations acquire some two-photon moment mainly from the first singlet excitation, with a near-resonance enhancement mechanism [112, 113]. The decomposition into molecular orbitals' excitations within a TD-DFT framework reveals that, analogously to the one-photon case, where the dominant excitation is HOMO $\rightarrow$ LUMO for all chromophores, a common molecular orbital composition is also found for these higher excitations (see Fig. 12).

Marchant et al. reported an early application of multiphoton excitation in the high-energy bands of DsRed [110]. Femtosecond-pulsed irradiation at $\lambda < 760$ nm readily changes the fluorescence of DsRed from red to green, presumably by bleaching the red-emitting chromophore within the DsRed tetramer. Thanks to the three-dimensional localization of multiphoton excitation, DsRed "greening" could be exploited to optically highlight subcellular compartments in living cells. More generally, the availability of additional two-photon excitation bands is quite handy, given that laboratories are only rarely equipped with more than one two-photon laser. Drobizhev and coworkers suggest an approach for dual color two-photon imaging, based on the simultaneous excitation of blue and red FPs at 780 nm, corresponding to the low-energy excitation band of the former and the high-energy excitation band of the latter [111].

7 Outlook

The range of spectral properties displayed by FPs is surprisingly broad. The factors determining such variety are the extension of the chromophore π-electron system and the interactions of the chromophore with the surrounding protein matrix. Using mutagenesis on the already wide set of natural FPs, the researcher can play with these two factors to achieve the desired absorption/emission characteristic. The mechanisms of spectral tuning are understood in general terms. Nonetheless, a predictive structure–property relationship has proven elusive. Due to the interconnections among the determinants of optical properties, two mutations known to, say, redshift DsRed fluorescence when included separately into the sequence may not give rise to additional redshifting when included simultaneously. Properties such as quantum yield or extinction coefficient are yet more unpredictable.

These difficulties have not prevented an array of remarkable accomplishments in manipulating FP excitation/emission, from the early generation of blue- and red-shifted variants of *av*GFP to more recent achievements of bright FPs in the near-infrared region. Despite the number and the variety of FPs available, there is much scope for advancement in further enlarging the available spectral window and improving FP brightness.

References

1. Shimomura O, Johnson FH, Saiga Y (1962) Extraction, purification and properties of Aequorin, a bioluminescent protein from the luminous hydromedusan, *Aequorea*. J Cell Comp Physiol 59:223–239
2. Prasher DC, Eckenrode VK, Ward WW, Prendergast FG, Cormier MJ (1992) Primary structure of the *Aequorea victoria* green fluorescent protein. Gene 111:229–233
3. Chalfie M, Tu Y, Euskirchen G, Ward W, Prasher D (1994) Green fluorescent protein as a marker for gene expression. Science 263(5148):802–805
4. Inouye S, Tsuji FI (1994) *Aequorea* green fluorescent protein: expression of the gene and fluorescent characteristics of the recombinant protein. FEBS Lett 341:277–280
5. Tsien RY (1998) The green fluorescent protein. Annu Rev Biochem 67:509–544
6. Matz M, Fradkov AF, Labas Y, Savitsky A, Zaraisky AG, Markelov AZ, Lukyanov SA (1999) Fluorescent proteins from non-bioluminescent Anthozoa species. Nat Biotechnol 17:969–973
7. Shaner NC, Steinbach PA, Tsien RY (2005) A guide to choosing fluorescent proteins. Nat Methods 2(12):905–909
8. Patterson GH, Knobel SM, Sharif WD, Kain SR, Piston DW (1997) Use of the green fluorescent protein and its mutants in quantitative fluorescence microscopy. Biophys J 73(5):2782–2790
9. Pakhomov AA, Martynov VI (2008) GFP family: structural insights into spectral tuning. Chem Biol 15(8):755–764
10. Brejc K, Sixma TK, Kitts PA, Kain SR, Tsien RY, Ormö M, Remington SJ (1997) Structural basis for dual excitation and photoisomerization of the *Aequorea victoria* green fluorescent protein. Proc Natl Acad Sci USA 94:2306–2311
11. Cubitt AB, Heim R, Adams SR, Boyd AE, Gross LA, Tsien RY (1995) Understanding, improving and using green fluorescent proteins. Trends Biochem Sci 20:448–455

12. Heim R, Prasher DC, Tsien RY (1994) Wavelength mutations and posttranslational autoxidation of green fluorescent protein. Proc Natl Acad Sci USA 91:12501–12504
13. Wachter RM (2007) Chromogenic cross-link formation in green fluorescent protein. Acc Chem Res 40(2):120–127
14. Wachter RM, Watkins JL, Kim H (2010) Mechanistic diversity of red fluorescence acquisition by GFP-like proteins. Biochemistry 49(35):7417–7427
15. Kummer AD, Kompa C, Lossau H, Pollinger-Dammer F, Michel-Beyerle ME, Silva CM, Bylina EJ, Coleman WJ, Yang MM, Youvan DC (1998) Dramatic reduction in fluorescence quantum yield in mutants of green fluorescent protein due to fast internal conversion. Chem Phys 237:183–193
16. Ward WW, Bokman SH (1982) Reversible denaturation of *Aequorea* green fluorescent protein: physical separation and characterization of the renatured protein. Biochemistry 21:4535–4540
17. Bell AF, He X, Wachter RM, Tonge PJ (2000) Probing the ground state structure of the green fluorescent protein chromophore using Raman spectroscopy. Biochemistry 39: 4423–4431
18. Chattoraj M, King BA, Bublitz GU, Boxer SG (1996) Ultra-fast excited state dynamics in green fluorescent protein: multiple states and proton transfer. Proc Natl Acad Sci USA 93:8362–8367
19. Bizzarri R, Nifosì R, Abbruzzetti S, Rocchia W, Guidi S, Arosio D, Garau G, Campanini B, Grandi E, Ricci F, Viappiani C, Beltram F (2007) Green fluorescent protein ground states: the influence of a second protonation site near the chromophore. Biochemistry 46(18):5494–5504
20. Heim R, Cubitt AB, Tsien RY (1995) Improved green fluorescence. Nature 373:663–664
21. Ai H-W, Shaner NC, Cheng Z, Tsien RY, Campbell RE (2007) Exploration of new chromophore structures leads to the identification of improved blue fluorescent proteins. Biochemistry 46(20):5904–5910
22. Mena MA, Treynor TP, Mayo SL, Daugherty PS (2006) Blue fluorescent proteins with enhanced brightness and photostability from a structurally targeted library. Nat Biotechnol 24(12):1569–1571
23. Rizzo MA, Springer GH, Granada B, Piston DW (2004) An improved cyan fluorescent protein variant useful for fret. Nat Biotechnol 22(4):445–449
24. Ai H-W, Hazelwood KL, Davidson MW, Campbell RE (2008) Fluorescent protein FRET pairs for ratiometric imaging of dual biosensors. Nat Methods 5(5):401–403
25. Zapata-Hommer O, Griesbeck O (2003) Efficiently folding and circularly permuted variants of the sapphire mutant of GFP. BMC Biotechnol 3:5
26. Ando R, Mizuno H, Miyawaki A (2004) Regulated fast nucleocytoplasmic shuttling observed by reversible protein highlighting. Science 306(5700):1370–1373
27. Wachter RM, Elsliger MA, Kallio K, Hanson GT, Remington SJ (1998) Structural basis of spectral shifts in the yellow-emission variants of green fluorescent protein. Structure 6:1267–1277
28. Nagai T, Ibata K, Park ES, Kubota M, Mikoshiba K, Miyawaki A (2002) A variant of yellow fluorescent protein with fast and efficient maturation for cell-biological applications. Nat Biotechnol 20(1):87–90
29. Nguyen AW, Daugherty PS (2005) Evolutionary optimization of fluorescent proteins for intracellular FRET. Nat Biotechnol 23:355–360
30. Shagin DA, Barsova EV, Yanushevich YG, Fradkov AF, Lukyanov KA, Labas YA, Semenova TN, Ugalde JA, Meyers A, Nunez JM, Widder EA, Lukyanov SA, Matz MV (2004) GFP-like proteins as ubiquitous metazoan superfamily: evolution of functional features and structural complexity. Mol Biol Evol 21(5):841–850
31. Karasawa S, Araki T, Nagai T, Mizuno H, Miyawaki A (2004) Cyan-emitting and orange-emitting fluorescent proteins as a donor/acceptor pair for fluorescence resonance energy transfer. Biochem J 381:307–312

32. Shaner N, Campbell RE, Steinbach PA, Giepmans BNG, Palmer AE, Tsien RY (2004) Improved monomeric red, orange and yellow fluorescent proteins derived from *Discosoma* sp red fluorescent protein. Nat Biotechnol 22:1567–1572
33. Gurskaya NG, Verkhusha VV, Shcheglov AS, Staroverov DB, Chepurnykh TV, Fradkov AF, Lukyanov S, Lukyanov KA (2006) Engineering of a monomeric green-to-red photoactivatable fluorescent protein induced by blue light. Nat Biotechnol 24(4):461–465
34. Nienhaus K, Nienhaus GU, Wiedenmann J, Nar H (2005) Structural basis for photo-induced protein cleavage and green-to-red conversion of fluorescent protein EosFP. Proc Natl Acad Sci USA 102(26):9156–9159
35. Mizuno H, Mal TK, Tong KI, Ando R, Furuta T, Ikura M, Miyawaki A (2003) Photo-induced peptide cleavage in the green-to-red conversion of a fluorescent protein. Mol Cell 12:1051–1058
36. Lukyanov KA, Fradkov AF, Gurskaya NG, Matz MV, Labas YA, Savitsky AP, Markelov ML, Zaraisky AG, Zhao X, Fang Y, Tan W, Lukyanov SA (2000) Natural animal coloration can be determined by a nonfluorescent green fluorescent protein homolog. J Biol Chem 275(34):25879–25882
37. Campbell RE, Tour O, Palmer AE, Steinbach PA, Baird GS, Zacharias DA, Tsien RY (2002) A monomeric red fluorescent protein. Proc Natl Acad Sci USA 99:7877–7882
38. Wiedenmann J, Schenk A, Rcker C, Girod A, Spindler K-D, Nienhaus GU (2002) A far-red fluorescent protein with fast maturation and reduced oligomerization tendency from entacmaea quadricolor (anthozoa, actinaria). Proc Natl Acad Sci USA 99(18):11646–11651
39. Prescott M, Ling M, Beddoe T, Oakley AJ, Dove S, Hoegh-Guldberg O, Devenish RJ, Rossjohn J (2003) The 2.2 Å crystal structure of a pocilloporin pigment reveals a nonplanar chromophore conformation. Structure 22:275–284
40. Wang L, Jackson WC, Steinbach PA, Tsien RY (2004) Evolution of new nonantibody proteins via iterative somatic hypermutation. Proc Natl Acad Sci USA 101(48):16745–16749
41. Lin MZ, McKeown MR, Ng H-L, Aguilera TA, Shaner NC, Campbell RE, Adams SR, Gross LA, Ma W, Alber T, Tsien RY (2009) Autofluorescent proteins with excitation in the optical window for intravital imaging in mammals. Chem Biol 16(11):1169–1179
42. Chan MCY, Karasawa S, Mizuno H, Bosanac I, Ho D, Priv GG, Miyawaki A, Ikura M (2006) Structural characterization of a blue chromoprotein and its yellow mutant from the sea anemone *Cnidopus japonicus*. J Biol Chem 281(49):37813–37819
43. Gross LA, Baird GS, Hoffman RC, Baldridge KK, Tsien RY (2000) The structure of the chromophore within dsred, a red fluorescent protein from coral. Proc Natl Acad Sci USA 97:11990–11995
44. Petersen J, Wilmann PG, Beddoe T, Oakley AJ, De-venish RJ, Prescott M, Rossjohn J (2003) The 2.0-angstrom crystal structure of eqFP611, a far red fluorescent protein from the sea anemone *Entacmaea quadricolor*. J Biol Chem 287:44626–44631
45. Shu X, Shaner NC, Yarbrough CA, Tsien RY, Remington SJ (2006) Novel chromophores and buried charges control color in mFruits. Biochemistry 45(32):9639–9647
46. Kikuchi A, Fukumura E, Karasawa S, Mizuno H, Miyawaki A, Shiro Y (2008) Structural characterization of a thiazoline-containing chromophore in an orange fluorescent protein, monomeric Kusabira Orange. Biochemistry 47(44):11573–11580
47. Remington SJ, Wachter RM, Yarbrough DK, Branchaud B, Anderson DC, Kallio K, Lukyanov KA (2005) zFP538, a yellow-fluorescent protein from *Zoanthus*, contains a novel three-ring chromophore. Biochemistry 44(1):202–212
48. Quillin ML, Anstrom DM, Shu X, O'Leary S, Kallio K, Chudakov DM, Remington SJ (2005) Kindling fluorescent protein from *Anemonia sulcata*: dark-state structure at 1.38 Å resolution. Biochemistry 44(15):5774–5787
49. Tretyakova YA, Pakhomov AA, Martynov VI (2007) Chromophore structure of the kindling fluorescent protein asFP595 from *Anemonia sulcata*. J Am Chem Soc 129:7748–7749
50. Yampolsky IV, Remington SJ, Martynov VI, Potapov VK, Lukyanov S, Lukyanov KA (2005) Synthesis and properties of the chromophore of the asFP595 chromoprotein from *Anemonia sulcata*. Biochemistry 44(15):5788–5793

51. Schäfer LV, Groenhof G, Klingen AR, Ullmann GM, Boggio-Pasqua M, Robb MA, Grubmüller H (2007) Photoswitching of the fluorescent protein asFP595: mechanism, proton pathways, and absorption spectra. Angew Chem 119:536–542
52. Weber W, Helms V, McCammon JA, Langhoff PW (1999) Shedding light on the dark and weakly fluorescent states of green fluorescent proteins. Proc Natl Acad Sci USA 96:6177–6182
53. He X, Bell AF, Tonge PJ (2002) Isotopic labeling and normal-mode analysis of a model green fluorescent protein chromophore. J Phys Chem B 106:6056–6066
54. Niwa H, Inouye S, Hirano T, Matsuno T, Kojima S, Kubota M, Ohashi M, Tsuji FI (1996) Chemical nature of the light emitter of the *Aequorea* green fluorescent protein. Proc Nat Acad Sci USA 93:13617–13622
55. Voliani V, Bizzarri R, Nifosì R, Abbruzzetti S, Grandi E, Viappiani C, Beltram F (2008) Cis–trans photoisomerization of fluorescent-protein chromophores. J Phys Chem B 112(34):10714–10722
56. Nielsen SB, Lapierre A, Andersen JU, Pedersen UV, Tomita S, Andersen LH (2001) Absorption spectrum of the green fluorescent protein chromophore anion in vacuo. Phys Rev Lett 87:228102
57. Dong J, Solntsev KM, Tolbert LM (2006) Solvatochromism of the green fluorescence protein chromophore and its derivatives. J Am Chem Soc 128(37):12038–12039
58. Yampolsky IV, Balashova TA, Lukyanov KA (2009) Synthesis and spectral and chemical properties of the yellow fluorescent protein zFP538 chromophore. Biochemistry 48: 8077–8082
59. Boy S, Krogh H, Nielsen IB, Nielsen SB, Pedersen SU, Pedersen UV, Andersen LH, Bell AF, He X, Tonge PJ (2003) Vibrationally resolved photoabsorption spectroscopy of red fluorescent protein chromophore anions. Phys Rev Lett 90(11):118103
60. Yampolsky IV, Kislukhin AA, Amatov TT, Shcherbo D, Potapov VK, Lukyanov S, Lukyanov KA (2008) Synthesis and properties of the red chromophore of the green-to-red photoconvertible fluorescent protein Kaede and its analogs. Bioorg Chem 36(2):96–104
61. Wachter RM, King BA, Heim R, Kallio K, Tsien RY, Boxer SG, Remington SJ (1997) Crystal structure and photodynamic behavior of the blue emission variant Y66H/Y145F of green fluorescent protein. Biochemistry 36:9759–9765
62. Forbes MW, Jockusch RA (2009) Deactivation pathways of an isolated green fluorescent protein model chromophore studied by electronic action spectroscopy. J Am Chem Soc 131 (47):17038–17039
63. Boyé S, Nielsen SB, Krogh H, Nielsen IB, Pedersen UV, Bell AF, He X, Tonge PJ, Andersen LH (2003) Gas-phase absorption properties of DsRed model chromophores. Phys Chem Chem Phys 5(14):3021–3026
64. Andersen L, Lapierre A, Nielsen S, Nielsen I, Pedersen S, Pedersen U, Tomita S (2002) Chromophores of the green fluorescent protein studied in the gas phase. Eur Phys J D 20:597–600
65. Lammich L, Petersen MA, Nielsen MB, Andersen LH (2007) The gas-phase absorption spectrum of a neutral GFP model chromophore. Biophys J 92(1):201–207
66. Filippi C, Zaccheddu M, Buda F (2009) Absorption spectrum of the green fluorescent protein chromophore: a difficult case for ab initio methods? J Chem Theory Comput 5(8):2074–2087
67. Drobizhev M, Tillo S, Makarov NS, Hughes TE, Rebane A (2009) Absolute two-photon absorption spectra and two-photon brightness of orange and red fluorescent proteins. J Phys Chem B 113(4):855–859
68. Mandal D, Tahara T, Meech SR (2004) Excited-state dynamics in the green fluorescent protein chromophore. J Phys Chem B 108(3):1102–1108
69. Voityuk AA, Michel-Beyerle M-E, Rosch N (1997) Protonation effects on the chromophore of green fluorescent protein. quantum chemical study of the absorption spectrum. Chem Phys Lett 272:162–167

70. Martin M, Negri F, Olivucci M (2004) Origin, nature, and fate of the fluorescent state of the green fluorescent protein chromophore at the CASPT2//CASSCF resolution. J Am Chem Soc 126(17):5452–5464
71. Nifosì R, Amat P, Tozzini V (2007) Variation of spectral, structural and vibrational properties within the intrinsically fluorescent proteins family: a density functional study. J Comput Chem 28:2366–2377
72. Yan W, Zhang L, Xie D, Zeng J (2007) Electronic excitations of green fluorescent proteins: modeling solvatochromatic shifts of red fluorescent protein chromophore model compound in aqueous solutions. J Phys Chem B 111(50):14055–14063
73. Olsen S, Smith SC (2008) Bond selection in the photoisomerization reaction of anionic green fluorescent protein and kindling fluorescent protein chromophore models. J Am Chem Soc 130(27):8677–8689
74. Epifanovsky E, Polyakov I, Grigorenko B, Nemukhin A, Krylov AI (2009) Quantum chemical benchmark studies of the electronic properties of the green fluorescent protein chromophore. 1. Electronically excited and ionized states of the anionic chromophore in the gas phase. J Chem Theory Comput 5(7):1895–1906
75. Ma Y, Rohlfing M, Molteni C (2010) Modeling the excited states of biological chromophores within many-body green's function theory. J Chem Theory Comput 6(1):257–265
76. Kowalski K, Krishnamoorthy S, Villa O, Hammond JR, Govind N (2010) Active-space completely-renormalized equation-of-motion coupled-cluster formalism: excited-state studies of green fluorescent protein, free-base porphyrin, and oligoporphyrin dimer. J Chem Phys 132(15):154103
77. Xie D, Zeng J (2005) Electronic excitations of green fluorescent proteins: protonation states of chromophore model compound in solutions. J Comput Chem 26(14):1487–1496
78. Wan S, Liu S, Zhao G, Chen M, Han K, Sun M (2007) Photoabsorption of green and red fluorescent protein chromophore anions *in vacuo*. Biophys Chem 129(2–3):218–223
79. Olsen S, Smith SC (2007) Radiationless decay of red fluorescent protein chromophore models via twisted intramolecular charge-transfer states. J Am Chem Soc 129(7): 2054–2065
80. Sanchez-Garcia E, Doerr M, Thiel W (2010) QM/MM study of the absorption spectra of DsRed.m1 chromophores. J Comput Chem 31(8):1603–1612
81. Toniolo A, Granucci G, Martinez TJ (2003) Conical intersections in solution: a QM/MM study using floating occupation semiempirical configuration interaction wave functions. J Phys Chem A 107:3822
82. Drobizhev M, Tillo S, Makarov NS, Hughes TE, Rebane A (2009) Color hues in red fluorescent proteins are due to internal quadratic stark effect. J Phys Chem B 113(39):12860–12864
83. Hasegawa J-Y, Ise T, Fujimoto KJ, Kikuchi A, Fukumura E, Miyawaki A, Shiro Y (2010) Excited states of fluorescent proteins, mKO and DsRed: chromophore–protein electrostatic interaction behind the color variations. J Phys Chem B 114(8):2971–2979
84. He X, Bell AF, Tonge PJ (2002) Synthesis and spectroscopic studies of model red fluorescent protein chromophores. Org Lett 4(9):1523–1526
85. Laino T, Nifosì R, Tozzini V (2004) Relationship between structure and optical properties in green fluorescent proteins: an *ab initio* study of the active site. Chem Phys 298:17–28
86. Malo GD, Wang M, Wu D, Stelling AL, Tonge PJ, Wachter RM (2008) Crystal structure and Raman studies of dsFP483, a cyan fluorescent protein from *Discosoma striata*. J Mol Biol 378(4):871–886
87. Henderson JN, Remington SJ (2005) Crystal structures and mutational analysis of amFP486, a cyan fluorescent protein from *Anemonia majano*. Proc Natl Acad Sci USA 102(36):12712–12717
88. Henderson JN, Ai H-W, Campbell RE, Remington SJ (2007) Structural basis for reversible photobleaching of a green fluorescent protein homologue. Proc Natl Acad Sci USA 104(16):6672–6677

89. Ai H-W, Henderson JN, Remington SJ, Campbell RE (2006) Directed evolution of a monomeric, bright and photostable version of *Clavularia* cyan fluorescent protein: structural characterization and applications in fluorescence imaging. Biochem J 400(3):531–540
90. Wood TI, Barondeau DP, Hitomi C, Kassmann CJ, Tainer JA, Getzoff ED (2005) Defining the role of arginine 96 in green fluorescent protein fluorophore biosynthesis. Biochemistry 44(49):16211–16220
91. Ai H-W, Campbell RE (2008) Teal fluorescent proteins: characterization of a reversibly photoswitchable variant. In: Proceedings of the SPIE – The International Society for Optical Engineering, pp 68680D1–7
92. Nakano H, Okumura R, Goto C, Yamane T (2002) In vitro combinatorial mutagenesis of the 65thand 222nd positions of the green fluorescent protein of *Aequorea victoria*. Biotechnol Bioprocess Eng 7:311–315
93. Delagrave S, Hawtin RE, Silva CM, Yang MM, Youvan DC (1995) Red-shifted excitation mutants of the green fluorescent protein. Biotechnology (N Y) 13(2):151–154
94. Labas Y, Gurskaya NG, Yanushevich YG, Fradkov AF, Lukyanov KA, Lukyanov SA, Matz MV (2002) Diversity and evolution of the green fluorescent protein family. Proc Natl Acad Sci USA 99:4256–4261
95. Nienhaus K, Renzi F, Vallone B, Wiedenmann J, Nienhaus GU (2006) Exploring chromophore-protein interactions in fluorescent protein cmFP512 from *Cerianthus membranaceus*: X-ray structure analysis and optical spectroscopy. Biochemistry 45(43):12942–12953
96. Suto K, Masuda H, Takenaka Y, Tsuji FI, Mizuno H (2009) Structural basis for red-shifted emission of a GFP-like protein from the marine copepod *Chiridius poppei*. Genes Cells 14(6):727–737
97. Kummer AD, Wiehler J, Rehaber H, Kompa C, Steipe B, Michel-Beyerle ME (2000) Effects of threonine 203 replacements on excited-state dynamics and fluorescence properties of the green fluorescent protein (GFP). J Phys Chem B 104:4791–4798
98. Nifosì R, Ferrari A, Arcangeli C, Tozzini V, Pellegrini V, Beltram F (2003) Photoreversible dark state in a tristable green fluorescent protein variant. J Phys Chem B 107:1679–1684
99. Stiel AC, Trowitzsch S, Weber G, Andresen M, Eggeling C, Hell SW, Jakobs S, Wahl MC (2007) 1.8 a bright-state structure of the reversibly switchable fluorescent protein Dronpa guides the generation of fast switching variants. Biochem J 402(1):35–42
100. Shinobu A, Palm GJ, Schierbeek AJ, Agmon N (2010) Visualizing proton antenna in a high-resolution green fluorescent protein structure. J Am Chem Soc 132:11093–11102
101. Habuchi S, Dedecker P, Ichi Hotta J, Flors C, Ando R, Mizuno H, Miyawaki A, Hofkens J (2006) Photo-induced protonation/deprotonation in the GFP-like fluorescent protein Dronpa: mechanism responsible for the reversible photoswitching. Photochem Photobiol Sci 5(6):567–576
102. Shu X, Kallio K, Shi X, Abbyad P, Kanchanawong P, Childs W, Boxer SG, Remington SJ (2007) Ultrafast excited-state dynamics in the green fluorescent protein variant S65T/H148D. 1. Mutagenesis and structural studies. Biochemistry 46(43):12005–12013
103. Baird GS, Zacharias DA, Tsien RY (2000) Biochemistry, mutagenesis, and oligomerization of DsRed, a red fluorescent protein from coral. Proc Natl Acad Sci USA 97:11984–11989
104. Shcherbo D, Murphy CS, Ermakova GV, Solovieva EA, Chepurnykh TV, Shcheglov AS, Verkhusha VV, Pletnev VZ, Hazelwood KL, Roche PM, Lukyanov S, Zaraisky AG, Davidson MW, Chudakov DM (2009) Far-red fluorescent tags for protein imaging in living tissues. Biochem J 418(3):567–574
105. Tubbs JL, Tainer JA, Getzoff ED (2005) Crystallographic structures of Discosoma red fluorescent protein with immature and mature chromophores: linking peptide bond trans–cis isomerization and acylimine formation in chromophore maturation. Biochemistry 44(29):9833–9840
106. Shu X, Wang L, Colip L, Kallio K, Remington SJ (2009) Unique interactions between the chromophore and glutamate 16 lead to far-red emission in a red fluorescent protein. Protein Sci 18(2):460–466

107. Shcherbo D, Merzlyak EM, Chepurnykh TV, Fradkov AF, Ermakova GV, Solovieva EA, Lukyanov KA, Bogdanova EA, Zaraisky AG, Lukyanov S, Chudakov DM (2007) Bright far-red fluorescent protein for whole-body imaging. Nat Methods 4(9):741–746
108. Blab GA, Lommerse PHM, Cognet L, Harms GS, Schmidt T (2001) Two-photon excitation action cross-sections of the autofluorescent proteins. Chem Phys Lett 350:71–77
109. Heikal AA, Hess ST, Webb WW (2001) Multiphoton molecular spectroscopy and excited-state dynamics of enhanced green fluorescent protein (EGFP): acid–base specificity. Chem Phys 274:37–55
110. Marchant JS, Stutzmann GE, Leissring MA, LaFerla FM, Parker I (2001) Multiphoton-evoked color change of DsRed as an optical highlighter for cellular and subcellular labeling. Nat Biotechnol 19:645–649
111. Tillo SE, Hughes TE, Makarov NS, Rebane A, Drobizhev M (2010) A new approach to dual-color two-photon microscopy with fluorescent proteins. BMC Biotechnol 10:6
112. Drobizhev M, Makarov NS, Hughes T, Rebane A (2007) Resonance enhancement of two-photon absorption in fluorescent proteins. J Phys Chem B 111(50):14051–14054
113. Nifosì R, Luo Y (2007) Predictions of novel two-photon absorption bands in fluorescent proteins. J Phys Chem B 111(50):14043–14050
114. Hosoi H, Yamaguchi S, Mizuno H, Miyawaki A, Tahara T (2008) Hidden electronic excited state of enhanced green fluorescent protein. J Phys Chem B 112(10):2761–2763
115. Cubitt AB, Woollenweber LA, Heim R (1999) Understanding structure-function relationships in the Aequorea Victoria Green Fluorescent Protein Method Cell Biol 58:19–30
116. Morise H, Shimomura O, Johnson FH, Winant J (1974) Intermolecular energy transfer in the bioluminescent system of Aequorea Biochemistry. American Chemical Society 13: 2656–2662
117. Chudakov DM, Belousov VV, Zaraisky AG, Novoselov VV, Staroverov DB, Zorov DB, Lukyanov S, Lukyanov KA (2003) Kindling fluorescent proteins for precise in vivo photolabeling. Nat Biotechnol 21:191–194
118. Yang TT, Sinai P, Green G, Kitts PA, Chen YT, Lybarger L, Chervenak R, Patterson GH, Piston DW, Kain SR (1998) Improved fluorescence and dual color detection with enhanced blue and green variants of the green fluorescent protein. J Biol Chem 273:8212–8216

Primary Photophysical Processes in Chromoproteins

Stephen R. Meech

Abstract In this chapter, the diverse range of photophysical phenomena exhibited by chromoproteins is reviewed. Experimental and theoretical studies of both the electronic spectra and the ultrafast radiationless decay of the chromophore of green fluorescent protein (GFP) in solution are described as a function of solvent, temperature and substituent. The relevance of these observations to photophysical phenomena observed in chromoproteins which undergo photoconversion is discussed. Next, the excited state proton transfer found in GFP is described. Its potential to probe the dynamics of proton-transfer reactions in proteins is illustrated. Finally, the photophysics underlying the phenomenon of photoswitching in chromoproteins is discussed.

Keywords Fluorescence · Fluorescent protein · GFP · Isomerization · Photoswitching · Proton transfer · Ultrafast

Contents

S.R. Meech
School of Chemistry, University of East Anglia, Norwich NR4 7TJ, UK
e-mail: S.Meech@uea.ac.uk

G. Jung (ed.), *Fluorescent Proteins I*, Springer Ser Fluoresc (2012) 11: 41–68,
DOI 10.1007/4243_2011_19, © Springer-Verlag Berlin Heidelberg 2011,
Published online: 31 March 2011

1 Introduction

The green fluorescent protein (GFP) was first discovered and isolated from the jellyfish *Aequorea victoria* [1, 2]. The subsequent sequencing, cloning and expression of GFP revealed its remarkable utility in fluorescence imaging [3], a discovery which has since lead to it becoming one of the major tools of cell biology [4–7]. Since the discovery of wild-type GFP (avGFP), a substantial family of homologous proteins has been discovered and characterised [8]. This family has in common the structural features of an 11 stranded β-barrel through which is threaded an α-helix. This α-helix contains the chromophore, which is formed after protein folding by cyclisation and oxidation reactions involving three adjacent amino acid residues – a serine, a tyrosine and a glycine. The mechanism of chromophore formation has been extensively reviewed [9–11]. Crucially, the final result is a chromophore with an extended pi electron structure which is covalently bound to the protein backbone, absorbs in the blue-green region of the spectrum and, in many cases, emits with a high fluorescence quantum yield at green-red wavelengths. The chemical structure of the chromophore was identified as *p*-hydroxybenzilideneimidazolinone (HBDI, Fig. 1) [12, 13]. The high quantum yield, irreversible chromophore binding and the ease with which GFP can be fused to a desired target protein make it (and its mutants) an ideal protein-specific fluorescence label for live cell imaging [6, 14].

The structure of the folded protein and the basic structure of the chromophore are largely conserved throughout the ever widening GFP family. Despite this fundamental similarity, fluorescent proteins display an enormous range of photophysical properties, which greatly extends their utility in imaging [15, 16]. avGFP itself exhibits two bands arising from the protonated and deprotonated form of the phenolic hydroxyl group, and their relative proportion can be modified by both pH and mutagenesis [17–19]. Excitation of the neutral form of avGFP results in an excited state proton-transfer (ESPT) reaction, which is unique in biology [20]. The energy of the electronic absorption and emission bands can be modified over a wide range through mutagenesis [6]. Changes to the three amino acids that make up the chromophore may lead to large spectral shifts, while more subtle but still substantial shifts can be achieved through mutations in the surrounding residues [21]. The variety of colours that result is critical in bioimaging applications. Multiple emission frequencies permit multicolour imaging allowing more than one protein to be labelled in any given cell. When the energy levels of two GFP mutants are properly aligned, fluorescence resonance energy transfer (FRET) may be observed between

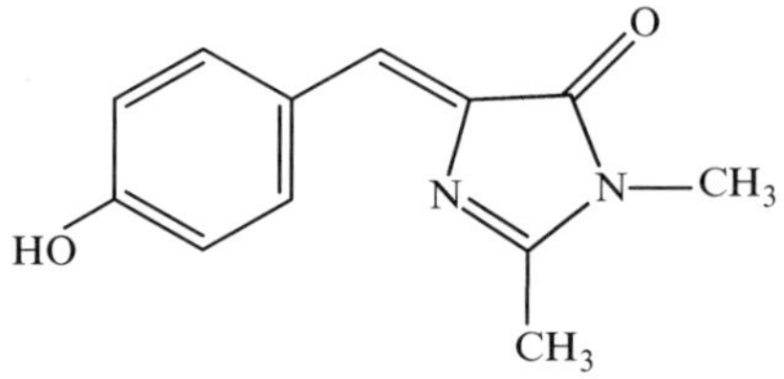

Fig. 1 Structure of the avGFP chromophore HBDI

them. Since FRET is strongly distance dependent on the nanometer scale, it is widely used to observe protein–protein interactions [14].

Thus, pH sensitivity and ESPT are photophysical phenomena exhibited by the original avGFP, to which may be added spectral tuning and FRET when GFP mutants are considered. However, the chromoproteins (CPs) now being discovered in a variety of marine organisms exhibit a still wider range of photophysical phenomena, all of which can in turn be modified by mutagenesis. Many CPs have a spectrum which is red-shifted compared to avGFP, because the original chromophore has undergone a further post-translational reaction leading to an extension of the pi delocalisation [10]. Some of these new CPs absorb strongly in the visible region but exhibit such fast radiationless decay as to be considered effectively non-fluorescent (hence CP is a preferred designation to FP; it is worth noting that CP has also been applied to coloured proteins which bind their chromophore non-covalently, it is used here as it is firmly embedded in the literature and little confusion arises, even though autochromic protein might be the more appropriate nomenclature) [22]. The mechanism is fast internal conversion (IC) and may be associated with non-planar forms of the chromophore [23]. In some CPs, the chromophore exists in either *trans* or *cis* forms, and under irradiation isomerisation between them may be observed [24–26]. Other CPs have been shown to exhibit photochemistry leading to large spectral shifts and changes in quantum yield, a phenomenon that can be used in 'optical highlighting' a specific protein population in vivo [27–29]. Finally some proteins can be photochemically and reversibly switched between emissive and non-emissive states, a phenomenon that forms the basis of a novel sub-diffraction-limited microscopy method currently yielding unprecedented 3D spatial resolution in bioimaging [30–32].

Thus, CPs add isomerisation, photochemistry and reversible photochromism to the photophysical phenomena already noted in avGFP. These new phenomena have already stimulated a number of new applications in life sciences. In this chapter, the primary photophysics of CPs will be reviewed, with a focus on excited state dynamics. First, the main time-resolved spectroscopic methods will be introduced. Next, the medium and substituent sensitivity of the GFP chromophore (HBDI), the mechanism of its fast radiationless decay and the relevance of this to photo-induced *cis–trans* isomerisation will be addressed. After that the ESPT reaction in avGFP and some of its mutants will be described. Finally, the mechanism of isomerisation and photochemistry in CPs will be discussed in relation to the CP photochromism described elsewhere in this volume.

2 Experimental Methods

2.1 Ultrafast Fluorescence Up-Conversion

Time-resolved fluorescence has been critical to understanding the photophysics of fluorescent proteins and is central to their application in fluorescence lifetime imaging. These aspects are covered elsewhere in this volume, particularly in the chapter by Jung. However, ultrafast fluorescence techniques (i.e., for decay processes faster than 100 ps) have proven especially useful in understanding the unusually rapid radiationless decay of the GFP chromophore (Sect. 3) and the kinetics of the unique ESPT reaction in avGFP (Sect. 4). Thus, it seems pertinent to outline the principles of the up-conversion method.

The starting point is a Kerr lens mode-locked titanium sapphire laser. These compact commercially available devices provide a train of ultrafast pulses (typically at a repetition rate of 70–100 MHz) with excellent stability [33]. The apparatus in our laboratory uses a Coherent Micra 10 source, which provides (following pulse compression, see below) pulses as short as 18 fs. The central wavelength is in the region 760–860 nm, the output power about 1 W and the bandwidth is typically 100 nm. Such a broad bandwidth is essential to support very short pulses, a consequence of the time–energy uncertainty relation.

The measurement of such short pulses is non-trivial, but well-established methods are available [34]. The most common is background-free autocorrelation. In this method, the output of the laser is split into two identical beams at a beam splitter. One beam is routed through a fixed delay, while the second is routed through a delay of variable path length. Typically a mirror is mounted on a delay stage which has a positional accuracy of better than 1 μm allowing control of the delay time between the two pulses with an accuracy of better than 6.7 fs. The two beams are made parallel and focussed to overlap spatially in a thin non-linear crystal, which generates second harmonic radiation (i.e., for an 800 nm centred input beam the output is at 400 nm, generated from two 800 nm photons). The 400 nm output from the crystal appears in three beams, the second harmonic of the two individual input beams plus a signal generated from both beam (each contributing one photon). These three signals emerge spatially separated due to the different phase matching conditions [34, 35]. Clearly, the two beam signal can only be generated when both pulses are overlapped in the crystal in time as well as in space. Thus, a measurement of the second harmonic intensity while scanning one pulse in time with the other is fixed yields a convolution of the temporal profile of one pulse with the other, the pulse autocorrelation. The autocorrelation serves as the instrument response function for deconvolution of the ultrafast fluorescence up-conversion traces (see below). However, the actual width of the laser pulse can be recovered from the autocorrelation if a pulse shape is assumed. Typically, a sech^2 profile is assumed for mode-locked titanium sapphire lasers, in which case the pulsewidth is 0.65 times the autocorrelation width.

Although the intensity autocorrelation is a simple and effective means of characterising laser pulses, additional information may be used to better characterise the pulse. At the simplest level the laser spectrum can also be measured and the time bandwidth product (pulsewidth times spectral width in frequency units, $\Delta\nu\ .\ \Delta\tau$) determined; for a sech^2 pulse this should take the value 0.315. Greatly superior methods of ultrafast pulse analysis are now available, which simultaneously characterise time structure, phase, spatial profile and amplitude of the laser pulse. These methods are described in more detail elsewhere [34].

The large spectral width required to support very short laser pulses gives rise to some problems. The refractive index of most optical materials is wavelength dependent (called the material's dispersion) with the consequence that the velocity of light in the medium is wavelength dependent. Specifically in the case of normal dispersion, blue light travels slower than red light, giving rise to positive group velocity dispersion (GVD) [33, 34]. Consequently, on transmission through any optical element, the spectrally broad pulse becomes temporally broader. Thus pulses emitted from the laser are typically temporally much broader than their bandwidth would suggest due to dispersion of intracavity elements and have to be compressed. Pulse compression is achieved by routing the plane-polarised pulse through a pair of prisms set at Brewster's angle (to minimise reflection losses) and retroreflecting the pulse back through the same prism pair. The dispersion of the spectrally broad beam and the different path lengths each wavelength follows in the (dispersive) glass of the prism contribute both positive and negative GVD. By controlling the separation and position of the prisms, it is possible to compensate the positive GVD of the cavity with negative GVD in the prism pair to yield the shortest pulse. Indeed negative GVD can be introduced to 'pre-compensate' positive GVD in optical elements, which the pulse must go through before reaching the sample.

For the up-conversion measurement of the fluorescence decay time, the compressed pulse is first routed to a thin non-linear crystal to generate the second harmonic [36, 37]. The ca. 800 nm fundamental and 400 nm second harmonic beams are separated at a dichroic mirror. The 400 nm beam is reflected by a series of mirrors and focused into the sample by a spherical mirror. Since the 400 nm beam is broadened by transmission through any optics it must be recompressed, since the shortest pulses are required *at the sample*. In our experiment, the bulky prism compressor is replaced by a compact dispersive mirror pair [38]. The 400 nm excitation pulse generates fluorescence from the sample, which is collected by a reflective Cassegranian microscope objective, which efficiently collects the emitted fluorescence without introducing a dispersive element (a lens) which would degrade time resolution (in a wavelength dependent manner). The objective focuses the light collected onto a non-linear crystal for sum-frequency generation (SFG). The residual 800 nm pulse is routed through a variable optical delay and passes through a half wave plate that sets the polarisation to the magic angle with respect to the excitation polarisation, to eliminate the contribution of orientational relaxation to the measured fluorescence kinetics. The beam then passes through a pair of dispersive mirrors to recompress the pulse before being focused into the non-linear SFG crystal.

In the SFG crystal, the train of fluorescence decays at wavelength λ_F (appearing at the laser repetition rate) is spatially overlapped with the train of short 800 nm pulses at λ_P. The SFG crystal is oriented to generate the sum frequency at a wavelength λ_{SFG}:

$$\frac{1}{\lambda_{SFG}} = \frac{1}{\lambda_F} + \frac{1}{\lambda_P}.$$

Since the SFG is generated by one fluorescence photon and one pulse photon, it can only be generated when both are present, and its intensity thus reflects the instantaneous intensity of the fluorescence (Fig. 2a) during the 800 nm pulse. Essentially the SFG samples the fluorescence intensity at the arrival time of the 800 nm pulse. By scanning the time delay between pump and probe, the SFG intensity maps out the fluorescence decay (Fig. 2b).

The SFG signal is detected in the UV region by a low noise photon counting photomultipler placed behind a monochromator, which selects the fluorescence wavelength (which is also optimised by angle tuning the SFG crystal). The time resolution is recorded by up-converting the instantaneous Raman radiation scattered

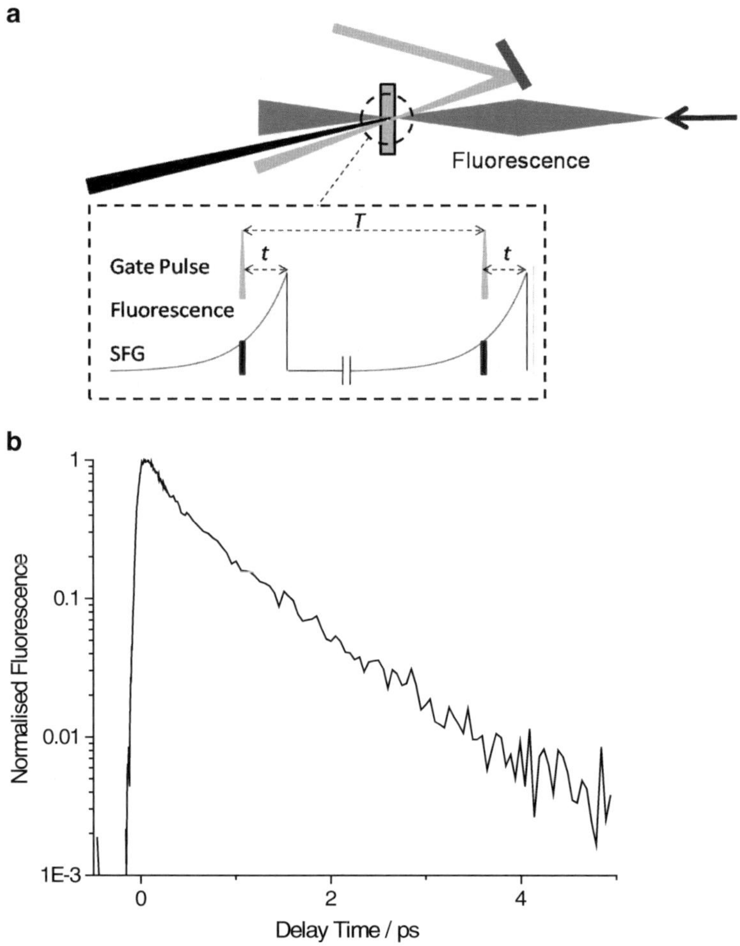

Fig. 2 (**a**) Schematic of the SFG process used for time resolution in up-conversion. (**b**) An example of an ultrafast fluorescence decay profile (anionic HBDI in ethanol) plotted on a log scale (excitation was at 400 nm with a sub 50 fs pulse and emission was detected at the peak of the fluorescence)

from the solvent. In the present case, the time resolution can be as short as 40 fs, although this time resolution is degraded by the need to use optical filters to block out scattered excitation radiation.

The up-conversion method is the best established and highest time resolution method for time-resolved fluorescence. It can, however, be time consuming, especially when fluorescence is detected at many wavelengths to generate a time-dependent fluorescence spectrum. Alternative broadband detection methods are available, including broadband up-conversion [39], Kerr gate detection [40–43] and streak camera measurements [44].

2.2 Transient Absorption

There are many different forms of the transient absorption experiment, however, in its essence it is a two pulse experiment [36, 45]. The first pulse excites the sample (the pump), while the second time delayed pulse (the probe) monitors the sample absorption as a function of time. In the simplest case, the pump excites a molecule from its ground state to an excited state. If the probe wavelength is at the same wavelength as the pump (and therefore the ground state absorption of the sample), its transmittance is increased after the pump and will decrease as a function of delay time as (or if) the ground state is repopulated. If the probe wavelength is set at a wavelength at which the newly formed excited state absorbs, then the transmittance will be decreased instantaneously by the pump pulse and will increase as the excited state decays. If the wavelength of the probe is set to a wavelength where the excited state emits fluorescence, then the probe pulse can stimulate emission from the excited state. This acts as a gain mechanism for the transmitted probe, so the apparent probe transmittance increases as the excited state is created and decreases as it decays. Of course in many important cases, including the CPs considered here, the excited state goes on to generate new species (e.g., by proton transfer or isomerisation), and the appearance and decay kinetics of these states can also be monitored. In general, these processes overlap one another in both time and wavelength requiring sophisticated analysis methods.

The laser apparatus used for transient absorption differs in some respects from that used in fluorescence, principally because to achieve a few per cent change in transmission, a moderately intense pump pulse is required (typically a 1 μJ pulse is focused into a 10^{-4} cm^2 cross-sectional area in the sample, compared to the nanojoule pulse energy used for background-free fluorescence up-conversion measurements). Thus, the preferred source is an amplified laser with a reduced repetition rate of a few kilohertz, to allow for sample recovery between pulses. In most cases, the change in absorbance as a result of the pump laser absorption should be kept small (a few per cent at most); hence, it is necessary to use difference methods normalised for the pump intensity and extensive signal averaging. Ideally this is done on a shot by shot basis with alternate pulses being used to record pumped and un-pumped sample transmission by the probe. With this methodology and a stable laser source, pump-induced changes in the sample optical density of 10^{-5} can be observed. In transient difference measurements, the increased transmission (or stimulated emission) appears as a negative signal (a 'bleach'), while induced absorption appears as a positive signal.

In some measurements, both pump and probe wavelengths are the same and the bleach and recovery of the ground state are observed. However, much more information is available from measurements at a range of wavelengths spanning the induced absorption and stimulated emission. To record such transient spectra, it is convenient to probe with an ultrafast broadband source of radiation. For intense laser sources, such broadband radiation can be obtained by white light continuum generation, in which an intense pulse of 800 nm radiation (for example) is focused into a CaF_2 or sapphire plate, which results in the generation of a short pulse of radiation with frequency components stretching between the UV and near IR through a third-order non-linear optical interaction [34]. This permits the simultaneous observation of the temporal evolution of a range of transients. Quite recently, it has proved possible to generate moderately broadband (approximately 500 cm^{-1}) radiation in the mid-IR spectral range. This permits the observation of photo-induced transients with both ultrafast time resolution and vibrational frequency resolution [46, 47]. Vibrational spectroscopy is in general more informative on and more sensitive to the molecular structure. An example showing the transient IR difference spectra of the HBDI chromophore is shown in Fig. 3. For either visible or IR probes, the entire 3D time–frequency–intensity surface can be simultaneously analysed by global analysis methods [48].

3 Photophysics in HBDI

Given its name, one initially surprising feature of the GFP chromophore (HBDI) is that it does not in fact fluoresce, or at least not very much; in room temperature aqueous solution, the quantum yield is on the order of 10^{-4}. This low quantum yield is also observed in the denatured protein and short lengths of peptide containing the chromophore [13]. These results are in sharp contrast to avGFP, where the quantum yield for emission is about 0.8 [6]; evidently the folded protein structure has a profound effect on the radiationless decay of the chromophore. An important challenge in GFP photophysics is to understand both the mechanism of the extremely rapid radiationless decay and the means by which the protein matrix can suppresses it so efficiently. Significantly a number of the more recently discovered CPs which contain the same basic chromophore as avGFP, and share essentially the same β-barrel structure, are non-fluorescent or only very weakly fluorescent [15, 49]. Thus, an understanding of radiationless decay in HBDI is central to understanding the mechanism of operation of photoswitchable CPs.

3.1 HBDI Spectroscopy

In Fig. 4, the absorption spectra of HBDI in the neutral and anionic states are compared with that of avGFP, which exists mainly as the neutral form but with

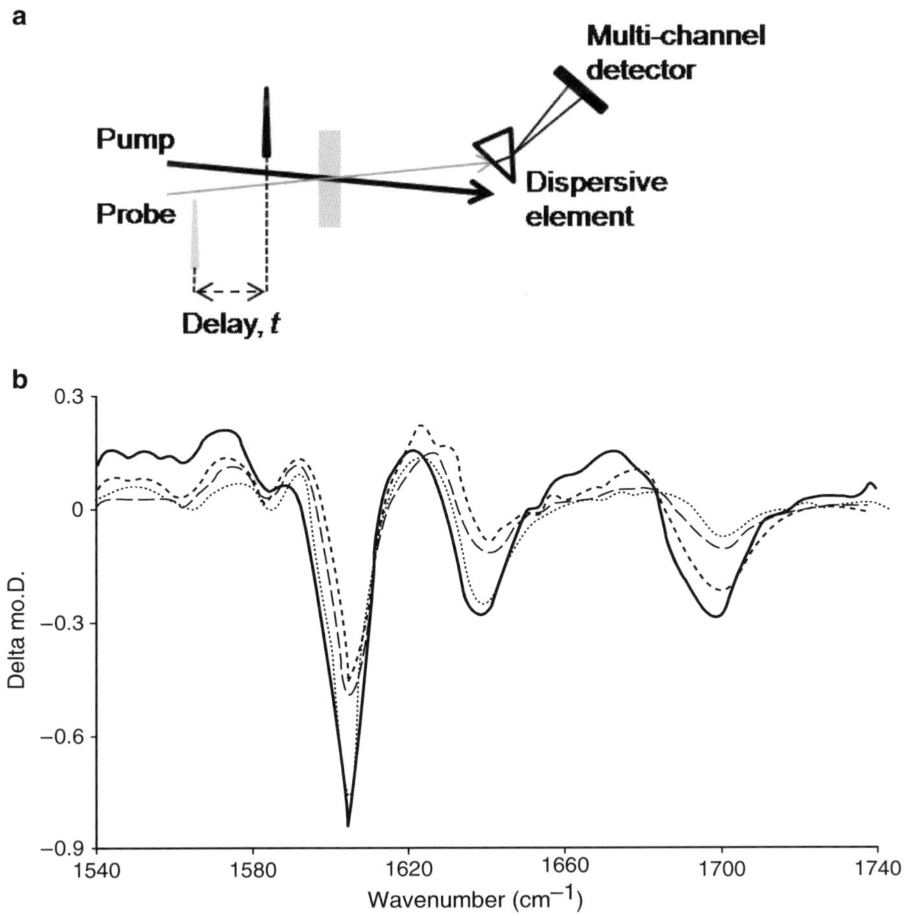

Fig. 3 (**a**) Simplified layout of pump-probe apparatus showing time delay between pump and multi-wavelength probe. (**b**) Transient IR transmission spectrum recorded for HBDI in DMSO, showing the bleach and recovery of the ground state accompanied by weak transient absorption. *Solid line* 2 ps delay, *short dash line* 4 ps, *long dash* 6 ps, *dot* 10 ps. The ground state recovery is fast, but included both excited state decay (see Sect. 3) and vibrational cooling in the ground state

a fraction in the anionic state (which can be formed photochemically [50, 51]). The neutral form absorbs to the blue of the anionic form, and in both cases the solution spectra are significantly blue-shifted from the corresponding protein spectra. This result points to a significant difference in electronic structure between the protein bound and the free forms of the chromophore. It has been shown that the peak absorption wavelength of GFP can be shifted significantly by mutations in the surrounding amino acid residues, suggesting a strong sensitivity to the environment. For example, a number of mutants with the S65T mutation, the yellow fluorescent proteins, show a large red-shift for the anionic form of the chromophore [52]. Intriguingly, the spectrum of HBDI was measured in the gas phase, where it

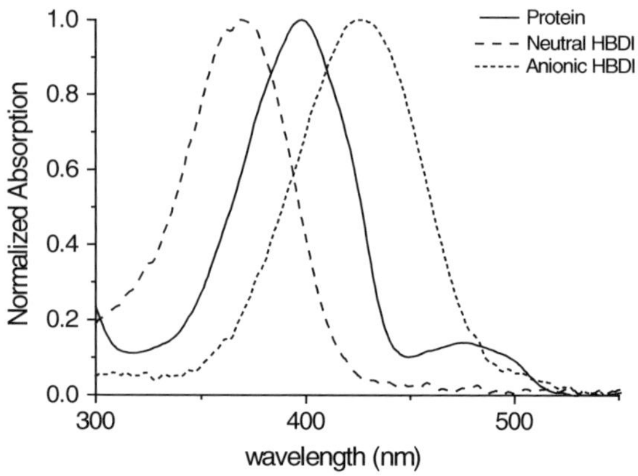

Fig. 4 Comparison of the spectrum of neutral and anionic HBDI in aqueous solution with the spectrum of avGFP

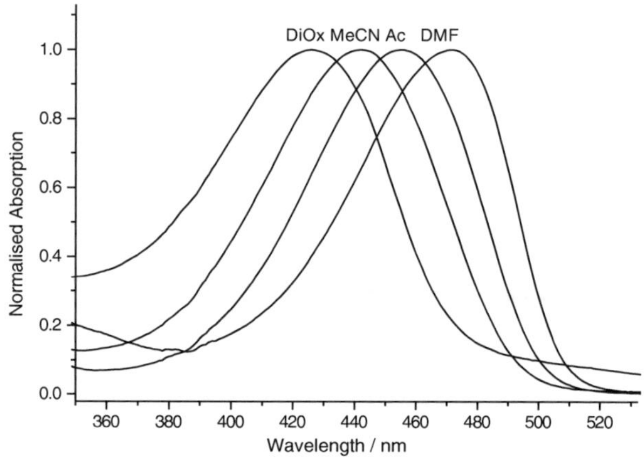

Fig. 5 Electronic absorption spectra of the HBDI anion in a range of non-H-bonding solvents

was reported that the spectrum matches the red-shifted protein absorption [53, 54]. One interpretation of this is that the protein provides a vacuum-like environment (although the medium must at least be polarisable). More recent measurements have suggested that the red-shifted transition in the gas phase might arise from a high multi-photon cross-section, and the one-photon absorption in fact peaks further to the blue, as seen in solution [55].

Solvatochromism in HBDI has been studied extensively, partly in an effort to reproduce the observed protein shift [56, 57]. The largest spectral shifts are observed for anionic HBDI, especially in polar non-H-bonding solvents such as DMSO and DMF (Fig. 5). The latter result is interesting because this represents the only solvent–charge combination for HBDI which comes close to reproducing the large red-shift seen in the protein environment. The significance of this result is unclear, as the solvent-induced shift is not accompanied by a significant increase in quantum yield. When the HBDI absorption spectra are measured in a range of solvents, no single solvent function (e.g., various polarity functions, π, E_T, etc. [30]) adequately fits all the data. However, in a detailed study, Dong et al. were able to obtain a good fit to the spectral shift data in a range of solvents using a combination of solvent acidity, basicity and polarity functions [57].

Recently, an attempt to reproduce the range of electronic transition energies found in CPs by tuning the electronic structure of HBDI with different substituents was reported [58]. Even very weak electron-donating substituents such as the methyl group were found to cause a dramatic shift in the spectrum of the HBDI anion when placed at the 3,5 positions on the phenyl ring (i.e., ortho to the OH group). The shifts for the neutral form are much smaller. The red-shift of the anion spectrum for the dimethyl derivative matches the protein shift between HBDI and avGFP even in the moderately polar solvent ethanol. Replacement of methyl with the somewhat more strongly electron-donating *t*-butyl substituent shifts the anion transition even further to the red, matching the transition energy seen in YFPs. These results suggest that even quite modest perturbations of the electronic structure of the chromophore are sufficient to cause large spectral shifts. The origin of this effect seems likely to lie in the charge transfer character of the transition, but confirmation will require quantum chemical calculations on these derivatives.

3.2 HBDI Photodynamics

The transient behaviour of HBDI following electronic excitation has been investigated in detail [59–65]. Single colour pump-probe polarisation spectroscopy and broadband transient absorption experiments showed that ground state repopulation occurs on an ultrafast timescale. This is consistent with the very weak fluorescence and suggests that the radiationless process is IC. The excited state fluorescence decay time has been measured in a range of solvents at room temperature. The lifetime of the neutral form is sub-picosecond and is slightly lengthened in non-polar solvents. The anionic form of HBDI has a slightly but consistently longer decay time than the neutral. In aqueous solution, both neutral and anionic forms have faster decay times than in other polar solvent, suggesting an enhanced quenching [62]. However, the solvent and charge effects are slight; the HBDI decay time is never longer than a few picoseconds, and it can be concluded that in all fluid solvents IC dominates the excited state decay. The observed excited state decay times are on the same order or slightly faster than the ground state recovery times, suggesting that a short-lived dark intermediate (perhaps simply a vibrationally hot ground state) may be involved in the relaxation pathway.

To provide information on the coordinate promoting IC, the effect of solvent viscosity on the decay rate has been investigated [61, 63, 65]. One plausible mechanism for radiationless decay in HBDI is excited state isomerisation. For a number of related molecules in solution, such behaviour is well characterised, for example stilbenes, azobenzenes and cyanine dyes (which are structurally somewhat similar to HBDI) all undergo fast isomerisation in the excited state [66–69]. The mechanism proposed involves a decreased bond order for the bridging double bond(s) in the excited state allowing nearly free rotation. The increase in the ground state energy during this excited state rotation causes ground and excited states to approach in energy or to cross at a conical intersection. At or near this point, rapid IC to the ground state occurs. Evidence supporting an isomerisation mechanism has been presented on the basis of stationary photochemical experiments, where the formation of a new ground state isomer under irradiation was observed [70–74]. The cross-section was dependent on the solvent as is the rate of the reverse isomerisation in the ground state [71, 73]. Interestingly, the solvent-dependent reverse process in the ground state (which is calculated to have a high energy barrier) may proceed by an addition elimination reaction [71, 75].

Excited state isomerisation involves large-scale structural reorganisation (e.g., a *cis–trans* isomerisation) on the upper potential energy surface, which may be opposed by solvent friction. Thus, measurements of the excited state lifetime as a function of solvent viscosity yield information about the nature of the coordinate promoting IC. The mean excited state lifetime for anionic HBDI increases by only a factor of 3 when the solvent is changed from methanol to glycerol, a 40-fold increase in viscosity [63]. This weak viscosity dependence suggests that the coordinate promoting IC in HBDI is not very sensitive to solvent friction (or is energetically sufficiently strongly downhill to provide a strong driving force to overcome the solvent friction). Thus, it seems unlikely that the coordinate promoting IC involves a large-scale structural change, such as a complete rotation about either of the exocyclic double bonds, as such a motion would have to displace a large volume of solvent. The weak viscosity dependence is consistent with a volume conserving motion involving a rearrangement localised on the bridging bonds – a hypothesis supported by studies of analogues of HBDI synthesised to lock the chromophore in the planar structure [76].

Further information can be obtained from the temperature dependence, which can also reveal the existence of barriers in the radiationless relaxation coordinate. Huppert and co-workers measured the fluorescence decay of HBDI in glycerol–water over a wide temperature range [59, 77]. They were able to model their data with a two-dimensional model involving phenyl ring torsion and a swinging motion in the bridging bonds [59]. Such an internal reorganisation could be volume conserving as suggested by the viscosity measurements. Litvinenko et al. presented an isoviscosity analysis of the fluorescence yield and ground state recovery time of HBDI in three charge states [61]. The data were similar in all charge states suggesting a common radiationless coordinate, and suggested a zero or negligible activation barrier along that coordinate. Mandal et al. proposed that the wavelength-independent non-exponential kinetics they observed in time-resolved fluorescence could be analysed in terms of a two-dimensional coordinate [62].

Further analysis of the coordinate leading to ultrafast IC is possible through quantum chemical calculation. Weber et al. considered the energetics of an excited state isomerisation reaction involving the bridging bonds of HBDI via a 90° rotation about three possible coordinates for three different charge states [78]. They found that only twisting about the imidazolinone double bond was barrierless for both neutral and ionic states (consistent with experiment), but it did not lead to a crossing of ground and excited states for the anion. They also considered the so-called hula twist mechanism. This is a volume conserving route to isomerisation, and thus consistent with experiment. However, it was found that this coordinate only leads to an S_0/S_1 crossing in the HBDI anion via a significant energy barrier.

More recent high-level quantum chemical calculations are consistent with a major contribution to the coordinate promoting IC arising from zero or low barrier single bond rotation, possibly coupled with a degree of pyramidalisation at the bridging carbon atom [79–81]. Pathways involving the volume conserving 'hula twist' motion are calculated to lead to fast IC but again only via an energy barrier [78, 80]. There are conflicting conclusions as to which single bond rotation is dominant in the excited state isomerisation. Olivucci and co-workers using solvent-free conditions found that two coordinates are important in achieving close approach of S_0 and S_1 in anionic HBDI – a fast stretching coordinate, corresponding to reduced bond order, and a slower rotation about the phenolic single bond [80]. This interpretation is consistent with the model proposed by Huppert and co-workers for the temperature-dependent HBDI fluorescence [59]. Gas phase calculations have recently been extended to neutral and cationic states, and in those case significant barriers are found to isomerisation [82]. This is inconsistent with experiment, where neutral HBDI has a shorter lifetime than the anion. This result suggests that details of the reactive potential surfaces may be modified in the condensed phase.

Altoe et al. included a polarisable continuum model of the solvent in their calculations for the HBDI anion and found that rotation about the imidazolinone double bond was the most significant in promoting IC, in contrast to the gas phase calculation [79]. Martinez and co-workers also identified an important role for the medium in their study of a molecule similar to HBDI [81, 83]. In vacuum, the excited state dynamics primarily involved twisting about the bridging double bond again accompanied by a large excursion in the phenyl torsion. These calculations, however, predicted a significant lifetime on the excited state surface, which is not consistent with recent gas phase measurements [55]. However, in the calculation for a water solvated chromophore, fast barrierless rotation about the double bond was predicted, leading to an S_1/S_0 conical intersection, and a sub-picosecond excited state lifetime, as found experimentally. These results therefore point to an important role for the medium in determining the coordinate leading to IC. This is in itself an important conclusion, as it suggests the means by which the protein can modulate the chromophore's photophysics. They also suggest that rotation about either or both bridging bonds may be important, with the mechanism being dependent on the environment. Significantly, Olsen [84, 85] has shown in calculations comparing HBDI and a model of the red-emitting CP chromophore (i.e., HBDI with an *N*-acylimine substitution) that a switch in the nature of the radiationless decay (and presumably isomerisation) coordinate from double bond to phenoxy bond rotation occurs on substitution. Recently, it was observed experimentally that the excited state decay of HBDI is somewhat dependent on the nature and location of substituents, as might have been predicted from the calculation [58]. However, in all cases the fluorescence quantum yield remains low.

The mechanism by which the protein renders the chromophore fluorescent has still not been definitively determined. Many calculations suggest that an almost 90° rotation about one or other of the bridging bonds is required for the closest approach of ground and excited states. It is easy to imagine that such a large volume excited state reorganisation could be sterically hindered by the protein, and molecular mechanics calculations suggest that many possible coordinates are so restricted [86, 87]. However, experiments in viscous media suggest that the coordinate promoting IC is volume conserving or at least has a large driving force, neither of which are in agreement with calculated coordinates requiring a large excited state reorganisation. Calculations do, however, suggest that the excited state relaxation pathway is sensitive to the chromophore's environment. It is also likely (e.g., on the basis of spectroscopic studies, section 3.1) that the electronic structure of the chromophore is modified on incorporation in the protein environment. It seems probable that some combination of steric and chemical (e.g., H-bond) effects can suppress excited state structure change, while environmental effects (charges on nearby residues, pi stacking, H-bonds) can modify the driving force, and that these factors operating together suppress IC. However, there is as yet no predictive model of the CP fluorescence enhancement mechanism.

While the mechanism suppressing IC remains to be determined, there is structural evidence to suggest that (at least some) CPs that are essentially non-fluorescent have a chromophore that exists in a non-planar ground state conformation [88, 89]. Conversely, it has been noted that both non-fluorescent and (weakly) fluorescent forms of a coral protein exhibit non-planar chromophore structures [90]. This suggests that more than one feature of the chromophore in the protein is responsible for controlling the fluorescence yield. Recently, a sterically crowded HBDI derivative formed by locating methyl groups in positions meta to the phenolic hydroxyl was studied. DFT calculations reveal a strongly twisted ground state. This derivative indeed exhibits extremely rapid fluorescence decay (ca. 100 fs even in the anionic form) [58]. This is consistent with the proposed correlation between non-planar ground state and fast radiationless decay. However, the analogy between synthetic chromophore and CP is not exact, as the non-planar derivative has a blue-shifted absorption relative to HBDI, while the non-planar non-emissive CPs are usually red absorbing.

4 ESPT in avGFP and Its Mutants

In the preceding section, it was shown that the photophysics of the GFP chromophore in solution are dominated by ultrafast structural changes resulting in IC. The effect of the protein matrix is so dramatic that IC can be neglected in avGFP, and the photophysics are instead dominated by a completely different mechanism, ESPT. Excitation into either neutral (usually labelled the A state) or anionic (B state) absorption bands of avGFP results in an intense green emission

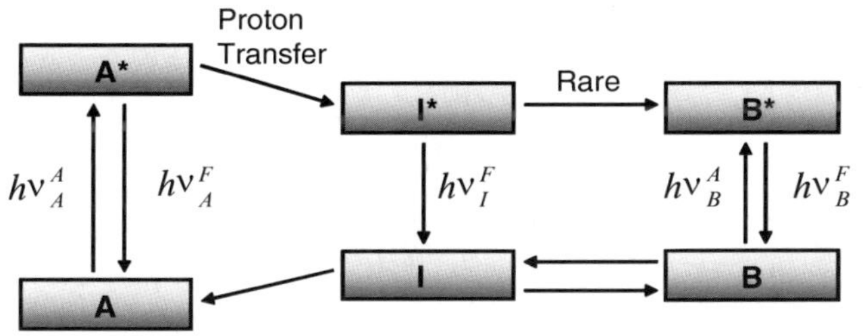

Fig. 6 Outline kinetic model of the ESPT mechanism in avGFP

arising almost exclusively from the anionic form. Boxer and co-workers measured the time-resolved fluorescence of avGFP with ultrafast time resolution [20]. The A* state decays in a non-single exponential fashion with a mean lifetime of 18 ps. The fluorescence spectrum of the anionic form was observed to grow in intensity on a similar timescale of a few tens of picoseconds, with no change in spectral profile [20, 91]. This result clearly points to the occurrence of an ESPT reaction. Such reactions are unique in biology, but have been well characterised in simpler molecular systems [92]. The assignment to ESPT was confirmed by the observation of a large deuterium isotope effect, which extended the A* state lifetime and correspondingly increased the rise time for the green emission [20]. Similar observations were made using transient absorption spectroscopy [93].

Since the population of the anionic (B) ground state does not increase rapidly as a result of irradiation, it is evident that the main fate of the deprotonated excited state is decay (mainly radiative) followed by re-protonation to recover the A ground state. Chattoraj and co-workers proposed a model which incorporates this behaviour (Fig. 6), where the emissive (deprotonated) state (called the I* state to distinguish it from the directly excited ground state, B) is formed in the geometry of the original ground state, and relaxes back to the A state. Ultrafast pump-dump-probe spectroscopy revealed fast I $\rightarrow$ A proton-transfer dynamics on the ground state surface which are sensitive to H/D isotope exchange [94]. It was proposed that the B state is populated by a reorganisation of the protein matrix about I* occurring with a low probability [20]. The X-ray structures of A and B states suggested that the reorganisation involves T203 reorientation [95], and steady-state photochemical measurements show that an irreversible A $\rightarrow$ B conversion can be effected photochemically, probably due to a low yield electron transfer and photodecarboxylation mechanism [50, 51].

The location of the proton acceptor was investigated by time-resolved vibrational spectroscopy [96–99]. The transient infrared difference spectrum was monitored following A state excitation with picosecond time resolution between 1500 and 1800 cm^{-1} (Fig. 7). The instantaneous appearance of four strong bleach bands (negative ΔOD) is associated with excitation of the chromophore ground state (cf. Fig. 3). This is accompanied by the immediate appearance of positive ΔOD signals due to vibrational modes in the excited state. These bands could be assigned to specific vibrational modes by isotope labelling and polarisation studies of HBDI in

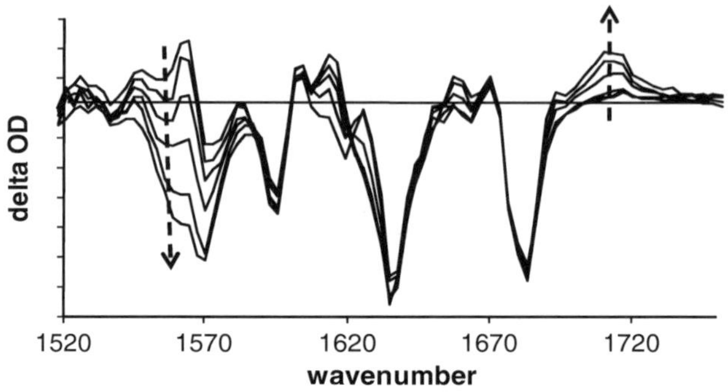

Fig. 7 Time-resolved IR spectroscopy of avGFP in D_2O recorded 2, 6, 10, 30, 100 and 200 ps after excitation at 400 nm. The direction of increasing/decreasing absorption is indicated

solution [97, 99]. Although the main bleach bands seen in avGFP can be assigned to the chromophore, the temporal evolution of the shape of the transient IR spectrum suggests additional underlying contributions from protein modes. The origin of some of these protein modes has been revealed by mutagenesis [99].

The proton-transfer dynamics and the site of the proton acceptor are revealed in two time-dependent bands – a bleach that develops as a function of time at 1560 cm^{-1} and a new transient absorption above 1700 cm^{-1} (Fig. 7). These bands evolve on the same picosecond timescale and are assigned to the conversion of a carboxylate to a carboxylic acid. Inspection of the structure of avGFP suggests that this transformation may be assigned to protonation of the residue E222, which, as first shown in the structural studies of Brecj et al. [95], is connected to the proton donor by a proton wire via a structural water molecule and the S205 residue. Thus, transient IR spectroscopy confirms the proposed assignment of the E222 residue as the proton acceptor. A comparison of the fluorescence decay time with the vibrational dynamics shows that donor decay and acceptor protonation occur simultaneously, suggesting a concerted mechanism for proton motion, or at least that any intermediates states are very short lived [96]. The role of the E222 residue has been supported by studies of mutants, polarisation-resolved measurements, isotopic labelling and observations over a wider spectral range [97, 100]. The non-exponential dynamics observed in both fluorescence and transient IR suggest dispersive kinetics, which observations over a wider spectral range were able to assign to side-chain disorder leading to different proton-transfer rates [100].

These studies of the structure and dynamics of the proton relay reaction have provided a rather detailed picture of the photophysics of avGFP. They also suggest the use of GFP as a model system for the study of proton-transfer reactions in proteins. Proton-transfer is one of the key steps in many biochemical reactions, and transfer along proton wires has often been invoked in biochemical mechanisms [101–103]. The unique ESPT reaction in GFP permits the measurement of the real-time dynamics of such proton-transfer reactions, following photoinitiation with an ultrafast laser pulse. The comparison between transient IR and fluorescence data already suggests that such long-range proton transport (along a preformed wire) can be very fast, as intermediate states in the chain were not detected. It has also proved possible to modify the proton wire through mutagenesis. Single mutations can disrupt the proton wire, trapping the chromophore in the A* state, which has a shorter lifetime than the I* state [104]. It has also been shown that a second mutation, H148D, which places a proton acceptor adjacent to the donor gives rise to a new proton-transfer reaction, presumably to the asparagine carboxylate group [105–108]. This redirected proton transfer has been shown to be extremely fast (<100 fs) [109]. It was suggested that this represents a case of proton transfer over a low barrier or barrierless hydrogen bond. Such low barrier H-bonds are implicated in a number of key enzyme reactions [110, 111].

Progress towards understanding the observed protein proton-transfer dynamics requires theoretical modelling of the observed transient behaviour. Although accurate quantum chemical calculations of excited states in proteins remain challenging, there has been significant progress in modelling ESPT in avGFP. In an early work, Lill and Helms used classical MD to simulate the proton transfer along the three-step proton wire [112]. They concluded that after transfer was triggered by ejection of the proton from the chromophore, the steps leading to protonation of E222 occur on the tens of femtosecond timescale. Subsequently, quantum chemical calculations have been reported by two groups using the geometry of the proton-transfer chain suggested by the protein structure, but in the absence of surrounding residues [113–116]. Both groups calculated that the proton transfer occurred in a single concerted step along a low barrier potential surface, with no stable intermediate states. These calculations also suggested that the first proton to move in the concerted process was the last in the chain (i.e., the proton protonating E222). Zhang et al. found that the H148 residue, which is not part of the proton-transfer chain but is H-bonded to the donor O atom, had a significant impact on the potential surface, suggesting an important role for the surrounding residues [116].

Lluch and co-workers used molecular dynamics to study the structure and stability of the proton relay chain in avGFP, and performed quantum chemical calculations on a reduced set of residues, using the geometries obtained from the MD simulations [117–119]. They investigated the proton-transfer surface for both ground and excited states of the chromophore, and found that the photoactive state was $\pi\pi^*$ rather than the $\pi\sigma^*$. The potential energy surfaces were calculated to have minima for the proton localised on the chromophore in the ground state and on E222 in the excited state, in agreement with experiment. The ESPT has a small barrier (ca. 2 kcal mol^{-1}) and is strongly downhill for the S203 to E222 step, as found in earlier calculations. This is consistent with a concerted but asynchronous proton transfer, with the last proton 'leading'.

The majority of simulations suggest an effectively barrierless ultrafast ESPT in the avGFP geometry, whereas experimental results show transfer on a picosecond or tens of picoseconds timescale. Fang et al. recently reported a femtosecond-stimulated Raman study of the first few picoseconds of ESPT in avGFP [120]. Their results suggest that fluctuations in the structure of the chromophore may modulate the proton-transfer rate, and thus the frequency at which such fluctuations result in the formation of a barrierless proton-transfer potential may control the overall rate of proton transfer. Such details are yet to be incorporated into theoretical calculations, but when they are a picture of long range proton transfer in proteins may emerge with significance beyond that of understanding ESPT in avGFP.

5 Photochemistry and Photochromism

Photochemistry may induce a permanent change in the absorption spectrum of a chromophore under irradiation. In some cases, such a photochemical change is reversible, either thermally on a ground state potential surface or photochemically through excitation of the product state, in which case the phenomenon is called photochromism [121]. Photochromism has been known for many years in organic photochemistry where the colour change, and the structure change which often accompanies it, have a number of important applications [121]. However, the observation of photochemistry and photochromism in CPs has been a source of immense excitement and intense research activity [49, 122–125]. That early excitement has been more than justified by the important applications which have been demonstrated. For example, the ability to induce photochemically a permanent change in the colour of a spatially localised population of labelled proteins (optical highlighting [28, 29]) permits the tracking of selected sub-cellular structures. The applications of CP photochromism have proven even more remarkable. The ability to switch a protein between bright and dark states at will forms the basis of a novel method of super-resolution optical microscopy. Typically, resolution in an optical microscope is limited by the Rayleigh criterion to be no better than 200 nm [31, 32]. The position of a single fluorescent molecule may, however, be located within a few tens of nanometers by straightforward fitting procedures, provided there are no interfering molecules close by. However, there is no improvement in resolution if the fluorophores must be kept so far apart. The advantage of photoswitchable or kindling protein labels is that a dense population of fluorophores may be switched between dark and bright states, which permits the accumulation of a series of isolated fluorophore images with tens of nanometre resolution which, when combined, give a complete image with ultrahigh resolution [30–32].

A comprehensive description of photoconversion in CPs has been presented by van Thor elsewhere in this volume. In this section, we focus only on the primary photophysics, and specifically on how the photoconversion mechanism may relate to the phenomena of proton transfer and ultrafast IC described above.

Miyawaki and co-workers reported the remarkable observation that a protein isolated from coral possessing a chromophore with structure and spectroscopy similar to avGFP underwent photoconversion to create a red light-absorbing and -emitting species upon irradiation in the blue region of the spectrum [27]. The protein was named kaede (maple leaf in Japanese) which elegantly suggests its behaviour. Shortly later, very similar observations were made in the protein EosFP also isolated from coral [124]. Structural studies showed that the chemical structure of the red-emitting chromophore differs markedly from that of the GFP

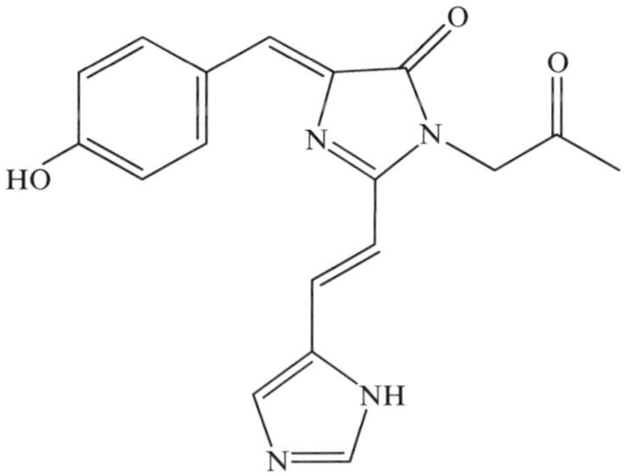

Fig. 8 Molecular structure of the red-emitting kaede chromophore

chromophore through the addition of a third ring conjugated to the original structure (Fig. 8). It was shown through mutagenesis that a histidine residue adjacent to the original chromophore is critical in forming this three ring system [29].

From the point of view of primary photochemical processes, the most intriguing result is that the photoconversion occurs exclusively from the neutral form of the chromophore, which has the same chemical structure (though not environment) as in avGFP [124]. This strongly suggests a role for ESPT, which accordingly is a feature of most of the proposed photoconversion mechanisms [10, 124, 126], though not all of them [127]. Specifically, it has been proposed that an initial ESPT step is followed by a reorganisation of the electronic structure of the anionic form of the chromophore in its excited state resulting in the extension of the chromophore (i.e., incorporation of the histidine residue) and a simultaneous bond scission of the main amino acid chain. The breaking of the chain was shown to be a feature of the photoconversion [29]. A number of further atom transfer and electronic structure changes result in the final red-absorbing chromophore. Related observations have been made on the engineered protein kikGR, which has a number of useful properties, including being readily titratable. Intriguingly, the red form of this protein has been shown to have a different chromophore structure compared to the kaede/EosFP, in which the third ring has a *cis* orientation with respect to the newly formed double bond, showing that rotation has occurred, which has important implications for the likely intermediate structures. Again, only excitation of the neutral state leads to photoconversion [128].

The role of ESPT appears critical at some point in the mechanism, because the photoconversion does not occur from the directly excited anionic chromophore even though it is readily excited from its ground state, and can be prepared in high yield by increasing slightly the pH. The action spectrum for photoconversion instead follows closely the absorption of the neutral form. Unfortunately the precise role of the ESPT is yet to be determined. One possibility is that the proton transferred is actively involved in the chemistry. It was suggested that protonation of the His residue via a proton wire, such as was observed in avGFP, might be an important step [124]. However, thus far no obvious route for such a long-range proton transfer has been established. A second possibility is that the proton itself is not involved in the subsequent chemical steps, but that the anionic excited state formed directly following ESPT is in the appropriate geometry to undergo further reaction, a geometry which is not directly accessible from the equilibrated anionic ground state. Most measurements suggest that the quantum yield for photoconversion is low, which is in contrast to the very facile ESPT of avGFP. It seems at least plausible that the mechanism involves ESPT as a primary step with the resultant anion either undergoing a second low cross-section step ultimately leading to photoconversion or a more facile reverse re-protonation in the ground state: such a mechanism recalls the A $\rightarrow$ I $\rightarrow$ B mechanism introduced for avGFP [20]. It seems likely that the role of the ESPT will be resolved through ultrafast and microsecond transient IR studies, although such measurements on these irreversibly photoactive proteins present a number of experimental difficulties. Some preliminary results for kikGR have been presented [129].

A still more complicated picture is presented by the primary photophysics of the photochromic CPs, of which Dronpa is the best-characterised example. Dronpa was engineered by Miyawaki and co-workers from a protein isolated from a coral [130, 131]. It was observed to be very weakly fluorescent following excitation in the neutral state, but on continued irradiation that state was converted with a relatively high cross-section to an anionic state which had a high fluorescence quantum yield. Critically, strong illumination of this emissive anionic state gives rise to a reverse photoconversion back to the dark state, with a low cross-section. The protein can be cycled many times through this photochromic cycle, which is clearly important for the ultraresolution microscopy application [131, 132]. Since the discovery of Dronpa, a number of other photoconvertible proteins have been reported [133], including some in which negative switching was achieved, where excitation of the neutral form switches fluorescence off and irradiation of the anionic form switches it on [24, 134]. Further mutagenesis has been used to control the rate of the on and off processes, which may vary quite dramatically [135].

The detailed mechanism of photochromic switching has yet to be fully elucidated. The mechanism is likely to involve a mixture of ESPT, *cis–trans* isomerisation, radiationless decay and ground state acid–base equilibria. It is not, however, clear that there is a common mechanism operating among the different photochromic CPs. In an NMR study, Mizuno et al. showed that the state of protonation of the chromophore was critical, and in particular that the anchoring of the anionic (fluorescent) form to the main β-barrel structure generates a rigid environment for this state [136]. In this model, ESPT is an important step [137], while *cis–trans* isomerisation between the off and on states is not critical. However, some X-ray structure measurements on dark- and light-adapted Dronpa suggested different isomers of the chromophore in the on and off states, a result that requires some excited (or ground state) structural reorganisation [138, 139]. This observation is consistent with the reported increased flexibility in the off state, which may well allow both fast radiationless decay and excited state isomerisation.

The interplay between isomerisation and fluorescence quantum yield is an interesting one. In a series of structural studies of photoswitchable proteins, it was shown that both planar *cis* and *trans* forms of the chromophore could be stabilised, and that both could be fluorescent. However, some mutants revealed a non-planar *trans* form, which may correspond to disorder or flexibility in the chromophore structure [26]. These disordered forms were found to be essentially non-fluorescent.

Quite recently, detailed quantum chemical calculations on the possible reaction pathways in Dronpa were performed by Li and co-workers [140]. They confirmed that a balance of proton-transfer and isomerisation steps may be important in the photoreversible process, and considered a number of possible mechanisms. Clearly, some detailed observations of the primary excited state processes in photoconvertible CPs will be critical in developing a unified picture of the mechanism(s) operating. In particular, a correlation of excited state dynamics and time-resolved vibrational spectroscopy will be highly informative, especially if it is accompanied by structure determinations. Such measurements are in progress.

6 Summary

Since their first discovery and application in bioimaging, the range and capabilities of the CP family have continued to grow. Recent years have seen dramatic progress particularly in the area of photoconvertible CPs, which have already proved to be of great utility in bioimaging. However, there is already evidence that this is by no means the end of the CP story. In the future, CPs may contribute to life sciences as much more than simply passive fluorescent probes. CPs with potential application in phototherapy have already been proposed [141]. Very recently, it was demonstrated that unnatural amino acids can be incorporated into GFP, modifying the properties of the chromophore [142, 143]. This exciting result suggests an entirely new range of engineered applications. At each new step in the CP story, the advances in protein science have been accompanied by detailed physical and chemical measurements of the underlying mechanism. Spectroscopic and kinetic studies have in turn been complemented by quantum chemical and molecular mechanics calculations. As the variety and range of CPs continues to grow, so will the need for the detailed understanding of the underlying photophysical phenomena that dynamics, spectroscopy, theoretical calculation and molecular dynamics simulation can supply.

Acknowledgements I am grateful to EPSRC for financial support (EP/H025715), to my students and postdoctoral fellows for their work and insights over the years, and to my collaborators for their generous advice and assistance.

References

1. Shimomura O, Johnson FH (1969) Properties of bioluminescent protein aequorin. Biochemistry 8:3991
2. Morise H, Shimomur O, Johnson FH, Winant J (1974) Intermolecular energy-transfer in bioluminescent system of Aequorea. Biochemistry 13:2656–2662
3. Chalfie M, Tu Y, Euskirchen G, Ward WW, Prasher DC (1994) Green fluorescent protein as a marker for gene-expression. Science 263:802–805

4. Cubitt AB, Heim R, Adams SR, Boyd AE, Gross LA, Tsien RY (1995) Understanding, improving and using green fluorescent proteins. Trends Biochem Sci 20:448–455
5. Shaner NC, Steinbach PA, Tsien RY (2005) A guide to choosing fluorescent proteins. Nat Methods 2:905–909
6. Tsien RY (1998) The green fluorescent protein. Annu Rev Biochem 67:509–544
7. Chudakov DM, Matz MV, Lukyanov S, Lukyanov KA (2010) Fluorescent proteins and their applications in imaging living cells and tissues. Physiol Rev 90:1103–1163
8. Alieva NO, Konzen KA, Field SF, Meleshkevitch EA, Hunt ME, Beltran-Ramirez V, Miller DJ, Wiedenmann J, Salih A, Matz MV (2008) Diversity and evolution of coral fluorescent proteins. PLoS One 3
9. Pouwels LJ, Zhang LP, Chan NH, Dorrestein PC, Wachter RM (2008) Kinetic isotope effect studies on the de novo rate of chromophore formation in fast- and slow-maturing GFP variants. Biochemistry 47:10111–10122
10. Wachter RM, Watkins JL, Kim H (2010) Mechanistic diversity of red fluorescence acquisition by GFP-like proteins. Biochemistry 49:7417–7427
11. Wachter RM (2007) Chromogenic cross-link formation in green fluorescent protein. Acc Chem Res 40:120–127
12. Shimomura O (1979) Structure of the chromophore of Aequorea green fluorescent protein. FEBS Lett 104:220–222
13. Niwa H, Inouye S, Hirano T, Matsuno T, Kojima S, Kubota M, Ohashi M, Tsuji FI (1996) Chemical nature of the light emitter of the Aequorea green fluorescent protein. Proc Natl Acad Sci USA 93:13617–13622
14. Frommer WB, Davidson MW, Campbell RE (2009) Genetically encoded biosensors based on engineered fluorescent proteins. Chem Soc Rev 38:2833–2841
15. Meech SR (2009) Excited state reactions in fluorescent proteins. Chem Soc Rev 38:2922–2934
16. van Thor JJ (2009) Photoreactions and dynamics of the green fluorescent protein. Chem Soc Rev 38:2935–2950
17. Bell AF, He X, Wachter RM, Tonge PJ (2000) Probing the ground state structure of the green fluorescent protein chromophore using Raman spectroscopy. Biochemistry 39:4423–4431
18. Elsliger MA, Wachter RM, Hanson GT, Kallio K, Remington SJ (1999) Structural and spectral response of green fluorescent protein variants to changes in pH. Biochemistry 38:5296–5301
19. Llopis J, McCaffery JM, Miyawaki A, Farquhar MG, Tsien RY (1998) Measurement of cytosolic, mitochondrial, and Golgi pH in single living cells with green fluorescent proteins. Proc Natl Acad Sci USA 95:6803–6808
20. Chattoraj M, King BA, Bublitz GU, Boxer SG (1996) Ultra-fast excited state dynamics in green fluorescent protein: Multiple states and proton transfer. Proc Natl Acad Sci USA 93:8362–8367
21. Sample V, Newman RH, Zhang J (2009) The structure and function of fluorescent proteins. Chem Soc Rev 38:2852–2864
22. Lukyanov KA, Fradkov AF, Gurskaya NG, Matz MV, Labas YA, Savitsky AP, Markelov ML, Zaraisky AG, Zhao XN, Fang Y, Tan WY, Lukyanov SA (2000) Natural animal coloration can be determined by a nonfluorescent green fluorescent protein homolog. J Biol Chem 275:25879–25882
23. Schuttrigkeit TA, von Feilitzsch T, Kompa CK, Lukyanov KA, Savitsky AP, Voityuk AA, Michel-Beyerle ME (2006) Femtosecond study of light-induced fluorescence increase of the dark chromoprotein asFP595. Chem Phys 323:149–160
24. Adam V, Lelimousin M, Boehme S, Desfonds G, Nienhaus K, Field MJ, Wiedenmann J, McSweeney S, Nienhaus GU, Bourgeois D (2008) Structural characterization of IrisFP, an optical highlighter undergoing multiple photo-induced transformations. Proc Natl Acad Sci USA 105:18343–18348

25. Kredel S, Nienhaus K, Oswald F, Wolff M, Ivanchenko S, Cymer F, Jeromin A, Michels FJ, Spindler KD, Heilker R, Nienhaus GU, Wiedenmannl J (2008) Optimized and far-red-emitting variants of fluorescent protein eqFP611. Chem Biol 15:224–233
26. Nienhaus K, Nar H, Heilker R, Wiedenmann J, Nienhaus GU (2008) Trans-cis isomerization is responsible for the red-shifted fluorescence in variants of the red fluorescent protein eqFP611. J Am Chem Soc 130:12578
27. Ando R, Hama H, Yamamoto-Hino M, Mizuno H, Miyawaki A (2002) An optical marker based on the UV-induced green-to-red photoconversion of a fluorescent protein. Proc Natl Acad Sci USA 99:12651–12656
28. Habuchi S, Tsutsui H, Kochaniak AB, Miyawaki A, van Oijen AM (2008) mKikGR, a monomeric photoswitchable fluorescent protein. PLoS One 3
29. Mizuno H, Mal TK, Tong KI, Ando R, Furuta T, Ikura M, Miyawakil A (2003) Photo-induced peptide cleavage in the green-to-red conversion of a fluorescent protein. Mol Cell 12:1051–1058
30. Vaziri A, Tang JY, Shroff H, Shank CV (2008) Multilayer three-dimensional super resolution imaging of thick biological samples. Proc Natl Acad Sci USA 105:20221–20226
31. Bates M, Huang B, Dempsey GT, Zhuang XW (2007) Multicolor super-resolution imaging with photo-switchable fluorescent probes. Science 317:1749–1753
32. Patterson G, Davidson M, Manley S, Lippincott-Schwartz J (2010) Annu Rev Phys Chem 61:345–367
33. Rulliere C (ed) (2003) Femtosecond laser pulses: principles and experiments. Springer, New York
34. Diels JC, Rudolphe W (2006) Ultrashort laser pulse phenomena. Academic, New York
35. Andrews DL, Allcock P (2002) Optical harmonics in molecular systems. Wiley-VCH, Weinheim
36. Fleming GR (1986) Chemical applications of ultrafast spectroscopy. Oxford University Press, Oxford
37. Rhee H, Joo T (2005) Noncolinear phase matching in fluorescence upconversion. Opt Lett 30:96–98
38. Heisler IA, Kondo M, Meech SR (2009) Reactive dynamics in confined liquids: ultrafast torsional dynamics of auramine O in nanoconfined water in aerosol OT reverse micelles. J Phys Chem B 113:1623–1631
39. Vengris M, van der Horst MA, Zgrablic G, van Stokkum IHM, Haacke S, Chergui M, Hellingwerf KJ, van Grondelle R, Larsen DS (2004) Contrasting the excited-state dynamics of the photoactive yellow protein chromophore: protein versus solvent environments. Biophys J 87:1848–1857
40. Arzhantsev S, Zachariasse KA, Maroncelli M (2006) Photophysics of trans-4-(dimethylamino)-4′-cyanostilbene and its use as a solvation probe. J Phys Chem A 110:3454–3470
41. Kwok WM, Ma C, Phillips D, Matousek P, Parker AW, Towrie M (2000) Picosecond time-resolved study of 4-dimethylaminobenzonitrile in polar and nonpolar solvents. J Phys Chem A 104:4188–4197
42. Schmidt B, Laimgruber S, Zinth W, Gilch P (2003) A broadband Kerr shutter for femtosecond fluorescence spectroscopy. Appl Phys B Lasers Opts 76:809–814
43. Takeda J, Nakajima K, Kurita S, Tomimoto S, Saito S, Suemoto T (2000) Femtosecond optical Kerr gate fluorescence spectroscopy for ultrafast relaxation processes. J Lumin 87–89:927–929
44. van Stokkum IHM, Gobets B, Gensch T, van Mourik F, Hellingwerf KJ, van Grondelle R, Kennis JTM (2006) (Sub)-picosecond spectral evolution of fluorescence in photoactive proteins studied with a synchroscan streak camera system. Photochem Photobiol 82:380–388
45. Schmitt M, Dietzek B, Hermann G, Popp J (2007) Femtosecond time-resolved spectroscopy on biological photoreceptor chromophores. Laser Photon Rev 1:57–78
46. Greetham G, Towrie M, Matousek P, Parker AW (2010) J Appl Spectrosc 64:1320

47. Towrie M, Grills DC, Dyer J, Weinstein JA, Matousek P, Barton R, Bailey PD, Subramaniam N, Kwok WM, Ma CS, Phillips D, Parker AW, George MW (2003) Development of a broadband picosecond infrared spectrometer and its incorporation into an existing ultrafast time-resolved resonance Raman, UV/visible, and fluorescence spectroscopic apparatus. Appl Spectrosc 57:367–380
48. van Stokkum IHM, Larsen DS, van Grondelle R (2004) Global and target analysis of time-resolved spectra. Biochim Biophys Acta Bioenergetics 1657:82–104
49. Remington SJ (2006) Fluorescent proteins: maturation, photochemistry and photophysics. Curr Opin Struct Biol 16:714–721
50. van Thor JJ, Pierik AJ, Nugteren-Roodzant I, Xie AH, Hellingwerf KJ (1998) Characterization of the photoconversion of green fluorescent protein with FTIR spectroscopy. Biochemistry 37:16915–16921
51. Bell AF, Stoner-Ma D, Wachter RM, Tonge PJ (2003) Light-driven decarboxylation of wild-type green fluorescent protein. J Am Chem Soc 125:6919–6926
52. Wachter RM, Elsliger MA, Kallio K, Hanson GT, Remington SJ (1998) Structural basis of spectral shifts in the yellow-emission variants of green fluorescent protein. Struct Folding Design 6:1267–1277
53. Andersen LH, Lapierre A, Nielsen SB, Nielsen IB, Pedersen SU, Pedersen UV, Tomita S (2002) Chromophores of the green fluorescent protein studied in the gas phase. Eur Phys J D 20:597–600
54. Lammich L, Petersen MA, Nielsen MB, Andersen LH (2007) The gas-phase absorption spectrum of a neutral GFP model chromophore. Biophys J 92:201–207
55. Forbes MW, Jockusch RA (2009) Deactivation pathways of an isolated green fluorescent protein model chromophore studied by electronic action spectroscopy. J Am Chem Soc 131:17038
56. Webber NM, Meech SR (2007) Electronic spectroscopy and solvatochromism in the chromophore of GFP and the Y66F mutant. Photochem Photobiol Sci 9:276–281
57. Dong J, Solntsev KM, Tolbert LM (2006) Solvatochromism of the green fluorescence protein chromophore and its derivatives. J Am Chem Soc 128:12038–12039
58. Conyard J, Kondo M, Heisler IA, Baldridge A, Tolbert LM, Solntsev KM, Meech SR (2011) J Phys Chem B 115:1863–1873
59. Gepshtein R, Huppert D, Agmon N (2006) Deactivation mechanism of the green fluorescent chromophore. J Phys Chem B 110:4434–4442
60. Vengris M, van Stokkum IHM, He X, Bell AF, Tonge PJ, van Grondelle R, Larsen DS (2004) Ultrafast excited and ground-state dynamics of the green fluorescent protein chromophore in solution. J Phys Chem A 108:4587–4598
61. Litvinenko KL, Webber NM, Meech SR (2003) Internal conversion in the chromophore of the green fluorescent protein: temperature dependence and isoviscosity analysis. J Phys Chem A 107:2616–2623
62. Mandal D, Tahara T, Meech SR (2004) Excited-state dynamics in the green fluorescent protein chromophore. J Phys Chem B 108:1102–1108
63. Mandal D, Tahara T, Webber NM, Meech SR (2002) Ultrafast fluorescence of the chromophore of the green fluorescent protein in alcohol solutions. Chem Phys Lett 358:495–501
64. Webber NM, Litvinenko KL, Meech SR (2001) Radiationless relaxation in a synthetic analogue of the green fluorescent protein chromophore. J Phys Chem B 105:8036–8039
65. Kummer AD, Kompa C, Niwa H, Hirano T, Kojima S, Michel-Beyerle ME (2002) Viscosity-dependent fluorescence decay of the GFP chromophore in solution due to fast internal conversion. J Phys Chem B 106:7554–7559
66. Sension RJ, Repinec ST, Szarka AZ, Hochstrasser RM (1993) Femtosecond laser studies of the cis-stilbene photoisomerization reactions. J Chem Phys 98:6291–6315
67. Kim SK, Fleming GR (1988) Reorientation and isomerization of trans-stilbene in alkane solutions. J Phys Chem 92:2168–2172

68. Nagele T, Hoche R, Zinth W, Wachtveitl J (1997) Femtosecond photoisomerization of cis-azobenzene. Chem Phys Lett 272:489–495
69. Sundstrom V, Gillbro T (1982) Viscosity-dependent isomerization yields of some cyanine dyes – a picosecond laser spectroscopy study. J Phys Chem 86:1788–1794
70. He X, Bell AF, Tonge PJ (2003) Ground state isomerization of a model green fluorescent protein chromophore. FEBS Lett 549:35–38
71. Dong J, Abulwerdi F, Baldridge A, Kowalik J, Solntsev KM, Tolbert LM (2008) Isomerization in fluorescent protein chromophores involves addition/elimination. J Am Chem Soc 130:14096
72. Voliani V, Bizzarri R, Nifosi R, Abbruzzetti S, Grandi E, Viappiani C, Beltram F (2008) Cis-trans photoisomerization of fluorescent-protein chromophores. J Phys Chem B 112: 10714–10722
73. Yang JS, Huang GJ, Liu YH, Peng SM (2008) Photoisomerization of the green fluorescence protein chromophore and the meta- and para-amino analogues. Chem Commun:1344–1346
74. Huang GJ, Yang JS (2010) The N-arylamino conjugation effect in the photochemistry of fluorescent protein chromophores and aminostilbenes. Chem Asian J 5:2075–2085
75. Wang DQ, Merz T, van Gunsteren WF (2010) The thermal isomerization of the GFP chromophore: a computational study. Phys Chem Chem Phys 12:11051–11061.
76. Wu LX, Burgess K (2008) Syntheses of highly fluorescent GFP-chromophore analogues. J Am Chem Soc 130:4089–4096
77. Stavrov SS, Solntsev KM, Tolbert LM, Huppert D (2006) Probing the decay coordinate of the green fluorescent protein: arrest of cis-trans isomerization by the protein significantly narrows the fluorescence spectra. J Am Chem Soc 128:1540–1546
78. Weber W, Helms V, McCammon JA, Langhoff PW (1999) Shedding light on the dark and weakly fluorescent states of green fluorescent proteins. Proc Natl Acad Sci USA 96: 6177–6182
79. Altoe P, Bernardi F, Garavelli M, Orlandi G, Negri F (2005) Solvent effects on the vibrational activity and photodynamics of the green fluorescent protein chromophore: a quantum-chemical study. J Am Chem Soc 127:3952–3963
80. Martin ME, Negri F, Olivucci M (2004) Origin, nature, and fate of the fluorescent state of the green fluorescent protein chromophore at the CASPT2//CASSCF resolution. J Am Chem Soc 126:5452–5464
81. Toniolo A, Olsen S, Manohar L, Martinez TJ (2004) Conical intersection dynamics in solution: the chromophore of green fluorescent protein. Farad Discuss 127:149–163
82. Polyakov IV, Grigorenko BL, Epifanovsky EM, Krylov AI, Nemukhin AV (2010) Potential energy landscape of the electronic states of the GFP chromophore in different protonation forms: electronic transition energies and conical intersections. J Chem Theory Comput 6:2377–2387
83. Olsen S, Manohar L, Martinez TJ (2002) Features of interest on the S-0 and S-1 potential energy surfaces of a model green fluorescent protein chromophore. Biophys J 82:359A–359A
84. Olsen S, Smith SC (2007) Radiationless decay of red fluorescent protein chromophore models via twisted intramolecular charge-transfer states. J Am Chem Soc 129:2054–2065
85. Olsen S, Smith SC (2008) Bond selection in the photoisomerization reaction of anionic green fluorescent protein and kindling fluorescent protein chromophore models. J Am Chem Soc 130:8677–8689
86. Baffour-Awuah NYA, Zimmer M (2004) Hula-twisting in green fluorescent protein. Chem Phys 303:7–11
87. Megley CM, Dickson LA, Maddalo SL, Chandler GJ, Zimmer M (2009) Photophysics and dihedral freedom of the chromophore in yellow, blue, and green fluorescent protein. J Phys Chem B 113:302–308
88. Wilmann PG, Petersen J, Pettikiriarachchi A, Buckle AM, Smith SC, Olsen S, Perugini MA, Devenish RJ, Prescott M, Rossjohn J (2005) The 2.1 angstrom crystal structure of the far-red fluorescent protein HcRed: inherent conformational flexibility of the chromophore. J Mol Biol 349:223–237

89. Henderson JN, Ai HW, Campbell RE, Remington SJ (2007) Structural basis for reversible photobleaching of a green fluorescent protein homologue. Proc Natl Acad Sci USA 104: 6672–6677
90. Prescott M, Ling M, Beddoe T, Oakley AJ, Dove S, Hoegh-Guldberg O, Devenish RJ, Rossjohn J (2003) The 2.2 A crystal structure of a pocilloporin pigment reveals a nonplanar chromophore conformation. Structure 11:275–284
91. Jaye AA, Stoner-Ma D, Matousek P, Towrie M, Tonge PJ, Meech SR (2006) Time-resolved emission spectra of green fluorescent protein. Photochem Photobiol 82:373–379
92. Arnaut LG, Formosinho SJ (1993) Excited-state proton-transfer reactions. 1. Fundamentals and intermolecular reactions. J Photochem Photobiol A Chem 75:1–20
93. Lossau H, Kummer A, Heinecke R, PollingerDammer F, Kompa C, Bieser G, Jonsson T, Silva CM, Yang MM, Youvan DC, MichelBeyerle ME (1996) Time-resolved spectroscopy of wild-type and mutant green fluorescent proteins reveals excited state deprotonation consistent with fluorophore-protein interactions. Chem Phys 213:1–16
94. Kennis JTM, Larsen DS, van Stokkum NHM, Vengris M, van Thor JJ, van Grondelle R (2004) Uncovering the hidden ground state of green fluorescent protein. Proc Natl Acad Sci USA 101:17988–17993
95. Brejc K, Sixma TK, Kitts PA, Kain SR, Tsien RY, Ormo M, Remington SJ (1997) Structural basis for dual excitation and photoisomerization of the Aequorea victoria green fluorescent protein. Proc Natl Acad Sci USA 94:2306–2311
96. Stoner-Ma D, Jaye AA, Matousek P, Towrie M, Meech SR, Tonge PJ (2005) Observation of excited-state proton transfer in green fluorescent protein using ultrafast vibrational spectroscopy. J Am Chem Soc 127:2864–2865
97. Stoner-Ma D, Melief EH, Nappa J, Ronayne KL, Tonge PJ, Meech SR (2006) Proton relay reaction in green fluorescent protein (GFP): polarization-resolved ultrafast vibrational spectroscopy of isotopically edited GFP. J Phys Chem B 110:22009–22018
98. van Thor JJ, Georgiev GY, Towrie M, Sage JT (2005) Ultrafast and low barrier motions in the photoreactions of the green fluorescent protein. J Biol Chem 280:33652–33659
99. van Thor JJ, Zanetti G, Ronayne KL, Towrie M (2005) Structural events in the photocycle of green fluorescent protein. J Phys Chem B 109:16099–16108
100. van Thor JJ, Ronayne KL, Towrie M, Sage JT (2008) Balance between ultrafast parallel reactions in the green fluorescent protein has a structural origin. Biophys J 95:1902–1912
101. Frank RAW, Titman CM, Pratap JV, Luisi BF, Perham RN (2004) A molecular switch and proton wire synchronize the active sites in thiamine enzymes. Science 306:872–876
102. Burykin A, Warshel A (2003) What really prevents proton transport through aquaporin? Charge self-energy versus proton wire proposals. Biophys J 85:3696–3706
103. Cui Q, Karplus M (2003) Is a "proton wire" concerted or stepwise? A model study of proton transfer in carbonic anhydrase. J Phys Chem B 107:1071–1078
104. Stoner-Ma D, Jaye AA, Ronayne KL, Nappa J, Tonge PJ, Meech SR (2008) Ultrafast electronic and vibrational dynamics of stabilized A state mutants of the green fluorescent protein (GFP): snipping the proton wire. Chem Phys 350:193–200
105. Shi X, Abbyad P, Shu X, Kallio K, Kanchanawong P, Childs W, Remington SJ, Boxer SG (2007) Ultrafast excited-state dynamics in the green fluorescent protein variant S65T/H148D. 2. Unusual photophysical properties. Biochemistry 46:12014–12025
106. Shu X, Kallio K, Shi X, Abbyad P, Kanchanawong P, Childs W, Boxer SG, Remington SJ (2007) Ultrafast excited-state dynamics in the green fluorescent protein variant S65T/H148D. 1. Mutagenesis and structural studies. Biochemistry 46:12005–12013
107. Leiderman P, Genosar L, Huppert D, Shu X, Remington SJ, Solntsev KM, Tolbert LM (2007) Ultrafast excited-state dynamics in the green fluorescent protein variant S65T/H148D. 3. Short- and long-time dynamics of the excited-state proton transfer. Biochemistry 46:12026–12036
108. Stoner-Ma D, Jaye AA, Ronayne KL, Nappa J, Meech SR, Tonge PJ (2008) An alternate proton acceptor for excited-state proton transfer in green fluorescent protein: rewiring GFP. J Am Chem Soc 130:1227–1235

109. Kondo M, Meech SR, Tonge PJ, Stoner-Ma D, Heisler IA (2010) Ultrafast dynamics of protein proton transfer on short hydrogen bond potential energy surfaces: S65T/H148D GFP. J Am Chem Soc 152:1452–1453
110. Cleland WW, Kreevoy MM (1994) Low-barrier hydrogen-bonds and enzymatic catalysis. Science 264:1887–1890
111. Cleland WW (2010) Adv Phys Org Chem 44:1–17
112. Lill MA, Helms V (2002) Proton shuttle in green fluorescent protein studied by dynamic simulations. Proc Natl Acad Sci USA 99:2778–2781
113. Wang SF, Smith SC (2006) Leading coordinate analysis of reaction pathways in proton chain transfer: application to a two-proton transfer model for the green fluorescent protein. Chem Phys 326:204–209
114. Wang SF, Smith SC (2007) Mechanistic aspects of proton chain transfer in the green fluorescent protein – Part II. A comparison of minimal quantum chemical models. Phys Chem Chem Phys 9:452–458
115. Zhang H, Smith SC (2007) Model real-time quantum dynamical simulations of proton transfer in the green fluorescent protein (GFP). J Theor Comput Chem 6:789–802
116. Zhang RB, Nguyen MT, Ceulemans A (2005) A concerted mechanism of proton transfer in green fluorescent protein. A theoretical study. Chem Phys Lett 404:250–256
117. Vendrell O, Gelabert R, Moreno M, Lluch JM (2008) A potential energy function for heterogeneous proton-wires. Ground end photoactive states of the proton-wire in the green fluorescent protein. J Chem Theory Comput 4:1138–1150
118. Vendrell O, Gelabert R, Moreno M, Lluch JM (2008) Exploring the effects of intramolecular vibrational energy redistribution on the operation of the proton wire in green fluorescent protein. J Phys Chem B 112:13443–13452
119. Vendrell O, Gelabert R, Moreno M, Lluch JM (2008) Operation of the proton wire in green fluorescent protein. A quantum dynamics simulation. J Phys Chem B 112:5500–5511
120. Fang C, Frontiera RR, Tran R, Mathies RA (2009) Mapping GFP structure evolution during proton transfer with femtosecond Raman spectroscopy. Nature 462:200–204
121. Irie M (2000) Chem Rev 100(5):1685–1716
122. Ando R, Mizuno H, Miyawaki A (2004) Regulated fast nucleocytoplasmic shuttling observed by reversible protein highlighting. Science 306:1370–1373
123. Gurskaya NG, Verkhusha VV, Shcheglov AS, Staroverov DB, Chepurnykh TV, Fradkov AF, Lukyanov S, Lukyanov KA (2006) Engineering of a monomeric green-to-red photo-activatable fluorescent protein induced by blue light. Nat Biotechnol 24:461–465
124. Wiedenmann J, Ivanchenko S, Oswald F, Schmitt F, Rocker C, Salih A, Spindler KD, Nienhaus GU (2004) EosFP, a fluorescent marker protein with UV-inducible green-to-red fluorescence conversion. Proc Natl Acad Sci USA 101:15905–15910
125. Shaner NC, Patterson GH, Davidson MW (2007) Advances in fluorescent protein technology. J Cell Sci 120:4247–4260
126. Hayashi I, Mizuno H, Tong KI, Furuta T, Tanaka F, Yoshimura M, Miyawaki A, Ikura M (2007) Crystallographic evidence for water-assisted photo-induced peptide cleavage in the stony coral fluorescent protein kaede. J Mol Biol 372:918–926
127. Lelimousin M, Adam V, Nienhaus GU, Bourgeois D, Field MJ (2009) Photoconversion of the fluorescent protein EosFP: a hybrid potential simulation study reveals intersystem crossings. J Am Chem Soc 131:16814–16823
128. Tsutsui H, Karasawa S, Shimizu H, Nukina N, Miyawaki A (2005) Semi-rational engineering of a coral fluorescent protein into an efficient highlighter. EMBO Rep 6:233–238
129. Lukacs A, Kondo M, Heisler IA, Miyawaki A, Tsutsui H, Towrie M, Greetham G, Tonge PJ, Stoner-Ma D, Meech SR (2010) In: Ultrafast phenomena XVII. Jonas D, Riedle E, Schoenlein R, Chergui M, Taylor A (eds) OUP: 511–513
130. Dedecker P, Hotta J, Ando R, Miyawaki A, Engelborghs Y, Hofkens J (2006) Fast and reversible photoswitching of the fluorescent protein Dronpa as evidenced by fluorescence correlation spectroscopy. Biophys J 91:L45–L47

131. Habuchi S, Ando R, Dedecker P, Verheijen W, Mizuno H, Miyawaki A, Hofkens J (2005) Reversible single-molecule photoswitching in the GFP-like fluorescent protein Dronpa. Proc Natl Acad Sci USA 102:9511–9516
132. Habuchi S, Dedecker P, Hotta JI, Flors C, Ando R, Mizuno H, Miyawaki A, Hofkens J (2006) Photo-induced protonation/deprotonation in the GFP-like fluorescent protein Dronpa: mechanism responsible for the reversible photoswitching. Photochem Photobiol Sci 5: 567–576
133. Subach FV, Zhang LJ, Gadella TWJ, Gurskaya NG, Lukyanov KA, Verkhusha VV (2010) Red fluorescent protein with reversibly photoswitchable absorbance for photochromic FRET. Chem Biol 17:745–755
134. Stiel AC, Andresen M, Bock H, Hilbert M, Schilde J, Schonle A, Eggeling C, Egner A, Hell SW, Jakobs S (2008) Generation of monomeric reversibly switchable red fluorescent proteins for far-field fluorescence nanoscopy. Biophys J 95:2989–2997
135. Ando R, Flors C, Mizuno H, Hofkens J, Miyawaki A (2007) Highlighted generation of fluorescence signals using simultaneous two-color irradiation on Dronpa mutants. Biophys J 92:L97–L99
136. Mizuno H, Mal TK, Walchli M, Kikuchi A, Fukano T, Ando R, Jeyakanthan J, Taka J, Shiro Y, Ikura M, Miyawaki A (2008) Light-dependent regulation of structural flexibility in a photochromic fluorescent protein. Proc Natl Acad Sci USA 105:9227–9232
137. Fron E, Flors C, Schweitzer G, Habuchi S, Mizuno H, Ando R, De Schryver FC, Miyawaki A, Hofkens J (2007) Ultrafast excited-state dynamics of the photoswitchable protein dronpa. J Am Chem Soc 129:4870
138. Andresen M, Stiel AC, Trowitzsch S, Weber G, Eggeling C, Wahl MC, Hell SW, Jakobs S (2007) Structural basis for reversible photoswitching in Dronpa. Proc Natl Acad Sci USA 104:13005–13009
139. Stiel AC, Trowitzsch S, Weber G, Andresen M, Eggeling C, Hell SW, Jakobs S, Wahl MC (2007) 1.8 Angstrom bright-state structure of the reversibly switchable fluorescent protein Dronpa guides the generation of fast switching variants. Biochem J 402:35–42
140. Li X, Chung LW, Mizuno H, Miyawaki A, Morokuma K (2010) A theoretical study on the nature of on- and off-states of reversibly photoswitching fluorescent protein Dronpa: absorption, emission, protonation, and Raman. J Phys Chem B 114:1114–1126
141. Bulina ME, Chudakov DM, Britanova OV, Yanushevich YG, Staroverov DB, Chepurnykh TV, Merzlyak EM, Shkrob MA, Lukyanov S, Lukyanov KA (2006) A genetically encoded photosensitizer. Nat Biotechnol 24:95–99
142. Kent KP, Childs W, Boxer SG (2008) Deconstructing green fluorescent protein. J Am Chem Soc 130:9664
143. Kent KP, Oltrogge LM, Boxer SG (2009) Synthetic control of green fluorescent protein. J Am Chem Soc 131:15988

Fluorescence Lifetime of Fluorescent Proteins

Gregor Jung, Andreas Brockhinke, Thomas Gensch, Benjamin Hötzer, Stefanie Schwedler, and Seena Koyadan Veettil

Abstract Fluorescence is a photophysical phenomenon, which obeys basic physical laws. The fluorescence of the autofluorescent proteins arises on the molecular level from chromophores, which are buried in the protein matrix. The three-dimensional, well-defined architecture of the surrounding is a prerequisite for their function. Excitation of the isolated chromophores leads only to a negligible light emission at room temperature. Several processes competing with the radiative decay are responsible for the quenching. To understand how nature has learned to suppress these alternative pathways from the excited state in autofluorescent proteins, the molecular dynamics as well as the influence of several amino acids in the interior of the protein has to be analysed. We review the current status of the understanding of the non-radiative decay mechanisms for the different fluorescent protein classes, i.e., colours. Furthermore, we address what can be learned from fluorescence lifetime measurements and how they can be exploited for analytical purposes such as fluorescence lifetime imaging microscopy. Finally, we sketch the needs of increased fluorescence quantum yields and present strategies to prolong the fluorescence lifetimes.

Keywords FLIM · Internal conversion · Photophysics · Protein dynamics · TCSPC

Contents

G. Jung (✉), B. Hötzer, and S.K. Veettil
Biophysical Chemistry, Saarland University, Campus B2 2, 66123 Saarbrücken, Germany
e-mail: g.jung@mx.uni-saarland.de

A. Brockhinke and S. Schwedler
Physical Chemistry 1, University of Bielefeld, Postfach 10 01 31, 33501 Bielefeld, Germany

T. Gensch
Institute of Structural Biology and Biophysics 1 (Cellular Signaling, ISB-1), Forschungszentrum Jülich, 52425 Jülich, Germany

G. Jung (ed.), *Fluorescent Proteins I*, Springer Ser Fluoresc (2012) 11: 69–98,
DOI 10.1007/4243_2011_14,
Published online: 2 April 2011

1 Principles of Fluorescence Decays

1.1 Introduction into the Photophysics of a Fluorophore: The Jablonski Scheme

In this chapter, we will provide a treatment of fluorescence, which is based on basic physical laws. Some aspects are also covered in other chapters; we will rely on the work, which is presented elsewhere in this compendium.

The treatise can start with a consideration of a simplified Jablonski scheme (Fig. 1). Such diagrams are often used to visualize the possible transition of an isolated conjugated system. With respect to fluorescent proteins (FPs), this situation corresponds to a chromophore taken out of the protein barrel and embedded in an inert solvent or, even more stringent, into the vacuum. Both kinds of conditions were experimentally established and gave insight into the photophysics of FPs [1–3].

We can divide the different photophysical pathways in those where radiation is involved and those which are purely radiationless. The first group comprises transitions such as absorption, fluorescence and phosphorescence, the second group mainly consists of intersystem crossing (ISC), internal conversion (IC) and, if we account for the interaction with other molecules, quenching. The latter processes are not restricted to the first excited singlet and triplet state, S_1 and T_1, respectively, but occur also for higher excited states. Indeed, IC is the main decay mechanism, which prevents to observe strong emission from higher excited states.

The Jablonski scheme in Fig. 1 shows more photophysical pathways, which connect the different states. However, little is known about higher excited states S_n and T_n, i.e., about their energies and their lifetimes. Absorption spectra of synthetic chromophores show the excitation into higher singlet states at wavelengths $\lambda < 300$ nm. They are weaker than the $S_0 \rightarrow S_1$ transition [4]. However, these absorption bands are hidden in the spectra of proteins since the large number

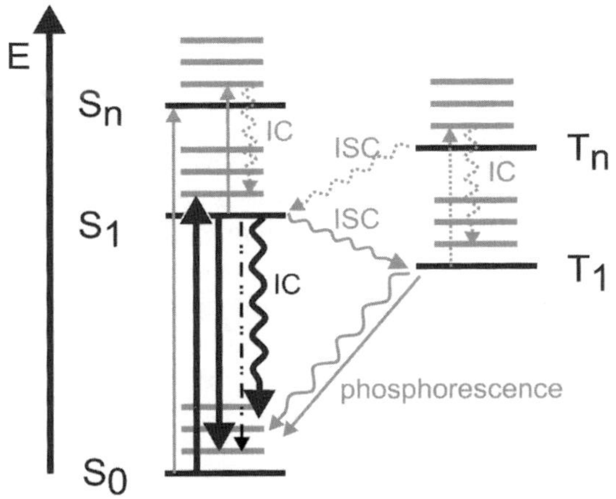

Fig. 1 Jablonski scheme for chromophores. With respect to fluorescent proteins, most spectroscopic data exist for absorption, emission and internal conversion (*black arrows*), i.e., transitions between the lowest two singlet-states S_0 and S_1. Experimental findings hint at some relevance of higher excited singlet-states S_n and the lowest triplet state T_1 (*grey, full line arrows*). Little is known about other possible transitions (*grey, dotted line arrows*)

of aromatic amino acids also absorbs strongly in the UV-range. Selective destruction of the chromophores in FPs, which would allow for verifying the energetic position in the natural environment by subtracting the spectra after some chemical treatment from the original spectra, is possible but likely not reliable enough [5]. More reliable is experimental and theoretical work on two-photon excitation, which gives insight into the energetics of higher excited states [6, 7]. That these higher excited singlet states are indeed of more than academic interest, can be inferred from two-photon excitation microscopy [8] and from the enhanced photoconversion upon excitation in these states [9, 10]: it was shown that both UV-excitation at 254 nm and excited-state absorption of the neutral chromophore lead to a rapid decarboxylation of a nearby glutamate at position 222. Based on these experiments, one could also hypothesize that other photochemical reactions such as photobleaching are accelerated upon excitation into higher excited states. Especially the characterization of excited-state absorption would be desirable since pump-dump schemes such as in stimulated-emission depletion (STED)-microscopy might suffer from such detrimental reactions [11, 12]. Excited-state absorption might also occur to some extent in microscopy within the diffraction-limited spot of a focussed laser beam, even with continuous-wave excitation. Due to the yet limited knowledge about these processes, we restrict the following discussion to the lowest excited states, i.e., S_1 and T_1.

The relevance of the lowest energetic triplet state is still under debate. In the past, fluorescence correlation spectroscopy (FCS) gave some evidence that the triplet state is significantly populated at high intensities in FPs [13]. Unfortunately, FCS only detects transient states due to their missing fluorescence emission, which is especially ambiguous in FPs: there are more photochemical reactions, which can lead to fluorescence intermittency. Only recently, some near-infrared (NIR) luminescence with a lifetime of tens of microseconds was detected, which was attributed to phosphorescence [14]. The experiments were originally designed to detect the weak emission of singlet oxygen $^1O_2$ upon photosensitization. When further spectroscopic characterization of the emitting species is provided, one could finally improve its impact on photochemical reactions such as the mentioned photosensitization as manifested e.g., in the so-called Killer Red proteins [15]. On the basis of the yet available data, we can neglect ISC as competitive process to fluorescence and, consequently, restrict the further discussion to the transitions between the two energetically lowest singlet states, i.e., S_1 and S_0.

1.2 Einstein Factors: Absorption, Spontaneous and Stimulated Emission

The kinetic formulation of Planck's law for the black-body radiation by Albert Einstein provided three molecular photophysical parameters, A_{21}, B_{12} and B_{21}. These three parameters are deduced when the spectrum of a thermal light source, such as the sun or a light bulb, is matched with the Boltzmann distribution. This so-called two-level system (TLS) is met in an approximation in the pair S_0 and S_1 of almost any good fluorophore. Therefore, the three Einstein factors A_{21} for spontaneous emission, B_{12} for absorption and B_{21} for stimulated emission are of *fundamental* validity and significance in fluorescence spectroscopy. For details, the interested reader is referred to more specialized articles, e.g., [16].

The complexity is reduced if two states of the same spin multiplicity, i.e., singlet or triplet, are treated. Due to the same degeneracy of both aforementioned TLS levels, the values of B_{21} and B_{12} become identical. The underlying processes are still absorption and stimulated emission. The only reason that the latter process is generally not observed in conventional fluorescence spectroscopy is the fact that the excited state is, at room temperature, much less populated than the electronic ground state. Stimulated emission, however, plays a role in pump-probe spectroscopy [17], in stimulated emission microscopy [18] and, also, in STED-microscopy [11, 12]. Furthermore, strong pumping of the excited state of FPs can lead to laser action which is, actually, a result of stimulated emission [19]. The importance of the parameter $B_{21} = B_{12}$ is its connection to the Einstein-factor A_{21} (1), where c is the speed of light and h is Planck's constant.

$$A_{21} = \left(\frac{8\pi h}{c^3}\right) \cdot \nu^3 \cdot B_{12}. \quad (1)$$

The unit of A_{21} is that of a frequency. In fact, A_{21} is the reciprocal value of the radiative lifetime τ_{rad} of the upper state in the TLS which is, here, the excited state S_1. In other words, the knowledge of the absolute absorbance, i.e., B_{12}, allows for calculating the fluorescence lifetime $\tau_{Fl} = \tau_{rad} = A_{21}^{-1}$. This equivalency, however, is only true if no fluorescence quenching mechanism takes place. This is another meaning of absorption spectroscopy for fluorescence spectroscopy. Equation (1) also shows that fluorescence, i.e., spontaneous emission, becomes more important at higher optical frequencies ν or shorter wavelengths λ. We will get back to this relation while comparing different proteins (Sect. 3).

One limitation of the early formulation by Einstein is that it accounts for only one wavelength λ or one optical frequency ν, respectively. Absorption as well as fluorescence spectra are characterized by band widths in the range of tens of nanometres (Fig. 2). This was solved in a careful reformulation of (1), (2) [23].

$$A_{21} = \left(\frac{8\pi \cdot n_0^2}{c^2}\right) \cdot \left(\frac{\ln 10}{N_A}\right) \cdot \frac{\langle \varepsilon_{abs}(\nu)/\nu \rangle}{\langle \nu_{fl}^{-3} \rangle}. \qquad (2)$$

The first term in parentheses takes the refractive index n_0 of the surrounding medium into account. The second term in parentheses, a constant, scales the molar decade extinction coefficient down to a molecular value (see Sect. 1.3). The brackets in the third term denote integrations over the frequency of the respective spectrum. The upper integral uses as argument the absorption spectrum $\varepsilon_{abs}(\nu)$, divided by the frequency ν. The lower integral mainly consists of the fluorescence

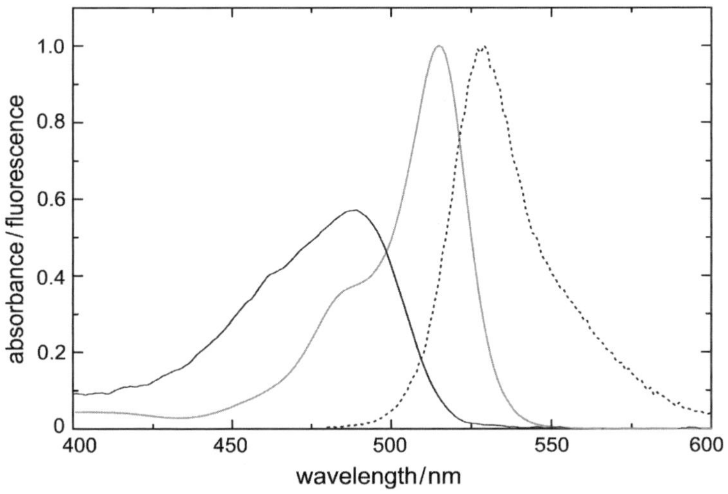

Fig. 2 Absorption spectra of eGFP (*black*) and eYFP (*grey*) and fluorescence spectrum of the latter (*black dotted*), normalized to the spectral area of eYFP. Taking the approximation of equal oscillator strengths f_{12} for eGFP and eYFP, ε_{max}(eGFP) can only be 0.57 × ε_{max}(eYFP) as the latter exhibits a distinctly smaller absorption spectrum. On the basis of ε_{max}(eYFP) = 83.400 M^{-1} cm^{-1}, $f_{12} = 0.48$ is calculated according to (4). Application of (2) leads to $\tau_{rad} = A_{21}^{-1} = 4.7$ ns [20]. The dependence of A_{21} on the spectral changes between GFPs and YFPs was observed to be lower than 7% [21], i.e., the radiative lifetime of eGFP can be as short as 4.4 ns which is very close to the value of [22]

spectrum $I_{fl}(\nu)$ and is used for calculating the centre frequency $\langle\nu\rangle$ of the fluorescence spectrum, weighted by ν^3 (3). In a rough approximation, one could also use the third power of the frequency at the maximum of the fluorescence spectrum, ν_{max}.

$$\left\langle \nu_{fl}^{-3} \right\rangle^{-1} = \frac{\int I_{fl}(\nu)d\nu}{\int \nu^{-3} I_{fl}(\nu)d\nu} \approx \nu_{max}^3. \tag{3}$$

One caveat has to be mentioned. Although the calculation according to (2) works reasonably well for good fluorophores, the absorbing ground state S_0 as well as the emitting excited state S_1 should possess a more or less similar electron distribution. This is satisfyingly fulfilled for the anionic chromophore species in FPs, i.e., most GFP (Green Fluorescent Protein), YFP (Yellow Fluorescent Protein) and, likely, red fluorescent proteins [21], although also structural changes due to vibrational relaxation may influence Einstein's relations. Equation (2) certainly fails when excited-state reactions such as excited-state proton transfer (ESPT) as in wild-type GFP (wt-GFP) are involved.

1.3 *Meaning of Absorption Spectra*

According to (2), there seem to be two ways to enhance the radiative rate constant A_{21}. The first approach is to enhance the refractive index n_0 of the surrounding. Its importance is experimentally proven and is used to map the refractive index within cells [24–26]. It also influences lifetime measurements at cryogenic temperatures where glycerol or sugars are added as cryoprotectors. Furthermore, the refractive index influence on A_{21} is surely one of the reasons for the, generally observed, shortened fluorescence lifetime in cells compared to aqueous solutions. Significantly higher refractive indices than in water were detected for the cytosol ($n_0 = 1.36$–1.38) and membranes ($n_0 = 1.45$) [27–29]. It is also especially important if τ_{Fl} is measured close to glass or, via a different mechanism, close to metal surfaces [30].

The other appealing approach would be to increase the extinction coefficient $\varepsilon_{abs}(\nu)$. However, the so-called oscillator strength f_{12} is given for a chromophoric system and can be hardly influenced by mutagenesis of the surrounding protein barrel. f_{12} is a dimensionless, molecular quantity for one electronic transition, e.g. absorption from S_0 to S_1. It is related to $\varepsilon_{abs}(\nu)$ which is experimentally accessible (4).

$$f_{12} = \left(\frac{4m_e\varepsilon_0 c}{e^2}\right) \cdot \left(\frac{\ln 10}{N_A}\right) \cdot \int \varepsilon_{abs}(\nu)d\nu. \tag{4}$$

The first term contains the electron mass m_e, the dielectric permittivity ε_0 of the vacuum and the elementary charge e. The conversion to the molecular quantity is, as in (2), provided by the second term in (4). Thus, this formula can be applied to almost any dye, but caution is advised when used for FPs.

A speciality of FPs is that $\varepsilon_{abs}(\nu)$ is related to 1 mole proteins but not to 1 mole of chromophoric units. The value neglects, on the one hand, that not every protein has a fully completed chromophore, i.e., incomplete chromophore maturation [31]. On the other hand, equilibria between different chromophore forms, e.g. the neutral and the anionic chromophore forms in wt-GFP, are also neglected. Mutations in FPs which do not alter the chromophore mainly influence these two parameters. Hence, a correction factor x might be introduced: x should denote the fraction of proteins of a certain mutant, which effectively contribute to the $S_0 \rightarrow S_1$ transition under investigation (5).

$$f_{12} = \left(\frac{4m_e\varepsilon_0 c}{e^2}\right) \cdot \left(\frac{\ln 10}{x \cdot N_A}\right) \cdot \int \varepsilon_{abs}(\nu)\mathrm{d}\nu. \tag{5}$$

The meaning of (4) is that, as f_{12} and both terms in the parentheses are constant, the integral over the frequency is also constant. In other words, the area under the absorption spectrum is constant for a specific transition of a certain chromophore form. A higher ε_{max} in an absorption spectrum results in a reduced width of the absorption band. In terms of quantum mechanics, the shape of the excitation as well as the emission spectrum is modulated by the Franck–Condon factors f_{FC} (see also Sect. 1.4). However, the natural lifetime τ_{rad} of the S_1 state is virtually constant if the spectral changes of the same transition upon mutation are minor. This relation was experimentally verified in a series of GFP and YFP variants; it turned out that the whole variation of A_{21} is less than 10% [21]. It is worth to emphasize that an accurate calculation of the extinction coefficients or the oscillator strength is absolutely necessary for a correct and unbiased determination of τ_{rad} (2).

1.4 Relevant Radiationless Processes

The decay of the upper state of a TLS like in Fig. 1 obeys first-order kinetics (6) with A_{21} as the decay rate constant. Sometimes, k_{rad} is used instead of A_{21}.

$$\frac{\mathrm{d}[S_1]}{\mathrm{d}t} = -A_{21}[S_1] \rightarrow [S_1](t) = [S_1] \cdot \exp(-A_{21} \cdot t). \tag{6}$$

The purely radiative lifetime, which could be measured if no fluorescence quenching took place, is $\tau_{rad} = A_{21}^{-1}$. Fluorescence lifetime measurements, however, always lead to τ_{fl}, which is shorter than τ_{rad}. A faster decay of the excited state S_1 is the result of additional decay channels from the excited state S_1 to the electronic ground state S_0. In the kinetic description, these exclusively non-radiative channels are accounted for by introducing additional rate constants. They are subsumed by k_{IC}, the rate constant for internal conversion (7).

$$\frac{\mathrm{d}[S_1]}{\mathrm{d}t} = -(A_{21} + k_{IC})[S_1] \rightarrow \frac{1}{\tau_{Fl}} = A_{21} + k_{IC}. \tag{7}$$

As we focus the discussion on the relaxation of S_1 to S_0, ISC and also all photochemical pathways are neglected, since, except from ESPT, the associated rate constants are too small to distort the conclusions drawn in this chapter for FPs.

Internal conversion is a generic term, which comprises different pathways, but the mechanism is similar for all of these relaxation channels. In any case, the large molecular excess energy, which corresponds to the energy difference ΔE between the ground and the excited state (between 160 and 300 kJ mol^{-1} in the visible range of the electromagnetic spectrum) has to be released. Although energy dissipation to the surrounding finally occurs by collisions, we will consider here only intramolecular processes, which refer to the chromophore itself. It should be noted, however, that the irreversibility of the transition $S_1 \rightarrow S_0$ is ensured by the mentioned energy dissipation. This is even true for ISC, which is, at first, an isoenergetic passage from one electronic state to another followed by rapid IC, as depicted in Fig. 1 in a simplified manner.

The most straightforward approach to IC is to consider a pair of potential functions, each for one electronic state (see Fig. 3). For the sake of simplicity, we

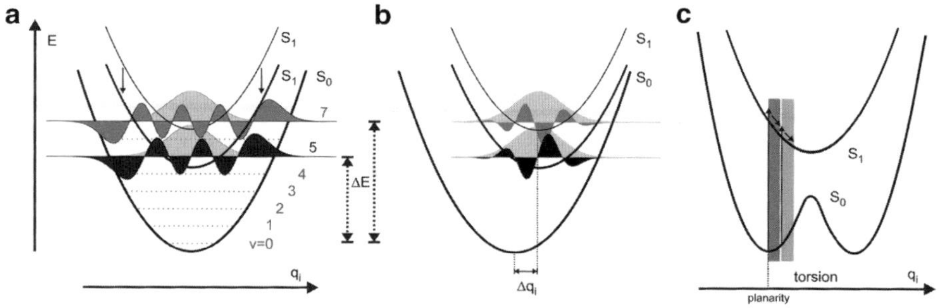

Fig. 3 Internal conversion and energy-gap law. The two singlet state S_0 and S_1 are described by harmonic oscillators along most coordinates q_i, as done in (**a**) and (**b**), but separated by the energy gap ΔE. (**a**) The radiationless transition from S_1 to S_0 depends on the overlap between the vibrational ground state wave function ($v = 0$) of S_1 (*light grey*) and the higher vibrational state ($v = 7$) of S_0. Upon decreasing ΔE by lowering e.g. the energy of S_1, overlap with a lower vibrational state (here $v = 5$) is established. (**b**) The overlap, i.e., the product of the S_1 wavefunction $v = 0$ with the S_0 wavefunctions $v = 5$ (*black*) or $v = 7$ (*dark grey*), respectively, is integrated over the whole coordinate yielding the Franck–Condon integral f_{FC}. Negative and positive values of the product cancel to a large extent. $f_{FC} = 0$ if there is no geometrical change upon excitation, i.e., $\Delta q_i = 0$, but increases with increasing Δq_i. Although both overlap functions look similar, the integral for $v = 5$ is ~10× larger than the integral for $v = 7$ in the depicted example. The rate constant for internal conversion, k_{ic}, is proportional to the square of the integrated overlap, i.e., the Franck–Condon factor f_{FC}^2 is 100× smaller for the larger change of v. (**c**) A reduction of ΔE can also occur by one-bond rotations. As the stiffness of the protein surrounding restricts the torsion to small movements – the *grey bars* symbolize the conformational latitude –, only twisted structures (*light grey*) of chromophores can reach regions in the excited state with smaller ΔE. Chromophores locked in planar structures (*dark grey*) maintain large energy gaps

assume an equal force constant k for both states. The likelihood of the crossing from the electronic S_1 state to the electronic ground state S_0, which is then, instantaneously, highly vibrationally excited, depends on the overlap of the vibrational wavefunctions of both states. This overlap is characterized by the Franck–Condon factor f_{FC}^2 [(8), see also Sect. 1.3] and depends in the case of radiationless transitions on several factors: (1) the difference of the vibrational quantum numbers Δv between the populated vibrational states of S_1 and S_0, (2) the force constant k of the vibration, (3) the energy of the vibration hc$\tilde{v}$ and (4) on the geometric displacement Δq_{i} of both potential curves [32–34].

$$f_{\mathrm{FC}}{}^2 = \frac{\exp(-\gamma) \cdot \gamma^{v_{SO}}}{v_{\mathrm{SO}}!}; \quad \gamma = \frac{k(\Delta q_i)^2}{2hc\tilde{v}}. \tag{8}$$

In (8) and Fig. 3, it is assumed that the radiationless transition occurs from the vibrational ground state in the first electronically excited state S_1, i.e., ($v_{S1} = 0$), to some vibrationally excited state v_{SO} in the electronic ground state S_0.

Any valence vibrations of hydrogen-atoms with $\tilde{v} \sim 3{,}000\ \mathrm{cm}^{-1}$ exhibit the strongest influence whenever they are coupled to the chromophore and undergo geometric changes upon excitation. Only 5–6 quanta of suchlike vibrations, i.e., $\Delta v = 5$–6, are required to surpass the energy gap ΔE between S_1 and S_0 in FPs. The effect becomes even faster if there is a larger geometric change Δq_{i} along the considered coordinate q_{i} between both electronic states. q_{i} is called a normal coordinate but could correspond e.g. to a bond length. It was experimentally found and reproduced by theoretical calculations that the rate constant k_{IC} roughly exponentially decreases with increasing ΔE [34]. The contribution of this so-called energy-gap law can be detected by a distinct change of τ_{Fl} upon H/D-isotope exchange: Δv is increased by a factor of $2^{1/2}$ upon deuteration and, concomitantly, k_{IC} is even smaller.

Although the importance of the energy-gap law is known for molecules, which exhibit fluorescence in the red region of the visible spectrum, little is found out about its importance in FPs. Figure 4 shows the only example so far (to our knowledge), where deuteration of the Gold Fluorescent Protein (GdFP) leads to a 1.5× longer τ_{Fl} [35]. However, the exact mechanism is still unknown. We propose that GdFP can be used to map the water exchange of cells by fluorescence lifetime microscopy once the incorporation of non-canonical amino acids into gene products of higher organisms will be realized.

Other, well-described decay mechanisms for different FP chromophores and their synthetic analogues are one-bond rotations [1, 36, 37]. Upon rotation, the energy of the ground state is raised whereas the energy of S_1 becomes reduced thus diminishing the energy gap ΔE. On the one hand, as the protein surrounding of the protein matrix is rather rigid, only small displacements of the chromophore from the adopted ground state conformation are possible and low energy-gap configurations in S_1 are hardly accessible. On the other hand, most chromophore surroundings fit better to a twisted chromophore [38]. It appears on the basis of multiple

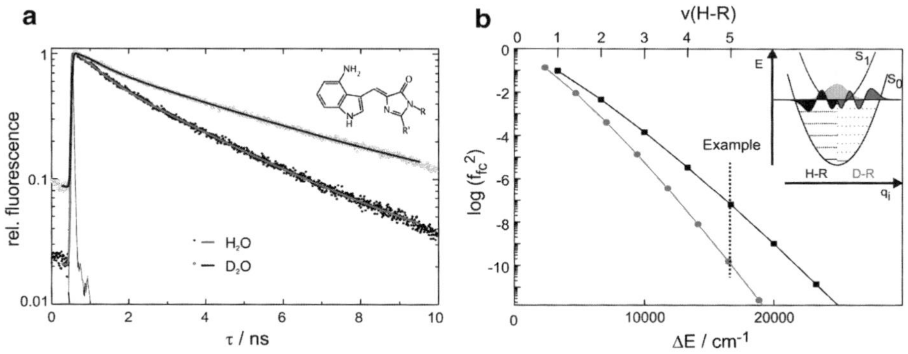

Fig. 4 Experimental isotope effect on the basis of the energy-gap law. (**a**) The longer component of the fluorescence decay in Gold fluorescent Protein (GdFP) [35] prolongs from $\tau_{\mathrm{Fl}} = 2.7$ ns to $\tau_{\mathrm{Fl}} = 4.2$ ns upon H_2O/D_2O exchange. The decay faster component is, probably, a result of the structural heterogeneity of GdFP. (**b**) The prolongation can be explained by the energy-gap law: ΔE of GdFP roughly corresponds to $\Delta v = 5$ quanta of an N–H vibration ($\tilde{\nu}_{\mathrm{N-H}} \sim 3{,}300\ \mathrm{cm}^{-1}$) but to $\Delta v = 7$ quanta of an N–D vibration ($\tilde{\nu}_{\mathrm{N-D}} \sim 2{,}300\ \mathrm{cm}^{-1}$). Calculations on the model of Fig. 3a, b and (8) (*inset*; wavefunctions are only half depicted) yield a reduction of k_{IC} by a factor 400 upon isotope exchange which is then too slow to compete with k_{rad}

structural analyses that the more planar a chromophore is in the crystallographic structure the longer is τ_{Fl}.

One-dimensional potential curves as depicted in Fig. 3 are only a rough approximation of the real circumstances of complex potential energy surfaces (PES). Even the chromophore itself exhibits a conformational space with more than 40 degrees of freedom. Displacements along some combinations of the underlying normal coordinates q_{i} lead to close approaches of the S_0 and S_1 PES, i.e., ΔE is small and efficient IC can occur. Even crossings of the different PES can exist in so-called conical intersections (CIs). However, isotope effects are unexpected for $\Delta E \sim 0$, and only quantum-mechanical treatments can give evidence for the corresponding radiationless transition from S_1 to S_0 [37]. It should be emphasized that only the knowledge of the coordinates q_{i}, i.e., geometric changes of the chromophore after excitation, gives a basis for the suppression of these movements and, thus, the fundament for a rational design of FPs with prolonged τ_{Fl}.

1.5 Other Mechanisms Which Compete with Fluorescence

There are other mechanisms which can lead to a reduction of τ_{Fl}. They are accounted for in (9) as additional rate constants.

$$\frac{\mathrm{d}[S_1]}{\mathrm{d}t} = -(A_{21} + k_{\mathrm{IC}} + k_{\mathrm{Iso}} + k_{\mathrm{PET}} + k_{\mathrm{FRET}} + \cdots)[S_1]$$
$$\rightarrow \frac{1}{\tau_{\mathrm{Fl}}} = A_{21} + k_{\mathrm{IC}} + k_{\mathrm{Iso}} + k_{\mathrm{PET}} + k_{\mathrm{FRET}} + \cdots. \quad (9)$$

ESPT is in FPs certainly one of the most evident and well-investigated mechanisms. Reversible cis-trans-isomerization can also occur when chromophoric systems, such as the chromophores of all FPs, contain exocyclic double-bonds. These isomerization reactions likely proceed via CI and are therefore ultrafast [38]. Photon-induced electron transfer (PET), in which transiently an electron is transferred between a nearby group and the chromophore, is hardly detectable if the back-reaction is faster than the forward reaction. Its prominent role can be concluded from the irreversible photoconversion of FPs. Here, PET can be deciphered as mechanism due to the irreversible release of carbon dioxide [10]. Thus, PET might generally show up in the photodynamics, e.g. in the fluorescence quenching of FP chromophores in the anionic form by halide anions, but experimental evidence is lacking. Theoretical calculations, however, struggle with the open-shell systems of transiently unpaired electrons.

Predictable changes of τ_{Fl} can be made in Förster resonance energy transfer (FRET). The good understanding of the theoretical background allows for designing energy donor–acceptor pairs with a sensing unit in between. FRET can be readout spectrally by taking the ratio of the donor and acceptor fluorescence, but also by lifetime changes of the donor [39]. The latter approach becomes especially promising if the acceptor is not fluorescent. Furthermore, lifetime changes do not depend on concentration changes or on fluctuations of the excitation intensity. Indeed, we recently developed a sensor for Cu^{2+}, where we exploited that Cu^{2+} exhibits a strong blue colour, i.e., broad absorbance of yellowish light, when bound to proteins [40].

Reduced τ_{Fl} can also result from aggregation. In the particular case of FPs, energy transfer on the picosecond time scale in homo- or hetero-dimers and oligomers distributes the energy among the individual chromophores (Schüttrigkeit et al. 2001) [20]. If one of the chromophores is in a dark state, then fluorescence vanishes. Self-quenching due to the formation of excimers, which can occur for organic dyes in close contact, is only a minor issue in FPs as the chromophores are buried deep within the protein barrel.

1.6 Importance of Long Fluorescence Lifetimes

The most important aspect of long τ_{Fl} is met when different FPs with chromophores of similar electronic structure are compared. In this case, the measured fluorescent lifetime is proportional to the fluorescent quantum yield Φ_{Fl} (10). A long τ_{Fl} therefore corresponds to a high Φ_{Fl}.

$$\Phi_{Fl} = \tau_{Fl} \cdot A_{21}. \tag{10}$$

Most GFPs and YFPs actually fulfil these requirements of similar electronic structures, and we verified this way of computing Φ_{Fl} by comparison with a classical approach [21]. High Φ_{Fl} are beneficial in FRET approaches, in single-molecule experiments, etc. as more fluorescence photons are generated from the same number of excitation cycles. An excitation cycle denotes the excitation from S_0 to S_1 with the subsequent decay back to S_0.

Longer τ_{Fl} are also preferential for analytical purposes. Also time-resolved anisotropy experiments would benefit from a longer τ_{Fl} as the rotation of FPs is already in the range of tens of nanoseconds [41]. As stated above, the actual limit is given by the radiative rate constant A_{21}. Hence, a prolongation of τ_{FL} can only be achieved if the extinction coefficient $\varepsilon_{abs}(\nu)$ or, more correctly, the oscillator strength f_{12} can be diminished. At present, this can be solely realized with the incorporation of small, non-canonical chromophores into FPs, where a trade-off between IC and the ν^3 dependence of A_{21} as limiting factors has to be found. It should also be emphasized that the total brightness of FPs will suffer from lowering f_{12} due to its connection with the extinction coefficients (5).

Finally, the sensitivity of lifetime measurements due to increased non-radiative rate constants k_j greatly improves as the lifetime change $\Delta\tau_{Fl}$ scales with the square of τ_{Fl} (11).

$$\tau_{FL} = \left(A_{21} + \sum k_i\right)^{-1} \rightarrow \frac{d\tau_{Fl}}{dk_j} = -\left(A_{21} + \sum k_i\right)^{-2} \rightarrow \Delta\tau_{FL} \approx -\tau_{FL}{}^2 \cdot \Delta k_j. \quad (11)$$

Examples for the changes of the mentioned rate constants, i.e., Δk_j, are altered FRET efficiencies (9), the accelerated decay due to a higher local refractive index n_0 according to (2).

2 Measurement of Fluorescence Decays

In the following, we shortly overview several experimental methods for the determination of τ_{Fl}. Recently, several monographs on this topic were released [39, 42, 43]. Also, original publications appeared in the past which compare the different methods in terms of photon economy, acquisition speed, etc [44–46]. Differences in the τ_{Fl} can be used as contrast mechanism, alternative to the emission colour, and allow for localizing different FPs [47]. All these more technical aspects are beyond the scope of this article, which focuses on lifetime variation of FPs.

2.1 Fluorescence Lifetime Measurements

Lifetime information can be obtained from various methods with different lifetime resolution. The fastest light-induced processes below 1 ps require pump-probe techniques. Excited-state lifetimes on this timescale lead to negligible few fluorescence photons with $\Phi_{Fl} < 10^{-4}$. The S_1 state is probed here either by stimulated emission or by upconversion of the few fluorescence photons. Although these methods can be applied, in principle, to lifetimes on the nanosecond timescale, the experiments are mostly designed, and thus more accurate, for faster processes. Length differences in the optical pathways between the excitation and the probe pulse are converted to a time delay. Path differences of 1 m, which are rarely exceeded, correspond to roughly 3 ns.

More relevant for fluorescence lifetime measurements are, therefore, setups where fluorescence photons are directly detected by some light-sensitive device. The time delay between the excitation pulse, i.e., the corresponding electronic signal, and the photon detector output is then recorded and analyzed for a large number of photons (see Fig. 5a, time-correlated single-photon counting (TCSPC)). Pulsed discharge lamps of some spectrometers have instrumental response functions (IRFs) with a width of ~1 ns and τ_{Fl} becomes trustworthy after careful reconvolution fitting. By using pulsed lasers, the time resolution can be as good as 50 ps and below, but, then, it critically depends on the photon counting detector. The most sensitive photon detectors such as avalanche photodiodes are limited to ~200 ps or even worse. They are especially used in fluorescence microscopy. The

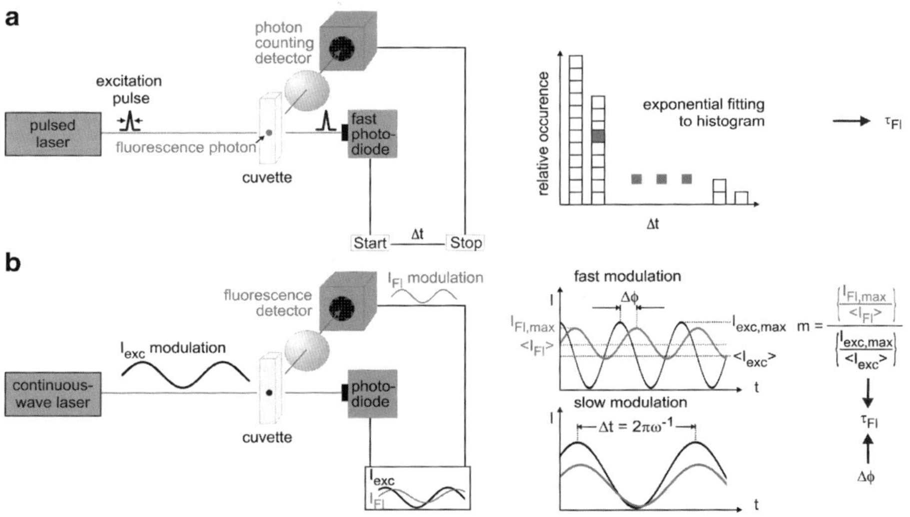

Fig. 5 Two widely used methods for the determination of τ_{Fl}, i.e., time-correlated single photon counting (**a**) and phase fluorimetry (**b**). (**a**) The time lag Δt between a single detected fluorescence photon and its excitation pulse is measured by fast electronics. A histogram is obtained from numerous fluorescence photons with individual Δt. Exponential fitting of the experimental histogram yields τ_{Fl}. (**b**) The sinusoidally modulated output I_{exc} of a laser generates a modulated fluorescence signal I_{Fl}. The phase shift $\Delta\phi$ between excitation (*black*) and emission (*grey*) depends on the modulation frequency ω and allows for calculating τ_{Fl}. Another quantity is the demodulation m, which is the ratio of the relative amplitude of the modulated I_{Fl} and the relative amplitude of I_{exc}. m decreases from 1 at slow modulation, i.e., ω is small, to minimally 0 at faster modulation, i.e., ω is large, and can be converted, as well, to τ_{Fl}

other extreme are streak cameras having the highest time resolution in this kind of experiments, even down to <1 ps. Here, photoelectrons are deflected by a rapidly changing electric field and then reconverted into light by a luminescence screen. A two-dimensional image of the luminescence then provides time resolution in the dimension along the electric field, i.e., the time axis. Often, the second axis is used for spectral information.

Phase fluorimetry is a completely different method, which works without pulsed excitation (Fig. 5b). Here, the intensity of the excitation light is modulated in time with a sine wave. The fluorescence arises with some retard, i.e., a phase-shift $\Delta\phi$, but it obeys the same modulation frequency ω. However, as the spontaneous fluorescence is a stochastic process of uncorrelated emitters, also the demodulation m of the fluorescence light drops as the inverse of the demodulation frequency, i.e., $2\pi\omega^{-1}$, approaches τ_{Fl}. Thus, both $\Delta\phi$ and m can be used for the determination of τ_{Fl}. In an ideal case, both values coincide [47–49].

2.2 *Fluorescence Lifetime Imaging Microscopy*

In principle, most of the aforementioned experiments can be used in imaging by performing lifetime experiments in a diffraction-limited spot. Consecutive recording of τ_{Fl} at neighbouring points in one dimension and then a line-by-line scan yields a two-dimensional representation of τ_{FL}, which can be extended to the third spatial dimension in confocal and multiphoton microscopy. Such experiments were performed with time-correlated single-photon counting, frequency modulation and even with streak-cameras; pump-probe experiments were hardly performed yet. As the recording of τ_{fl} at each pixel needs at least a fraction of an ms, a whole image is not acquired faster than several seconds. Faster processes cannot be visualized by this technique.

On the contrary, there are imaging experiments where the fluorescence is collected from a large area. The lifetime information is provided by some electronics, mostly an image intensifier, in front of a charge-coupled device (CCD). The fluorescence intensity in dependence of an adjustable time-gate, which is applied to the high-voltage of the image intensifier, enables extraction of τ_{Fl} from at least two images. Alternatively, modulation of the sensitivity of the detector array allows the combination of phase fluorimetry with microscopy.

2.3 *Lifetime Differences Between Microscopic and Cuvette Experiments*

Although τ_{Fl} measured in cuvette experiments should be identical to the values obtained in microscopy, several reasons can be identified why deviations are frequently observed. First of all, higher values of the refractive index n_0 of a cell enhance A_{21} [see (2)] and therefore lead to shorter τ_{Fl} [25, 41]. This explanation is mostly chosen when a clear dependence on other parameters is lacking, although experiments show that τ_{Fl} in the cytoplasm is not very different from aqueous solutions [50]. Second, local changes of the pH-value or the presence of a (yet unknown) quencher might lead to changes of τ_{Fl}. It was believed in the past that the fluorophore is well shielded from the solvent by the protein barrel. The finding that especially YFPs are sensitive to anions shows that this assumption is not generally valid. Indeed, we observed that even zwitterionic buffer molecules such as CAPS (*c*-hexyl-aminopropane sulfonic acid) of considerable size can quench τ_{Fl} of YFP at 10 mM concentrations by ~10%. Cellular systems contain a wealth of unidentified metabolites, and it is imaginable that some of these are locally so abundant that they can act as quenchers. Systematic investigations which address this issue are lacking so far, but would be of great importance. Third, temperature variation are certainly an issue [35, 51], at least for some mutants.

Finally, some FPs can undergo reversible or irreversible photochemistry, eventually with a concomitant colour change. Those processes are the basis of some recent microscopy applications of FPs allowing for intelligent timing and super-resolution imaging in cells. Focusing the radiation of 1 mW continuous wave, visible laser light to a diffraction limited spot can locally generate intensities I_{exc} up to 1 MW cm^{-2}. Such high intensities I_{exc} saturate the $S_0 \rightarrow S_1$ transition, i.e., more excitation cycles per second (k_{12}) cannot be performed. In other words, the first excited state S_1 is populated at least as strongly as the ground state S_0 (12). Here, k_{em} denotes the number of fluorescence photons per second which can be obtained from an individual emitter.

$$k_{\mathrm{em}} = \frac{k_{12}}{1 + k_{12} \cdot \tau_{Fl}} \cdot \Phi_{\mathrm{Fl}}; \quad k_{12} = \frac{\varepsilon(\nu_{\mathrm{exc}}) \cdot \ln 10}{N_{\mathrm{A}}} \cdot \frac{I_{\mathrm{exc}}}{h\nu_{\mathrm{exc}}}. \tag{12}$$

It is imaginable that excitation into higher excited states can occur under these conditions. However, in the limit that k_{12} reaches A_{21}, the observation of rare events is favoured. Rare events are processes with quantum yields below 0.1% and are detectable in cuvette experiments only under favourable circumstances such as colour-changes, accumulation techniques, etc. [9]. These processes, however, occur in microscopy with high probability due to the (orders of magnitude) higher excitation intensities and reduce the brightness [52]. Changes of τ_{Fl} in microscopy were actually observed for CFPs, some photoactivatable proteins, dsRed and also for enhanced GFP (eGFP) under the conditions of single-molecule spectroscopy [53–56]. Early evidence was provided that the change of τ_{Fl} is due to photoconversion (see Fig. 6) [21]. Also, preferential bleaching of one conformer in CFP (see below) with a longer lifetime was discussed [57], and one could also argue that light-driven isomerization leads to a rotamer with reduced τ_{Fl} [58, 59]. It is not clear whether there is a general mechanism and, if so, whether it always leads to a significant reduction of τ_{Fl}. A clear pattern, when a reduction of τ_{Fl} is likely to be observed, is still missing.

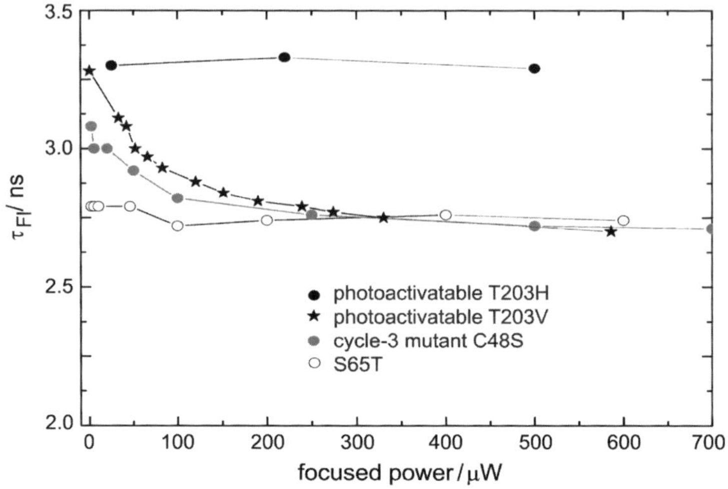

Fig. 6 Intensity-dependent change of τ_{Fl} due to diffraction-limited focusing of a pulsed laser ($\lambda_{exc} = 470$ nm, 63× NA 1.4 objective lens). The FPs were freely diffusing in a buffered solution at micromolar concentrations. This phenomenon was described for CFP [55, 57], dsRed [Bowen and Woodbury 2003] and even for eGFP [54]. Some GFP mutants such as T203V (see [56]) and C48S exhibit a change of τ_{Fl}, whereas the widely used mutants T203H and S65T appear stable in τ_{Fl}. The reason for this finding is unclear: preferential bleaching of one conformer, photoconversion or the light-induced population of some metastable isomer are discussed

2.4 Search for Proteins with Improved Photophysical Properties

As will be shown in Sect. 3, τ_{Fl} depends on the mutation pattern, and the understanding of the decay mechanisms allows for designing FPs with altered τ_{Fl}. Despite this rational proceeding, the progress is limited as protein folding, chromophore maturation and other biochemical parameters are also influenced by mutations. A rather elegant and straightforward approach unifies the rapid amplification of DNA by polymerase-chain reaction (PCR), the rapid conversion of DNA into proteins by bacteria and, finally, the screening capabilities of fluorescence lifetime imaging microscopy (FLIM) [48, 49, 60]. Accidental or specific mismatches during PCR actually create a multitude of sequences, which can subsequently result in a huge library of FPs with altered photophysical properties. On the one hand, any disturbance of the chromophore formation immediately will abort fluorescence. On the other hand, since τ_{Fl} can be measured independent from the fluorescence intensity, FP mutants with prolonged τ_{Fl} can be found as long as some visible protein fluorescence is detected. The expression yield can be optimized afterwards. The success of this approach, however, strongly relies on the way how the DNA library is generated. So far, most random mutagenesis only lead to a reduction of τ_{Fl}, whereas restriction of the combinatorial approach to few important positions enabled, quite recently, the generation of CFPs with $\Phi_{Fl} > 90\%$ and long τ_{Fl} (~4 ns) near τ_{rad} [61].

3 Fluorescence Lifetime of FPs

Most attempts to improve FPs were focused so far on fluorescence intensities, the spectral characteristics and the factors which have a strong impact on them. Fluorescence lifetimes were measured, if at all, as just one parameter among other spectroscopic data. It is therefore not astonishing that the values for τ_{Fl}, provided by different research groups, frequently vary more than the stated accuracy of the respective analysis. One prototypic example is τ_{Fl} of Cerulean, a CFP variant. It finds application as improved donor in FRET-constructs. Published values of τ_{Fl} span the range from 2.9 to 3.8 ns (Hoffmann et al. 2008) [58, 61, 62]. It should be noted that similar examples are widespread and can be found for YFPs as well as for dsRed. These ubiquitous deviations can be, at least in part, due to the alterations (temperature, refractive index) mentioned in the preceding sections; they could also hint at hidden quenchers or even at the proteins' history. Finally, even the mathematical treatment, i.e., monoexponential vs. biexponential fitting vs. fitting with a distribution of lifetimes can influence the outcome [59]. Hence, we prefer systematic investigations, where the effects of mutations are compared and from which qualitative conclusions are drawn.

3.1 Blue Fluorescent Proteins

The so-called blue fluorescent proteins (BFPs) denote FPs where the central amino acid Y66 in the wild-type GFP of the jellyfish *Aequorea victoria* is replaced by histidine (Y66H) [63] (Fig. 7, left). Only few time-resolved experiments were performed so far on this family, and it seems that the low Φ_{Fl} and other detrimental

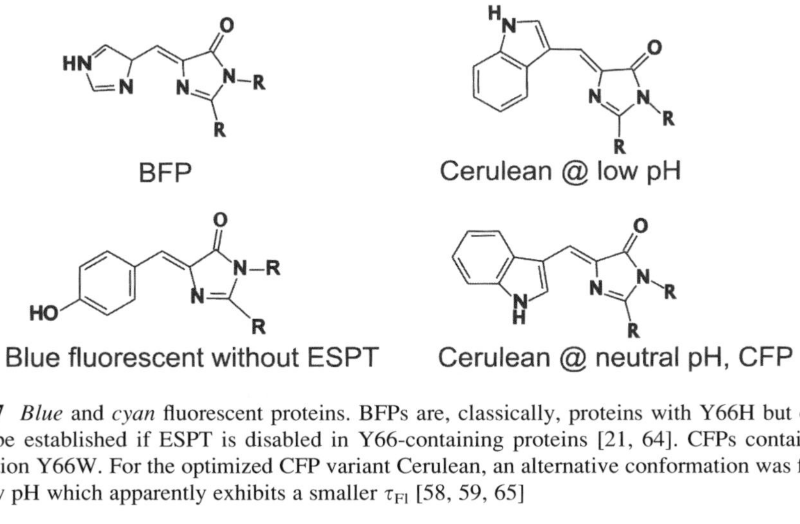

Fig. 7 *Blue* and *cyan* fluorescent proteins. BFPs are, classically, proteins with Y66H but could also be established if ESPT is disabled in Y66-containing proteins [21, 64]. CFPs contain the mutation Y66W. For the optimized CFP variant Cerulean, an alternative conformation was found at low pH which apparently exhibits a smaller τ_{Fl} [58, 59, 65]

properties prevented researchers from thorough investigations [66]. The introduction of H66 yields a protein with an averaged $\tau_{Fl} < 1$ ns at room temperature [67, 68]. Already, the additional mutations Y145F and S65T lead to proteins with distinctly prolonged τ_{Fl}, which could be further increased up to 3.65 ns by screening methods [49, 51, 64, 66]. Here, especially the mutation V224R appears beneficial for a long τ_{Fl}. A reduced conformational freedom or a stiffer hydrogen-bonding network is made responsible [69]. In combination with $\Phi_{Fl} \sim 0.5$, one could argue on the basis of (9) that the radiative lifetime τ_{rad} should be around 7 ns. However, all time-resolved experiments on BFPs have shown so far multiexponential decay behaviour for τ_{Fl} which suggests a strong heterogeneity of the chromophore surrounding. According to this model, some protein conformations lock the conjugated system in weakly or non-fluorescent states where rotations in the excited-state around the exocyclic C–C bonds are facilitated. These movements appear to be thermally activated and low-temperature experiments indeed show that the short lifetime components vanish [51, 67]. It should be noted that the longer lifetime component, despite a distinct temperature dependence, levels off at a maximum value $\tau_{Fl} \sim 3.5$ ns, which appears to be τ_{rad} in cryo-protectants.

Another way of disabling rotations around the exocyclic double bonds is the coordination of metal ions by the nitrogen-atoms of the imidazolinone and the histidine. It would be interesting to see whether binding of Zn^{2+}, which is not redox active and which cannot act as FRET quencher, to both heterocycles of the chromophore would result in similar values of τ_{Fl} at room and at low temperatures [70].

If we define BFP just by its fluorescence colour, then two further types of proteins should be mentioned: First, Y66F variants which exhibit $\tau_{Fl} << 1$ ns at room temperature but $\tau_{rad} \sim 4.5$ ns at cryogenic temperatures [67]; second, proteins containing Y66 where the ESPT is suppressed. This effect was observed in several proteins containing the mutation E222Q, and τ_{Fl} up to 1.5 ns at room temperature was observed [21]. However, further prolongation of τ_{Fl} and systematic investigations are missing yet owing to the low expression yields of E222Q containing proteins. In an alternative approach, suppression of ESPT was be achieved by the mutations S205V/T203V in enhanced GFP (eGFP) leading to the bright BFP mKalama [71]. In the same work, other bright BFPs, partially derived from other coral proteins, were presented which all exhibit $\Phi_{Fl} \sim 0.5$.

3.2 Cyan Fluorescent Proteins

Cyan Fluorescent Proteins (CFPs), where the central amino acid of the chromophore triad is replaced by tryptophane [63] (Fig. 7, right), were disregarded in terms of its photophysical characterization for almost one decade. The fluorescence maximum below 500 nm overlaps with strong cellular autofluorescence, and therefore, application of CFPs is mostly restricted to its use as FRET-donor [72]. A high-resolution crystallographic analysis of eCFP initiated further improvements: two different conformations are found and are correlated with two different lifetimes as found by TCSPC [35]. The major conformation with the longer $\tau_{Fl} = 2.8$ ns is associated with a conformation, where Y145 is buried within the chromophore pocket and H148 is exposed to the solvent. In the minor conformation, however, the occupation of the interior by both hydrophobic amino acids is reversed leading to a reduced $\tau_{Fl} = 0.6$ ns. Thus, by the substitution H148D, where the hydrophilic aspartate is exposed to the solvent at neutral pH, a CFP called Cerulean with a distinctly longer but not monoexponential τ_{Fl} is developed [62]. Molecular dynamics simulations are performed to understand the effect of various amino acid replacements on the lifetime [58, 73]. The lifetime differences between CFP and Cerulean are associated with the preferential stabilization of a completely planar structure in the latter (see Fig. 3c).

At lower pH values, however, a conformation of D148 is observed where this residue is bound through a hydrogen-bridge to the chromophore and, thus, stabilizing the latter in an isomeric rotamer [65]. Concomitantly, a distinct reduction of τ_{Fl} by 40% under acidic conditions is observed (Fig. 8a) [59]. CFPs can thus be used as FLIM-indicator revealing cellular acidification with spatial resolution (Fig. 8b).

The occurrence of multiple conformational states in CFPs challenges the assignment of individual lifetime components to crystallographically defined structures and backs lifetime heterogeneity [48, 59]. However, it should be mentioned that a higher Φ_{Fl} due to H148D was confirmed. Most recently, introduction of the smallest amino acid at position 148, i.e., mutation H148G, as well as bulkier amino acids at V224 (V224L, V224R) leads to prolonged $\tau_{Fl} > 4$ ns thus approaching τ_{rad} [61]. This goes together with a maximal fluorescence quantum yield near 1.

3.3 *Green and Yellow Fluorescent Proteins*

The most exhaustive investigations of the dependence of τ_{Fl} on structural effects are performed on proteins with Y66 as central aromatic amino acid as part of the

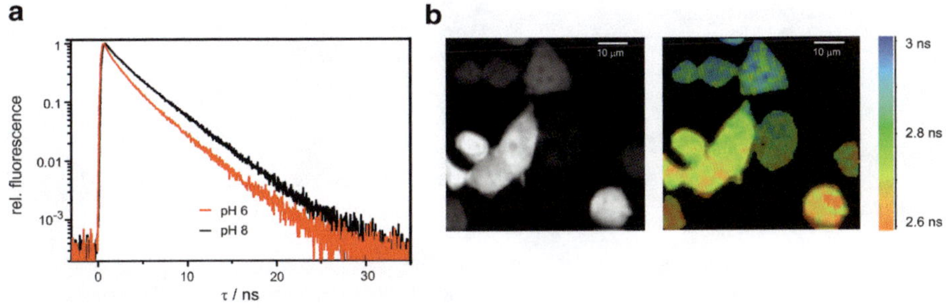

Fig. 8 ECFP as pH-sensor for FLIM. (**a**) ECFP fluorescence decay ($\lambda_{exc} = 440$ nm; $\lambda_{obs} = 480$ nm) is shorter at lower pH (apparent $pK_a = 6.5$) (**b**) Twophoton fluorescence images ($\lambda_{exc} = 820$ nm) of HEK293 cells expressing ECFP (*left*: intensity; *right*: fluorescence lifetime). The extracellular solution has a pH of 6.5 and 10 μM of the K^+/H^+ ionophore nigericin leading to lowering of intracellular pH in part of the cells

chromophore. In the meantime, many further natural sources of FPs are found but fundamental work on these proteins is scarcely performed. We therefore focus on avFP, which are derived from wild-type GFP from the pacific jellyfish *Aequorea victoria* and where general tendencies are provided, but similar considerations might also hold for the other proteins with identical chromophores. The common feature of GFP and YFP is that fluorescence of both stems from anionic chromophores, i.e., the phenolate form of Y66, and that is the reason why they are discussed together.

3.3.1 B-State Versus I-State: Electronic Effects

Already the first thorough, spectroscopic investigations of wt-GFP shows that the anionic chromophore R^- consists of two different forms, i.e., the intermediate state I and the stable ground state form B [68, 74]. It is believed that the I-state, of which the fluorescence is slightly red-shifted compared to that of the B-state, is formed only after ESPT. Moreover, the higher thermodynamic stability of the B-state is mainly attributed to a hydrogen-bonding between T203 and the chromophore, which stabilizes the negative charge of the phenolate (see structure at the top in Fig. 9a). Later on, it was shown that a minor population of the I-state exists even in the equilibrium with the neutral chromophore state and could be exclusively populated by point mutations [77]. Pump-dump-probe spectra verify this finding [17]. Independently, careful spectrally resolved lifetime measurements unravel two lifetime components in wt-GFP as well as in eGFP: the faster component with $\tau_{Fl} = 2.7–2.8$ ns is attributed to the predominant B-state whereas the second

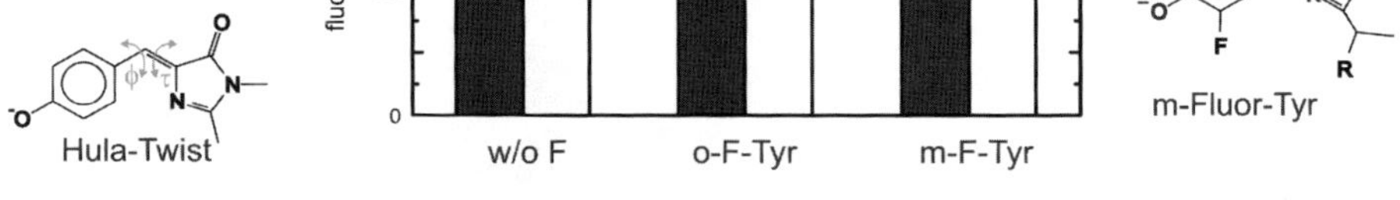

Fig. 9 Electronic effects on the fluorescence lifetime. (**a**) The electronic structure is roughly described by two mesomeric structures following the oxonol model [75]. The stronger the benzoidal character (*top*) is the more effective is internal conversion via rotation around ϕ. The hula-twist is described by the simultaneous rotation around ϕ and τ and should be less susceptible for preponderance of one mesomeric structure. (**b**) This phenomenon is also exemplified in the comparison of fluorinated chromophores in eGFP (*black bars*) and eYFP (*white*) with their native counterparts: The benzoidal character is enhanced upon fluorination leading to a reduced fluorescence lifetime τ_{Fl}. Structural effects might also play some role [76]

component $\tau_{Fl} = 3.3–3.4$ ns stems from the I-state [78, 79]. Subsequent comparative investigations confirm that the missing hydrogen-bond prolongs τ_{Fl} in FPs [21].

The reason for this, at first glance, unexpected behaviour is of purely electronic nature. The molecular mechanism, which is presumably responsible for the lifetime shortening, is the one-bond rotation around the angle ϕ. The anionic chromophore can be described by two mesomeric forms, i.e., the quinoidal and the benzoidal form. The double-bond character of the relevant exocyclic C–C bond is higher in the quinoidal form, thereby stiffening the chromophore. In contrast, the benzoidal form is strengthened when the negative charge is stabilized on the phenolate, e.g. by hydrogen-bonding. Here, the chromophore can accommodate easier to the torque pushing it away from planarity [38]. The same effect is observed when fluorinated tyrosines are incorporated into the chromophore [76]: the negative charge is stabilized because of the electron-withdrawing capability of fluorine atoms (−I effect), which then leads to a reduction of τ_{Fl} despite strong interactions with the surrounding in the protein (see Fig. 9b).

3.3.2 Structural Effects of the Protein Matrix

The above described electronic effects certainly are most concrete and point to the torsional deformation around ϕ as major decay mechanism (see also Sect. 1.4). This movement is made responsible for the fast excited-state decay of the synthetic GFP chromophore and of BFPs [1, 3, 67]. Also, sophisticated quantum-mechanical calculations confirm that an excited-state minimum is reached along this single-bond rotation of the isolated anionic chromophore, thus reducing the energy gap between S_0 and S_1 [37]. The effect of the protein matrix is to impose a strong counterforce to this rotation [80]. A temperature dependence, which apparently hints at an activated process, can be explained by the strong friction exerted by the protein [81]. Therefore, lifetime measurements allow for conclusions about protein dynamics [51].

There are also experimental data, which supports the importance of bulky aromatic amino acids such as phenylalanine or tyrosine at 203 for long τ_{Fl}. It seems that a $\tau_{Fl} > 3.5$ ns cannot be achieved in GFPs, i.e., proteins without aromatic residues at 203. In contrast, several authors reported τ_{Fl} close or even >4 ns for YFPs, thereby exceeding all published values for GFPs [21, 82, 83]. These values are close to the theoretical limit τ_{rad}, which probably is ~4.6 ns [22] (see Fig. 2). One could argue that the phenyl- or tyrosyl-moiety maintains the planarity of the chromophore.

Another important decay mechanism is the concerted two-bond flip around τ and ϕ, i.e., the so-called Hula-twist (HT) [84]. It comprises at least four different collective movements of atoms, with graduated importance [85]. That HT can actually occur in FPs, can be inferred from some switching reactions, especially of red FPs [86, 87]. It is the only way how cis-trans isomerization can proceed therein [38]. There is, however, no need of a completed isomerization reaction for efficient fluorescence quenching [88].

Experimental data provide evidence for this mechanism since only in this case changes in the vicinity of the imidazolinone heterocycle, e.g. by photoconversion should influence τ_{Fl} [21]. According to recent theoretical results, it seems that enough spatial freedom is given for efficient radiationless processes even in YFPs [69]. This finding could explain why τ_{Fl} of many YFPs, despite the steric constraints of Y203, is not superior to eGFP, i.e., around 3.0 ns [48]. One should mention at this point that the most widely used YFPs contain the mutation S65G for enhanced maturation, which is shown to distinctly reduce Φ_{Fl} as well as τ_{Fl} [21].

3.3.3 Some Peculiarities

Although the preceding paragraphs might evoke the impression that especially the photophysics of anionic chromophore species is completely understood, some astonishing facts should be listed, which deserve further investigations:

1. Red-edge excitation of wt-GFP yielded $\tau_{Fl} = 4.5$ ns which is, obviously, close to τ_{rad} [79]. There is hope that even with GFPs, higher τ_{Fl} like for CFPs can be achieved.
2. Even considerably longer τ_{Fl} of 6.2 ns (averaged value) of green emitting forms of dsRed were observed in single-molecule experiments [89]. This value is the longest τ_{Fl} ever measured for GFP-like molecules, and immediately raises the question whether the green emitting form of dsRed is different from avFP in terms of its electronic properties. Another issue is whether energy migration, i.e., homo-FRET [20], can explain such a strong prolongation as it is observed for ordinary reabsorption [90]. Many FPs form oligomers (dsRed is only a prominent example), so this process may have practical relevance when applying these FPs in cell imaging applications.
3. The preceding example highlights the analytical power of single-molecule experiments. Such investigations are also successful in showing that intense illumination leads to a reduction of τ_{Fl} of eGFP [54], whereas ensemble experiments on the very similar S65T failed (see Fig. 6). Single-molecule experiments are beneficial when the disappearance of some intermediate is faster than its generation: under these kinetic conditions, a small population with atypical behaviour can be easily overseen in ensemble investigations. It is still the question why some lifetime measurements in cellular environments and microscopy yield τ_{Fl}, which are considerably smaller than in cuvette experiments [47, 50].

3.4 Orange and Red Fluorescent Proteins

With the examination of more animals, from which FPs could be isolated afterwards, FPs with emission maxima $\lambda > 530$ nm became available [91, 92]. At the beginning, especially dsRed is investigated and the basis for further developments.

Owing to its tetrameric nature and its (incomplete) chromophore formation through a green intermediate, lifetime experiments are performed which aimed at the elucidation of intramolecular hetero-FRET [72, 89, 93, 94]. Most experiments show $\tau_{Fl} = 3.7$ ns as longest lifetime component. It is again the question whether energy migration can prolong τ_{Fl}. It appears, however, that monomerization and subsequent conversion into the mFruit-family is predominantly detrimental to Φ_{Fl}. [95–99]. Hardly any lifetime experiments of either oligomeric or monomeric proteins have been performed as tunable pulsed excitation sources are still limited in this spectral range [6, 100]. Comparative studies are vastly missing. In the theoretical treatment, which only recently began [101, 102], even more volume-conserving mechanisms should be taken into account [103].

Some general statements can be made: the extinction coefficient of some members of the mFruit-family exceeds a value of $10^5\ M^{-1}\ cm^{-1}$ at the wavelength of maximal absorbance [95], resulting in a reduction of τ_{rad} according to (2). The red-shifted fluorescence of these FPs acts in the opposite direction owing to the ν^3 dependence. As a first approximation, we expect that τ_{rad} of the new FPs is, hence, similar to that of other FPs. The limited number of yet available experimental data, however, points to a longer $\tau_{rad} \sim 5$–11(!) ns [6]. The achievable Φ_{Fl} and τ_{Fl} are not higher than those of other FPs, which might be traced back to the increasing importance of the energy-gap law (Fig. 4). Further comparative experiments, which also consider the incomplete chromophore maturation, are strongly desired for a better understanding.

4 Outlook

Long τ_{Fl} are an indication of high Φ_{Fl} and are advantageous for most applications in fluorescence technology. For avFPs, a good, but incomplete understanding of the factors, which influence τ_{Fl} has been achieved in the past decade. On the one hand, amino acid substitutions especially at positions 65, 145, 148, 203, 224 of the protein sequence were found to influence τ_{Fl}. In some of these cases, satisfying explanations how these mutations affect τ_{Fl} are lacking. On the other hand, comparative structural investigations revealed beneficial protein conformations and pointed to improvement strategies. The combination of both approaches by screening methods appears promising in finding advanced proteins [61]. Whereas for CFPs and YFPs proteins now exist where τ_{Fl} is close to the theoretical limit of τ_{rad}, more efforts have to be spent for the design of correspondingly better BFPs and GFPs. Lifetime experiments on these proteins, where broad distribution of τ_{Fl} showed up, indicate that the aim of τ_{Fl} close to τ_{rad} is challenging but not hopeless. Indeed, mutations might be introduced which favour stiffer protein conformations, thereby providing an optimal geometry for a planar chromophore and a reduced lifetime heterogeneity while maintaining other mandatory biophysical properties such as high expression and maturation yields etc.

The situation is slightly different for the red- and orange-fluorescent proteins. Comparative studies are widely missing and, therefore, the mutations which influence τ_{Fl} are almost unknown. Even the general decay mechanisms are not yet described. The timescale of their biotechnological evolution can be roughly estimated to be a decade, based on the previous developments with avFPs. We expect that orange and red FPs will be improved with respect to τ_{Fl} and Φ_{Fl} once the most appropriate candidates for imaging purposes are identified.

Acknowledgements This work was supported by the German Science Foundation (DFG JU650/2-2). We are indebted to Nediljko Budisa for the generous gift of proteins with unnatural amino acids. We also thank Dagmar Auerbach for careful proofreading.

References

1. Kummer AD, Kompa C, Niwa H, Hirano T, Kojima S, Michel-Beyerle ME (2002) Viscosity-dependent fluorescence decay of the GFP chromophore in solution due to fast internal conversion. J Phys Chem B 106:7557–7559
2. Lammich L, Petersen MA, Brøndsted NM, Andersen LH (2007) The gas-phase absorption spectrum of a neutral GFP model chromophore. Biophys J 92:201–207
3. Litvinenko KL, Webber NM, Meech SR (2003) Internal conversion in the chromophore of the green fluorescent protein: temperature dependence and isoviscosity analysis. J Phys Chem A 107:2616–2623
4. Kojima S, Hirano T, Niwa H, Ohashi M, Inouye S, Tsuji FI (1997) Mechanism of the redox reaction of the *Aequorea* green fluorescent protein (GFP). Tetrahedron Lett 38:2875–2878
5. Inouye S, Tsuji FI (1994) Evidence for redox forms of the *Aequorea* green fluorescent protein. FEBS Lett 351:211–214
6. Drobizhev M, Tillo S, Makarov NS, Hughes TE, Rebane A (2009) Absolute two-photon absorption spectra and two-photon brightness of orange and red fluorescent proteins. J Phys Chem 113:855–859
7. Nifosi R, Luo Y (2007) Predictions of novel two-photon absorption bands in fluorescent proteins. J Phys Chem B 111:14043–14050
8. Tillo SE, Hughes TE, Makarov NS, Rebane A, Drobizhev M (2010) A new approach to dual-color two-photon microscopy with fluorescent proteins. BMC Biotechnol 10:6
9. Langhojer F, Dimler F, Jung G, Brixner T (2009) Ultrafast photoconversion of the green fluorescent protein studied by accumulative femtosecond spectroscopy. Biophys J 96:2763–2770
10. van Thor JJ, Gensch T, Hellingwerf KJ, Johnson LN (2002) Phototransformation of green fluorescent protein with UV and visible light leads to decarboxylation of glutamate 222. Nat Struct Biol 9:37–41
11. Hell SW (2007) Far-field optical nanoscopy. Science 316:1153–1158
12. Klar TA, Jakobs S, Dyba M, Egner A, Hell SW (2000) Fluorescence microscopy with diffraction resolution limit broken by stimulated emission. Proc Natl Acad Sci USA 97:8206–8210
13. Widengren J, Mets Ü, Rigler R (1999) Photodynamic properties of green fluorescent proteins investigated by fluorescence correlation spectroscopy. Chem Phys 250:171–186
14. Jiménez-Banzo A, Nonell S, Hofkens J, Flors C (2008) Singlet oxygen photosensitization by EGFP and its chromophore HBDI. Biophys J 94:168–172
15. Bulina ME, Chudakov DM, Britanova OV, Yanushevich YG, Staroverov DB, Chepurnykh TV, Merzlyak EM, Shkrob MA, Lukyanov S, Lukyanov KA (2006) A genetically encoded photosensitizer. Nat Biotechnol 24:95–99

16. Hilborn RC (1982) Einstein coefficients, cross sections, f values, dipole moments and all that. Am J Phys 50:982–986
17. Kennis JTM, Larsen DS, van Stokkum IHM, Vengris M, van Thor JJ, van Grondelle R (2004) Uncovering the hidden ground state of green fluorescent protein. Proc Natl Acad Sci USA 101:17988–17993
18. Min W, Lu S, Chong S, Roy R, Holtom GR, Xie XS (2009) Imaging chromophores with undetectable fluorescence by stimulated emission microscopy. Nature 461:1105–1109
19. Pikas DJ, Kirkpatrick SM, Tewksbury E, Brott LL, Naik RR, Stone MO, Dennis WM (2002) Nonlinear saturation and lasing characteristics of green fluorescent protein. J Phys Chem B 106:4831–4837
20. Jung G, Ma Y, Prall BS, Fleming GR (2005) Ultrafast fluorescence depolarisation in the yellow fluorescent protein due to its dimerisation. Chem Phys Chem 6:1628–1632
21. Jung G, Wiehler J, Andreas Z (2005) The photophysics of green fluorescent protein: influence of the key amino acids at positions 65, 203, and 222. Biophys J 88:1932–1947
22. Heikal AA, Hess ST, Webb WW (2001) Multiphoton molecular spectroscopy and excited-state dynamics of enhanced green fluorescent protein (EGFP): acid-base specifity. Chem Phys 274:37–55
23. Strickler SJ, Berg RA (1962) Relationship between absorption intensity and fluorescence lifetime of molecules. J Chem Phys 37:814–822
24. Suhling K, Siegel J, Phillips D, French PMW, Lévêque-Fort S, Webb SED, Davis DM (2002) Imaging the environment of green fluorescent protein. Biophys J 83:3589–3595
25. Tregidgo C, Levitt JA, Suhling K (2008) Effect of refractive index on the fluorescence lifetime of green fluorescent protein. J Biomed Opt 13:031218
26. Van Manen H, Verkuijlen P, Wittendorp P, Subramaniam V, van den Berg TK, Roos D, Otto C (2008) Refractive index sensing of green fluorescent proteins in living cells using fluorescence lifetime imaging microscopy. Biophys J 94:L67–L69
27. Beuthan J, Minet O, Helfmann J, Herrig M, Müller G (1996) The spatial variation of the refractive index in biological cells. Phys Med Biol 41:369–382
28. Curl CL, Bellair CJ, Harris T, Allman BE, Harris PJ, Stewart AG, Roberts A, Nugent KA, Delbridge LMD (2005) Refractive index measurement in viable cells using quantitative phase-amplitude microscopy and confocal microscopy. Cytometry A 65:88–92
29. Yamauchi T, Iwai H, Miwa M, Yamashita Y (2008) Low-coherent quantitative phase microscope for nanometer-scale measurement of living cells morphology. Opt Express 16:12227–12238
30. Fu Y, Zhang J, Lakowicz JR (2008) Metal-enhanced fluorescence of single green fluorescent protein (GFP). Biochem Biophys Res Commun 376:712–717
31. Ormö M, Cubitt AB, Kallio K, Gross LA, Tsien RY, Remington SJ (1996) Crystal structure of the Aequorea victoria green fluorescent protein. Science 273:1392–1395
32. Englman R, Jortner J (1970) The energy gap law for radiationless transitions in large molecules. Mol Phys 18:145–164
33. Freed KF (1978) Radiationless transitions in molecules. Acc Chem Res 11:74–80
34. Henry BR, Siebrand W (1974) Radiationless transitions. In: Birks J (ed) Organic molecular photophysics, vol 1. Wiley, London
35. Bae JH, Rubini M, Jung G, Wiegand G, Seifert MHJ, Azim MK, Kim J, Zumbusch A, Holak TA, Moroder L, Huber R, Budisa N (2003) Expansion of the genetic code enables design of a novel "gold" class of green fluorescent proteins. J Mol Biol 328:1071–1081
36. Kummer AD, Kompa C, Lossau H, Pöllinger-Dammer F, Michel-Beyerle ME, Silva CM, Bylina EJ, Coleman WJ, Yang MM, Youvan DC (1998) Dramatic reduction in fluorescence quantum yield in mutants of green fluorescent protein due to fast internal conversion. Chem Phys 237:183–193
37. Martin ME, Negri F, Olivucci M (2004) Origin, nature, and fate of the fluorescent state of the green fluorescent protein chromophore at the CASPT2//CASSCF resolution. J Am Chem Soc 126:5452–5464

38. Maddalo S, Zimmer M (2006) The role of the protein matrix in green fluorescent protein fluorescence. Photochem Photobiol 82:367–372
39. Lakowicz JR (2006) Principles of fluorescence spectroscopy, 3rd edn. Springer, Berlin
40. Hötzer B, Ivanov R, Brumbarova T, Altmeier S, Kappl R, Bauer P, Jung G (2010) Determination of copper(II) ion concentration by lifetime measurements of green fluorescent protein. unpublished results
41. Borst JW, Hink MA, van Hoek A, Visser AJWG (2005) Effects of refractive index and viscosity on fluorescence and anisotropy decays of enhanced cyan and yellow fluorescent proteins. J Fluoresc 15:153–160
42. Becker W (2005) Advanced time-correlated single photon counting technique. Springer series in chemical physics, vol 81. Springer, Heidelberg
43. Gadella TWJ (ed) (2009) FRET and FLIM techniques. Elsevier Science & Technology, Amsterdam
44. Gerritsen HC, Asselbergs MAH, Agronskaia AV, Van Sark WGJHM (2002) Fluorescence lifetime imaging in scanning microscopes: acquisition speed, photon economy and lifetime resolution. J Microsc 206:218–224
45. Suhling K, French PMW, Phillip D (2005) Time-resolved fluorescence microscopy. Photochem Photobiol Sci 4:13–22
46. Yasuda R (2006) Imaging spatiotemporal dynamics of neuronal signalling using fluorescence resonance energy transfer and fluorescence lifetime imaging microscopy. Curr Opin Neurobiol 16:551–561
47. Pepperkok R, Squire A, Geley S, Bastiaens PIH (1999) Simultaneous detection of multiple green fluorescent proteins in live cells by fluorescence lifetime imaging microscopy. Curr Biol 9:269–272
48. Kremers G, Goedhart J, van Munster EB, Gadella TWJ Jr (2006) Cyan and yellow super fluorescent proteins with improved brightness, protein folding, and FRET Förster radius. Biochemistry 45:6570–6580
49. Kremers G, Goedhart J, van den Heuvel DJ, Gerritsen HC, Gadella TWJ Jr (2007) Improved green and blue fluorescent proteins for expression in bacteria and mammalian cells. Biochemistry 46:3775–3783
50. Hess ST, Sheets ED, Wagenknecht-Wiesner A, Heikal AA (2003) Quantitative analysis of the fluorescence properties of intrinsically fluorescent proteins in living cells. Biophys J 85:2566–2580
51. Mauring K, Deich J, Rosell FI, McAnaney TB, Moerner WE, Boxer SG (2005) Enhancement of the fluorescence of the blue fluorescent proteins by high pressure or low temperature. J Phys Chem B 109:12976–12981
52. Jung G, Zumbusch A (2006) Improving autofluorescent proteins: comparative studies of the effective brightness of Green Fluorescent Protein (GFP) mutants. Microsc Res Tech 69:175–185
53. Bowen B, Woodbury N (2003) Single-molecule fluorescence lifetime and anisotropy measurements of the red fluorescent protein, DsRed, in solution. Photochem Photobiol 77:362–369
54. Düser M, Zarrabi N, Bi Y, Zimmermann B, Dunn S, Börsch M (2006) 3D-localization of the a-subunit of FoF1-ATP synthase by time resolved single-molecule FRET. Proc SPIE 6092:60920
55. Hoffmann B, Zimmer T, Klöcker N, Kelbauskas L, König K, Benndorf K, Biskup C (2008) Prolonged irradiation of enhanced cyan fluorescent protein or Cerulean can invalidate Förster resonance energy transfer measurements. J Biomed Opt 13:031205
56. Jung G, Werner M, Schneider M (2008) Efficient photoconversion distorts the fluorescence lifetime of GFP in confocal microscopy: a model kinetic study on mutant Thr203Val. ChemPhysChem 9:1867–1874
57. Tramier M, Zahid M, Mevel J, Masse M, Coppey-Moisan M (2006) Sensitivity of CFP/YFP and GFP/mCherry pairs to donor photobleaching on FRET determination by fluorescence lifetime imaging microscopy in living cells. Microsc Res Tech 69:933–939

58. Lelimousin M, Noirclerc-Savoye M, Lazareno-Saez C, Paetzold B, Le Vot S, Chazal R, Macheboeuf P, Field MJ, Bourgeois D, Royant A (2009) Intrinsic dynamics in ECFP and Cerulean control fluorescence quantum yield. Biochemistry 48:10038–10046
59. Villoing A, Ridhoir M, Cinquin B, Erard M, Alvarez L, Vallverdu G, Pernot P, Grailhe R, Fe M, Pasquier H (2008) Complex fluorescence of the cyan fluorescent protein: comparisons with the H148D variant and consequences for quantitative cell imaging. Biochemistry 47:12483–12492
60. Scruggs AW, Flores CL, Wachter R, Woodbury NW (2005) Development and characterization of green fluorescent protein mutants with altered lifetimes. Biochemistry 44:13377–13384
61. Goedhart J, van Weeren L, Hink MA, Vischer NOE, Jalink K, Gadella TWJ Jr (2010) Bright cyan fluorescent protein variants identified by fluorescence lifetime screening. Nat Methods 7:137–139
62. Rizzo MA, Springer GH, Granada B, Piston DW (2004) An improved cyan fluorescent protein variant useful for FRET. Nat Biotechnol 22:445–449
63. Heim R, Tsien RY (1996) Engineering green fluorescent protein for improved brightness, longer wavelengths and fluorescence resonance energy transfer. Curr Biol 6:178–182
64. Wachter RM, King BA, Heim R, Kallio K, Tsien RY, Boxer SG, Remington SJ (1997) Crystal structure and photodynamic behavior of the blue emission variant Y66H/Y145F of green fluorescent protein. Biochemistry 36:9759–9765
65. Malo GD, Pouwels LJ, Wang M, Weichsel A, Montfort WR, Rizzo MA, Piston DW, Wachter RM (2007) X-ray structure of Cerulean GFP: a tryptophan-based chromophore useful for fluorescence lifetime imaging. Biochemistry 46:9865–9873
66. Mena MA, Treynor TP, Mayo SL, Daugherty PS (2006) Blue fluorescent proteins with enhanced brightness and photostability from a structurally targeted library. Nat Biotechnol 24:1569–1571
67. Kummer AD, Wiehler J, Schüttrigkeit TA, Berger BW, Steipe B, Michel-Beyerle ME (2002) Picosecond time-resolved fluorescence from blue-emitting chromophore variants Y66F and Y66H of the green fluorescent protein. Chem Bio Chem 3:659–663
68. Lossau HS, Kummer A, Heinecke R, Pöllinger-Dammer F, Kompa C, Beiser G, Johnsson T, Silva CM, Yang MM, Youvan DC, Michel-Beyerle ME (1996) Time-resolved spectroscopy of wild-type and mutant green fluorescent proteins reveals excited state deprotonation consistent with fluorophore-protein interactions. Chem Phys 213:1–16
69. Megley CM, Dickson LA, Maddalo SL, Chandler GJ, Zimmer M (2009) Photophysics and dihedral freedom of the chromophore in yellow, blue, and green fluorescent protein. J Phys Chem B 113:302–308
70. Barondeau DP, Kassmann CJ, Tainer JA, Getzoff ED (2002) Structural chemistry of a green fluorescent protein Zn biosensor. J Am Chem Soc 124:3522–3524
71. Ai HW, Shaner NC, Cheng Z, Tsien RY, Campbell R (2007) Exploration of new chromophore structures leads to the identification of improved blue fluorescent proteins. Biochemistry 46:5904–5910
72. Tramier M, Gautier I, Piolot T, Ravalet S, Kemnitz K, Coppey J, Durieux C, Mignotte V, Coppey-Moisan M (2002) Picosecond-hetero-FRET microscopy to probe protein-protein interactions in live cells. Biophys J 83:3570–3577
73. Demachy I, Ridard J, Laguitton-Pasquier H, Durnerin E, Vallverdu G, Archirel P, Lévy B (2005) Cyan fluorescent protein: molecular dynamics, simulations, and electronic absorption spectrum. J Phys Chem B 109:24121–24133
74. Chattoraj M, King BA, Bublitz GA, Boxer SG (1996) Ultra-fast excited state dynamics in green fluorescent protein: multiple states and proton transfer. Proc Natl Acad Sci USA 93:862–8367
75. Olson S, McKenzie RH (2010) A dark excited state of fluorescent chromophores, considered as Brooker dyes. Chem Phys Lett 492:150–156
76. Pal PP, Bae JH, Azim MK, Hess P, Friedrich R, Huber R, Moroder L, Budisa N (2005) Structural and spectral response of *Aequorea victoria* green fluorescent proteins to chromophore fluorination. Biochemistry 44:3663–3672

77. Wiehler J, Jung G, Seebacher C, Zumbusch A, Steipe B (2003) Mutagenic stabilization of the photocycle intermediate of green fluorescent protein (GFP). ChemBioChem 4:1164–1171
78. Cotlet M, Hofkens J, Maus M, Gensch T, Van der Auweraer M, Michiels J, Dirix G, Van Guyse M, Vanderleyden J, Visser AJWG, De Schryver FC (2001) Excited-state dynamics in the enhanced green fluorescent protein mutant probed by picosecond time-resolved single photon counting spectroscopy. J Phys Chem B 105:4999–5006
79. Striker G, Subramaniam V, Seidel CAM, Volkmer A (1999) Photochromicity and fluorescence lifetimes of green fluorescent protein. J Phys Chem B 103:8612–8617
80. Stavrov SS, Solntsev KM, Tolbert LM, Huppert D (2006) Probing the decay coordinate of the green fluorescent protein: arrest of cis-trans isomerization by the protein significantly narrows the fluorescence spectra. J Am Chem Soc 128:1540–1546
81. Bagchi B, Fleming GR, Oxtoby DW (1983) Theory of electronic relaxation in solution in the absence of an activation barrier. J Chem Phys 78:7375–7385
82. Kummer AD, Wiehler J, Rehaber H, Kompa C, Steipe B, Michel-Beyerle ME (2000) Effects of threonine 203 replacements on excited-state dynamics and fluorescence properties of the green fluorescent protein (GFP). J Phys Chem B 104:4791–4798
83. Schwille P, Kummer S, Heikal AA, Moerner WE, Webb Watt W (2000) Fluorescence correlation spectroscopy reveals fast optical excitation-driven intramolecular dynamics of yellow fluorescent proteins. Proc Natl Acad Sci USA 97:151–156
84. Weber W, Helms V, McCammon JA, Langhoff PW (1999) Shedding light on the dark and weakly fluorescent states of green fluorescent proteins. Proc Natl Acad Sci USA 96:6177–6182
85. Baffour-Awuah NYA, Zimmer M (2004) Hula-twisting in green fluorescent protein. Chem Phys 303:7–11
86. Andresen M, Wahl MC, Stiel AC, Gräter F, Schäfer LV, Trowitzsch S, Weber G, Eggeling C, Grubmüller H, Hell SW, Jakobs S (2005) Structure and mechanism of the reversible photoswitch of a fluorescent protein. Proc Natl Acad Sci USA 102:13070–13074
87. Nifosi R, Tozzini V (2006) Cis-trans photoisomerization of the chromophore in the green fluorescent protein variant E^2GFP: a molecular dynamics study. Chem Phys 323:358–368
88. Mizuno H, Mal TK, Wälchi M, Kikuchi A, Fukano T, Ando R, Jeyakanthan TJ, Shiro Y, Ikura M, Miyawaki A (2008) Light-dependent regulation of structural flexibility in a photochromic fluorescent protein. Proc Natl Acad Sci USA 105:9227–9232
89. Schleifenbaum F, Blum C, Elgass K, Subramaniam V, Meixner AJ (2008) New insights into the photophysics of DsRed by multiparameter spectroscopy on single proteins. J Phys Chem B 112:7669–7674
90. Bateman RJ, Chance RR, Hornig JF (1974) Fluorescence reabsorption in anthracene single crystals: lifetime variations with emission wavelength and temperature. Chem Phys 4:402–408
91. Chudakov DM, Matz MV, Lukyanov S, Lukyanov KA (2010) Fluorescent proteins and their applications in imaging living cells and tissues. Physiol Rev 90:1103–1163
92. Müller-Taubenberger A, Anderson KI (2007) Recent advances using green and red fluorescent protein variants. Appl Microbiol Biotechnol 77:1–12
93. Heikal AA, Hess ST, Baird GS, Tsien RY, Webb WW (2000) Molecular spectroscopy and dynamics of intrinsically fluorescent proteins: coral red (dsRed) and yellow (Citrine). Proc Natl Acad Sci USA 97:11996–12001
94. Schüttrigkeit TA, Zachariae U, von Feilitzsch T, Wiehler J, von Hummel J, Steipe B, Michel-Beyerle ME (2001) Picosecond time-resolved FRET in the fluorescent protein from Discosoma Red (wt-DsRed). ChemPhysChem 2:325–328
95. Nienhaus GU, Wiedenmann J (2009) Structure, dynamics and optical properties of fluorescent proteins: perspectives for marker development. Chem Phys Chem 2009:1369–1379
96. Schmid JA, Neumeier H (2005) Evolutions in science triggered by green fluorescent protein (GFP). Chem Bio Chem 6:1149–1156
97. Seefeldt B, Kasper R, Seidel T, Tinnefeld P, Dietz K, Heilemann M, Sauer M (2008) Fluorescent proteins for single-molecule fluorescence applications. J Biophoton 1:74–82

98. Shaner N, Campbell RE, Steinbach PA, Giepmans BNG, Palmer AE, Tsien RY (2004) Improved monomeric red, orange and yellow fluorescent proteins derived from Discosoma sp. red fluorescent protein. Nat Biotechnol 22:1567–1572
99. Shaner NC, Steinbach PA, Tsien RY (2005) A guide to choosing fluorescent proteins. Nat Methods 2:905–909
100. Cox G, Matz M, Salih A (2007) Fluorescence lifetime imaging of coral fluorescent proteins. Microsc Res Tech 70:243–251
101. Hasegawa J, Ise T, Fujimoto KJ, Kikuchi A, Fukumura E, Miyawaki A, Shiro Y (2010) Excited states of fluorescent proteins, mKO and DsRed: chromophore-protein electrostatic interaction behind the color variations. J Phys Chem B 114:2971–2979
102. Yan W, Zhang L, Xie D, Zeng J (2007) Electronic excitations of green fluorescent proteins: modeling solvatochromatic shifts of red fluorescent protein chromophore model compound in aqueous solutions. J Phys Chem B 111:14055–14063
103. Liu RSH, Yang L, Liu J (2007) Mechanisms of photoisomerization of polyenes in confined media: from organic glasses to protein binding cavities. Photochem Photobiol 83:2–10

Synthetic Biology of Autofluorescent Proteins

Michael Georg Hoesl, Lars Merkel, and Nediljko Budisa

Abstract Autofluorescent proteins (FPs), which to date are predominately used as tools in cell biology and spectroscopy, have arrived in the focus of synthetic biology. Thereby, the intention is to supplement classically used protein design methods such as site-directed mutagenesis or guided evolution by expanding the scope of protein synthesis. This is achieved by the co-translational introduction of novel noncanonical amino acids (NCAAs) into proteins. In the following chapter, we present current applications of an expanded amino acid repertoire for the design of spectral and folding properties of FPs. We will show that NCAAs are not only useful tools to study fundamental aspects of photophysics but also have great potential to generate novel FP tools for cell biology applications. On the one hand, aromatic amino acids other than the naturally occurring His, Tyr, Phe, and Trp were used to create novel spectral classes of FPs by direct chromophore modification. On the other hand, NCAAs were also applied for "FP protein matrix engineering" to influence chromophore fluorescence and overall folding. We also illustrate a practical application of these principles by presenting "golden annexin A5" as a novel apoptosis detection tool designed by synthetic biology methods. Finally, we describe a potential route to convert any protein of interest into a chromo-protein by introduction of novel synthetic autofluorescent amino acids.

Keywords Chromophore variants, ECFP, EYFP, GFP, GFP structure, non-canonical amino acid, synthetic biology

Contents

M.G. Hoesl, L. Merkel, and N. Budisa (✉)
Department of Chemistry, Berlin Institute of Technology/TU Berlin, Biocatalysis Group, Franklinstraße 29, 10587 Berlin, Germany
e-mail: nediljko.budisa@tu-berlin.de

G. Jung (ed.), *Fluorescent Proteins I*, Springer Ser Fluoresc (2012) 11: 99–130,
DOI 10.1007/4243_2011_23,
Published online: 14 April 2011

1 Introduction

As explained in detail in the preceding chapters, a great variety of fluorescent protein spectral classes was gained in the last years by protein engineering of *Aequorea victoria* GFP (*av*GFP, e.g., generation of ECFP) or "natural" isolation from other species, mainly reef corals from the species *Anthozoa,* e.g., *ds*Red [1]. In most naturally occurring fluorescent proteins, the chromophore is 4-(*p*-hydroxy-benzylidene)-5-imidazolinone (*p*-HBI) or a derivative with an extended π-electron system obtained by further autocatalytic reactions from a *p*-HBI intermediate [1, 2].

While the *p*-HBI structure is obtained from a reaction between aa65, Tyr66, and Gly67, the key feature of engineered *av*GFP variants is an exchange of the aromatic Tyr residue by other canonical aromatic amino acids. The occurrence of an aromatic residue at position 66 is crucial for chromophore fluorescence. Thereby, enhanced blue fluorescent protein (EBFP) was designed by introduction of His and enhanced cyan fluorescent protein (ECFP) by introduction of Trp at position 66. Y66F also gave rise to a functional chromophore; however, the fluorescence characteristics were largely reduced in comparison to the other mutants and not much effort was made to develop the mutant further. The mutants Tyr66 → Phe/His/Trp show a characteristic blue emission since the excited state proton transfer (ESPT), responsible for the characteristic green fluorescence in native *av*GFP, is prohibited [3]. The only *av*GFP derivative with a chromophore identical to *av*GFP but significantly different spectral properties is enhanced yellow fluorescent protein (EYFP). It was designed by the mutation T203Y leading to juxtaposition of Y203 and the phenolate anion of the *p*-HBI chromophore. This leads to π–π stacking interactions stabilizing the excited state dipole moment and thus red-shifting the excitation and emission maxima about 20 nm when compared to *av*GFP. According to these findings, Roger Tsien subdivided the *av*GFP mutants into seven classes [4]. These were already discussed in great detail in the chapters by Nifosi and Tozzini, Wiedenmann et al. and Jung et al.

It is often not considered that beyond the key mutations in the chromophore, most *av*GFP derivates contain other crucial mutations responsible for lower aggregation tendencies, better folding (e.g., GFPuv [5]; F99S, M153T, V163A) and higher fluorescence intensities (e.g., EGFP [6], ECFP, EYFP [7, 8]). The importance of such "secondary" mutations is best illustrated in case of ECFP. A sole mutation of Y66W in *av*GFP indeed resulted in the characteristic absorbance and fluorescence maxima of ECFP but the fluorescence intensity was negligible [9]. Only by introduction of three additional mutations (N146I, M153T and V163A), which optimize the Trp chromophore accommodation within the GFP scaffold, good fluorescence intensities were achieved [7].

In general, *av*GFP derivates are monomers when expressed heterologously since the wild-type *av*GFP has an only weak tendency to dimerize. In contrast, most other FPs are naturally occurring as multimers, which strongly limit their applications in cell and molecular biology. To make these proteins available for general use, a great amount of engineering had to be performed. For example, the first isolated red FP *ds*Red [10] had to be mutated at not less than 45 positions to obtain the now widely used *ds*Red-monomer (Clontech).

The need of vast protein optimization in case of many FPs demonstrates the need to explore other strategies for the development of novel spectral classes of FPs as well. One strategy is the use of noncanonical amino acids (NCAAs) because classical protein engineering methods are restrained to the standard genetic code. Aromatic noncanonical residues are especially interesting for FP engineering because a higher amount of variation can be introduced at position 66 (with the standard genetic code only His, Tyr, Phe, and Trp are available). A greater variety of substituted aromatic residues with interesting spectral and other biophysical properties are nowadays available for FP *in vivo* expression, mainly in bacterial cells (see Fig. 1). This chapter will try to give an overview of the works exploring the effects of NCAAs on FP structure and biophysical properties. All abbreviations and structures of the used NCAAs are summarized in Fig. 1.

2 Chromophore Redesign

2.1 Chromophore with 4-Aminotryptophan

Among all noncanonical indole moieties tested for Trp66 substitution in ECFP to date, the greatest change in spectral properties was delivered by the indole → 4-aminoindole substitution (see Fig. 1) in the chromophore of ECFP [11]. The resulting protein was named gold fluorescent protein (GdFP) and exhibited maximal absorbance and emission at 466 and 574 nm, respectively. This makes GdFP the most red-shifted *av*GFP variant with the largest Stokes shift (108 nm) known to date. Its fluorescence is 69 nm red-shifted when compared to the parent ECFP and 47 nm when compared to the emission maximum of EYFP (see Fig. 2). In addition, GdFP is also characterized by increased thermostability and decreased aggregation tendency.

2.2 Spectroscopic Features of GdFP

2.2.1 Red-Shift in Absorbance and Fluorescence

From the different classes of neutral *av*GFP chromophores, it is known that an increase in size of the delocalized π-system results in a corresponding increase in excitation and emission wavelengths (see Table 1).

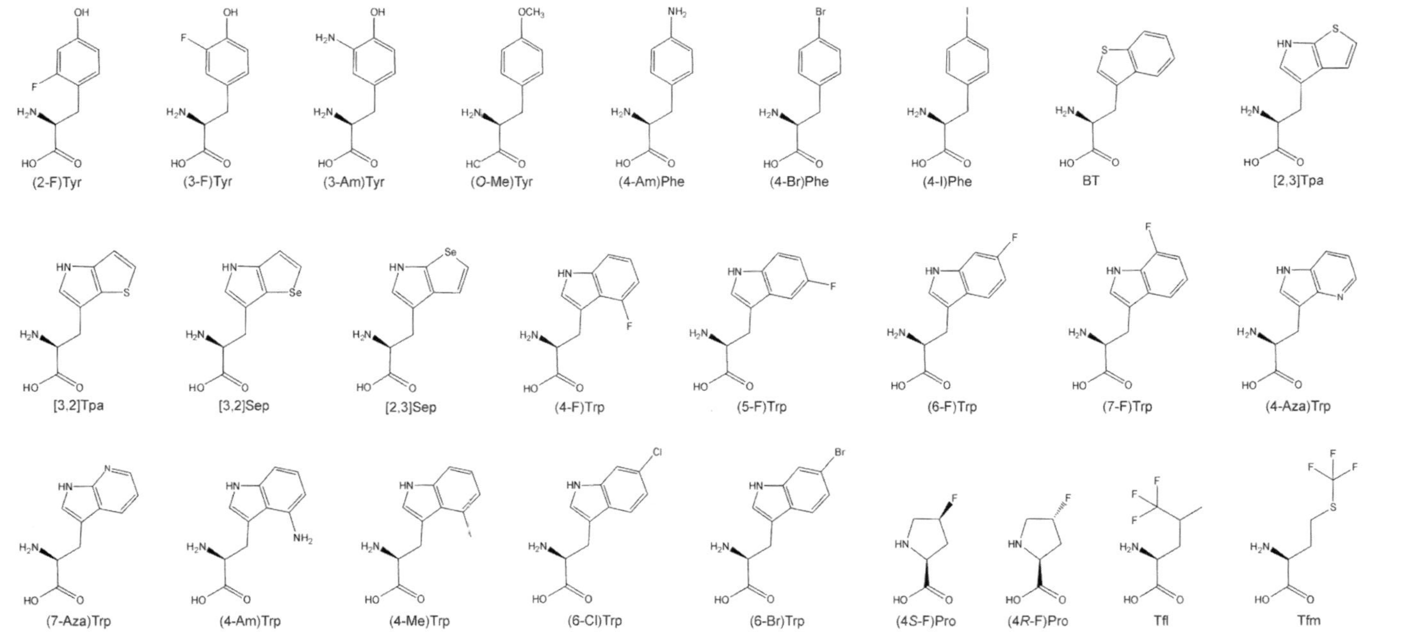

Fig. 1 Overview of noncanonical amino acids incorporated into *av*GFPs. (2-F)Tyr, 2-fluorotyrosine; (3-F)Tyr, 3-fluorotyrosine; (3-Am)Tyr, 3-aminotyrosine; (*O*-Me)Tyr, *O*-methyltyrosine; (4-Am)Phe, 4-aminophenylalanine, (4-Br)Phe, 4-bromophenylalanine; (4-I)Phe, 4-iodophenylalanine; BT, benzothienylalanine; [2,3]Tpa, β-thieno[2,3-*b*]pyrrolylalanine; [3,2]Tpa, β-thieno[3,2-*b*]pyrrolylalanine; [3,2]Sep, β-selenolo[3,2-*b*]pyrrolylalanine; [2,3]Sep, β-selenolo[2,3-*b*]pyrrolylalanine; (4-F)Trp, 4-fluorotryptophan; (5-F)Trp; 5-fluorotryptophan; (6-F)Trp, 6-fluorotryptophan; (7-F)Trp, 7-fluorotryptophan; (4-Aza)Trp, 4-azatryptophan; (7-Aza)Trp, 7-azatryptophan; (4-Am)Trp, 4-aminotryptophan; (4-Me)Trp, 4-methyltryptophan; (6-Cl)Trp, 6-chlorotryptophan; (6-Br)Trp, 6-bromotryptophan; (4*S*-F)Pro, 4(*S*)-fluoroproline; (4*R*-F)Pro, 4(*R*)-fluoroproline; Tfl, 5′,5′,5′-trifluoroleucine and Tfm, trifluoromethionine

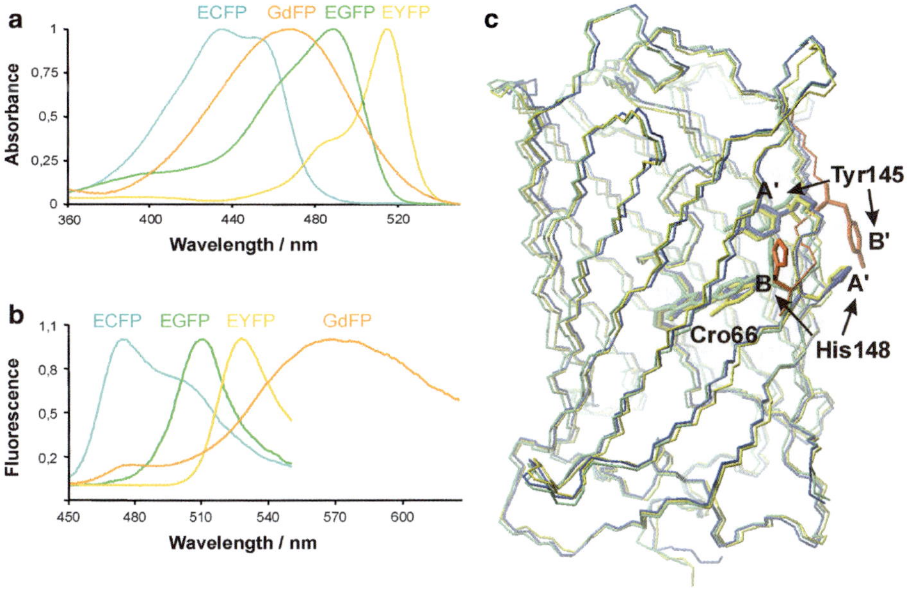

Fig. 2 Photophysical properties and structure of GdFP in comparison with common *av*GFP mutants. Normalized absorbance (**a**) and fluorescence (**b**) spectra of GdFP, ECFP, EGFP and EYFP. While ECFP (Trp66) has two characteristic absorbance maxima at 434 and 452 nm, GdFP ((4-Am)Trp66) has only one, but red-shifted maximum at 466 nm. The absorbance spectra of EGFP and EYFP are further red-shifted to 488 and 512 nm, respectively. In contrast, GdFP shows the most red-shifted fluorescence ($\lambda_{max} = 574$ nm) when compared to all *av*GFP variants produced by traditional DNA technologies. ECFP, EGFP and EYFP only have maximum fluorescence wavelength of 476 (505), 509 and 527 nm, respectively. Detailed information about maximum absorbance and fluorescence wavelengths of the different FPs can be found in Table 1 and in the text. (**c**) Structural overlay of GdFP (*yellow*) with EGFP (*green*) and ECFP (*blue*). The chromophore and residues Tyr145 and His148 are represented as sticks in the same colors with the exception of ECFP, where *blue* (conformation A′ of ECFP) and *red* (conformation B′ of ECFP) are used

The same effect is observed if the chromophore is additionally increased by conjugation with a fourth amino acid as present for example in *ds*Red [18]. GdFP, however, maintains the basic *p*-HBI structure of the *av*GFP chromophores but also exhibits an emission wavelength comparable to the more extended *ds*Red chromophore. One possible explanation for such a significant change in spectral properties was the emergence of intramolecular charge transfer processes (ICTs): the introduced amino group is acting as an additional electron donor and is conjugated to the delocalized π-system of the tryptophan moiety. This leads to a resonance stabilization of the excited state explaining the vast red-shift in the emission spectrum [11].

2.2.2 Stokes Shift

The chromophore of GdFP is not only characterized by a large red-shift in absorbance and fluorescence spectra when compared to ECFP but also by a Stokes shift

Table 1 Overview of the spectral properties of the noncanonical FP variants. Additional values in brackets were taken from [12]

Variant	Absorbance maximum/ nm	Extinction coefficient/ M^{-1} cm^{-1}	Emission maximum/ nm	Quantum yield	Relative fluorescence	Ref.
CFP6	449	25,900 ± 1,000	488	0.140	–	[13]
CFP6[(6-Cl)Trp]	430	19,800 ± 6,400	474	0.050	–	[13]
CFP6[(6-Br)Trp]	430	20,600 ± 3,800	474	0.058	–	[13]
CFP6[BT]	415	8,100 ± 1,100	472	0.048	–	[13]
GFPuv	397/475	25,000 ± 2,000/ 6,000 ± 500	506	0.760	–	[14]
GFPuv[(4-Am)Phe]	435	31,000 ± 4,000	498	0.43	–	[14]
GFPuv[(4-Me)Tyr]	394	27,000 ± 1,000	460	0.37	–	[14]
GFPuv[(4-Br)Phe]	375	20,000 ± 1,000	428	0.014	–	[14]
GFPuv[(4-I)Phe]	381	16,000 ± 1,000	438	0.013	–	[14]
ECFP	434	24,800 (20,000)	476	(0.15)	1	[15]
ECFP[(4-F)Trp]	426	14,810 ± 697	470	–	0.4	[15]
ECFP[(5-F)Trp]	426	19,172 730	466	–	0.7	[15]
ECFP[(6-F)Trp]	432	21,762 ± 238	474	–	0.5	[15]
ECFP[(7-F)Trp]	426	16,650 ± 837	471	–	0.4	[15]
ECFP[(4-Me)Trp]	438	23,450 ± 561	480	–	1	[15]
ECFP[(4-Am)Trp]	466	23,700	574	0.15	–	[11], [70]
ECFP[[2,3]Sep]	435	4,533 ± 239	–	–	–	[15]
ECFP[[3,2]Sep]	388	11,650 ± 975	–	–	–	[15]
ECFP[[2,3]Tpa]	385	6,840 ± 421	–	–	–	[15]
ECFP[[3,2]Tpa]	388	6,350 ± 717	–	–	–	[15]
EGFP	488	35,570 ± 1,316 (23,000)	510	(0.7)	1.00	[15]
EGFP[[2,3]Sep]	488	40,520 ± 1,261	509	–	0.9	[15]
EGFP[[3,2]Sep]	488	49,930 ± 1,597	509	–	0.83	[15]
EGFP[[2,3]Tpa]	488	30,715 ± 1,076	509	–	1.25	[15]
EGFP[[3,2]Tpa]	488	43,715 ± 1,455	509	–	0.83	[15]
*ds*Red	556	59,000 (27,300)	603 (592)	0.04 (0.14)	–	[16]
*ds*Red[(3-Am)Tyr]	556	59,000	615	0.16	–	[16]
*ds*Red[(3-F)Tyr]	556	47,000	591	0.25	–	[16]
EGFP[(2-F)Tyr]	482	22,190 ± 1,174	504	–	0.6	[17]
EGFP[(3-F)Tyr]	485	30,200 ± 1,200	514	–	0.9	[17]
EYFP	514	48,770 ± 1,250 (45,000)	527	(0.54)	1	[17]
EYFP[(2-F)Tyr]	504	55,460 ± 2,180	520	–	1.2	[17]
EYFP[(3-F)Tyr]	518	48,790 ± 1,440	533	–	1	[17]

of >100 nm. A comparable Stokes shift could only be detected in class I and III *av*GFPs when the chromophore is excited in its neutral form at 395 nm and fluoresces from its anionic form at 511 nm. However, the high Stokes shift of *av*GFP occurs due to ESPT [3, 19]. In GdFP, time-resolved experiments do not give any indication for ESPT [11].

A possible explanation for the large Stokes shift could be that, in contrast to ECFP, the dipole moment of the GdFP chromophore is considerably larger in the excited state than in the ground state. The additional amino group attached to the aromatic chromophore moiety has a great influence on the nature of the excited state and leads to a remarkable charge separation in the excited state of the GdFP chromophore. In contrast to ICT where a negative charge is (partially) relocated, charge separation means that a neutral state is split up into a positive and a negative charge which then are located at different positions in the chromophore. This intramolecular charge separation causes changes in the electrostatic potential around the chromophore. As a consequence, the protein matrix relaxes toward new, more favorable electrostatic contacts between the chromophore and its surrounding. Due to this structural relaxation, the energy level of the excited chromophore is lowered, leading to a red-shift in fluorescence but not absorbance, thereby causing the detected large Stokes shift (personal communication of Prof. M. Michel-Beyerle).

2.2.3 Appearance of GdFP Absorbance and Fluorescence Spectra

Within the *av*GFP family, the spectral features of ECFP are distinct because the absorbance as well as the fluorescence spectrum consists of two clearly distinguishable maxima (i.e., spectral bands). In contrast, absorption and emission spectra of GdFP consist of broad bands with no resolvable structures. In this context, it was speculated that the double absorbance and fluorescence peaks of ECFP arise from ground-state structural heterogeneity within the region between residues 145 and 149 or the chromophore itself (see Fig. 2c). In the vicinity of the ECFP chromophore, two structural conformations, named A′ and B′, were indeed detectable in the crystal structure. In contrast, GdFP did not show any heterogeneity in this respect; this was postulated to be the main reason for its spectral homogeneity. However, recent investigations on Cerulean GFP, an engineered mutant without ground-state heterogeneity, gave exactly the same double-humped spectra [20], and quantum mechanical calculations revealed that chromophore spectra in conformation A′ and B′ would be identical [21]. In the context of these observations, Malo et al. gave a possible explanation for the double-humped peaks. Taking into account that indole is in the core of the chromophore structure, they argued that the complex indole photophysics have been transmitted to the chromophore itself [20]. Namely, indole shows two overlapping $\pi \rightarrow \pi^*$ electronic transitions $1L_a$ and $1L_b$. Although this argumentation sounds plausible, it does not explain the vanishing of the second absorbance and fluorescence band in GdFP because (4-Am)Trp is a derivative of indole with the same underlying electronic transitions [22].

2.2.4 Influence of the Structural Context on Spectral Properties

The three-dimensional structures of ECFP and GdFP were found to be almost identical (see Fig. 2c). The most prominent novel interaction, which could be identified from crystallographic distance considerations, was a slight shift of the amino-chromophore of GdFP toward Phe165 (see Fig. 3). The two residues

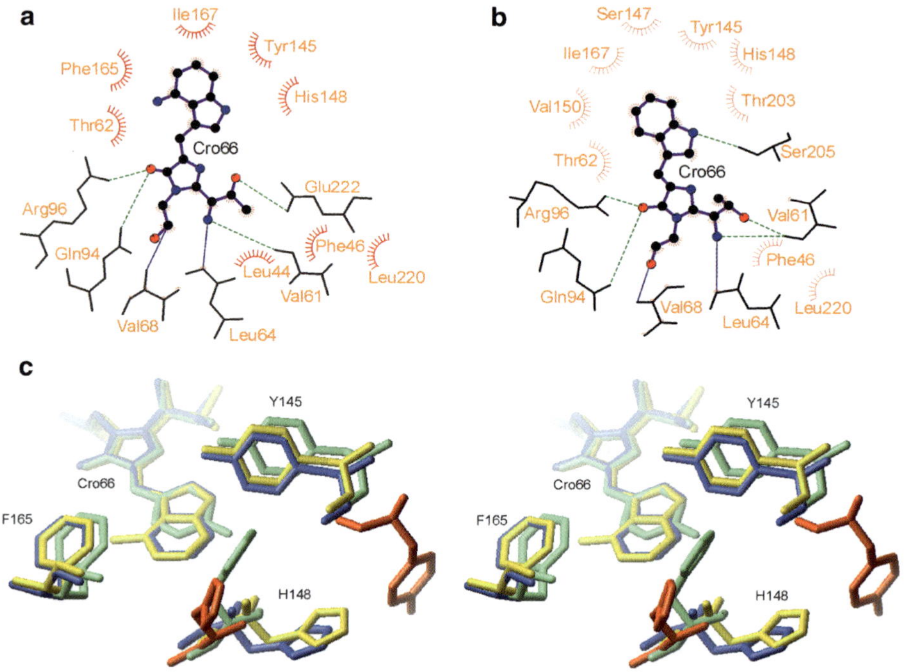

Fig. 3 Chromophore-surrounding structure of GdFP in comparison to EGFP and ECFP. The schematic representation of interactions between the protein matrix and the chromophore of GdFP (**a**) and ECFP (**b**; major configuration A′) are plotted. Note the additionally appearing possible interaction between Phe165 and the amino group of the chromophore in GdFP. (**c**) Stereo-view of the chromophores in EGFP (*green*) ECFP (*blue*) and GdFP (*yellow*) and their environments. Tyr145 and His148 of the ECFP minor-form are shown in *red*. Note a slight shift of the amino indole moiety of the GdFP chromophore toward the Phe165

are 3.2–4.5 Å apart, which makes interactions between the amino group and the aromatic ring of Phe165 possible. This novel interaction was assigned as an additional factor that might further contribute to more effective intramolecular charge transfer in the excited state of the GdFP chromophore when compared to the ground state. Thus, it cannot be excluded that this novel interaction is also involved in the occurrence of highly red-shifted absorbance and fluorescence as well as the large Stokes shift.

Interestingly, careful inspection of the residue configurations in the chromophore vicinity revealed that the chromophore of GdFP has less hydrophobic contacts with other amino acids in its proximity than the ECFP chromophore (Fig. 3a, b). This could be associated with higher flexibility of the chromophore substantially contributing to the increase in radiationless fluorescence decay [23]. Indeed, this was found in GdFP since the quantum yield was lower when compared to ECFP.

2.3 *Chromophore Halogenations*

2.3.1 Trp and Phe Halogenations

David Tirrell and co-workers incorporated position 6 halogenated Trp analogs in a CFP variant (CFP6; F64L, S65T, Y66W, V163A) [13], the laboratory of Peter Schultz incorporated *para* substituted Br- and I-phenylalanine [14] (see Fig. 1) in GFPuv and our group studied the effects of fluorinated Trp analogs on the chromophore spectral properties of ECFP [15].

Generally, halogenation of Trp and Phe in the *av*GFP chromophore leads to a blue shift in fluorescence. Tirrell and associates reported a 14 nm blue shifted fluorescence for CFP6[(6-Cl)Trp] and CFP6[(6-Br)Trp] when compared to CFP6 (see Table 1). Schultz and co-workers found exactly the same blue shift in fluorescence for GFPuv [(4-Br)Phe] and GFPuv[(4-I)Phe] when these two variants are compared to a class 7 GFP[4] with phenylalanine in the chromophore. A direct comparison to GFPuv is not reasonable since a high red-shift in fluorescence wavelength occurs because of inhibition of ESPT for (Br)Phe and (I)Phe substituted chromophores. Monofluorination of Trp at position 4–7 also caused slight blue shifts in fluorescence between 2 and 10 nm in ECFP [15]. The same tendency was also found when the Tyr residue in *ds*Red was exchanged to (3-F)Tyr (See Sect. 2.4.2.) [16].

Quantum yields (QYs) of the halogenated *av*GFP variants were determined as well. While incorporation of (6-Cl)Trp and (6-Br)Trp into CFP6 yielded ~60% reduction in quantum yield when compared to CFP6, (4-Br)Phe and (4-I)Phe caused ~98% reduction when introduced into GFPuv. This effect was explained by the influence of halogens on the π-electron system according to the spin-orbital coupling theory [24]. In other words, halogens promote intersystem crossing, which leads to radiationless decay of the excitation energy. QYs were not measured for the fluorinated ECFP variants. However, the same effect can also be seen in the relative fluorescence when compared to ECFP, where all variants exhibit significantly lower fluorescence intensities. Noteworthy, the strength of the effect was dependent on the position of the indole ring substitution, which was found in halogenated porphyrins as well [25].

Finally, the low quantum yields of the brominated and iodinated GFPuv and CFP6 might also be attributed to sterical effects caused by halogen presence since Br and I have very high van der Waals volumes [26], which might not be well accommodated in the *av*GFP structure. In line with this argumentation, CFP6 does not have the mutations N146I and M153T, which lead to better spectral properties by enhancing the accommodation of the Trp-containing chromophore (*vide supra*).

2.3.2 Tyr Fluorination

EGFP and EYFP

In the context of GFP spectroscopy, the use of Tyr analogs is not only a suitable tool to study the influence of certain functional groups on the GFP's spectral properties, e.g., intensity of absorbance and fluorescence, and blue/red-shifts in the emission maxima. It also bears the potential to modify the anion/neutral equilibrium of the Tyr's hydroxyl group because electron donating or withdrawing groups have an influence on the acidity of the –OH group. In EYFP, it has to be considered that a global replacement of Tyr by its analogs not only leads to a modified chromophore but also to a modified Tyr203 in the vicinity (stacking position), which also will have influence on the spectral properties of the chromophore.

Spectral Properties

Global substitution of all Tyr residues by (2-F)Tyr (see Fig. 1) in both, EGFP and EYFP, resulted in protein variants with slight blue shifts (6–10 nm) in the absorbance as well as in the fluorescence maxima of their chromophores (see Table 1). This is consistent with the findings of Trp fluorinations in ECFP (see above). In case of EGFP[(3-F)Tyr], however, the absorbance spectrum is 3 nm blue shifted, whereas the fluorescence maximum is slightly red-shifted (4 nm). Conversely, in EYFP[(3-F)Tyr] both, a red-shifted absorbance and fluorescence was found. Although it is difficult to unambiguously identify a possible origin of this spectral behavior, one has to take into account the regioisomeric differences between the fluorinated Tyr residues. For example, different positioning of fluorine atoms in the phenol ring of Tyr certainly influences electron delocalization, the orientation of transition- and ground-state dipoles. Studying the role of protonation on protein spectral properties, we found uniform behavior of all fluorinated variants of EGFP and EYFP at low (pH 1) or high (pH 11) pH. At low pH, when only the protonated form of the chromophore is present and the protein is acid denatured, all variants exhibited the same absorbance as EGFP or EYFP ($\lambda_{max} = 378$ nm) [17]. At high pH (only deprotonated form of the chromophore is present), the protein variants had equally blue-shifted absorbance maxima when compared to ECFP or EYFP (blue shift of 3 nm to a λ_{max} of 444 nm). Obviously, the main reason for the spectral differences between these variants in the folded state are sterical effects caused by the fluorination at different positions of the phenol side chain of the Tyr residues.

Influence of Tyr Fluorination on Chromophore pKa

The influence of the ring substituents on the pK_a of the Tyr side chain could indeed be detected in variants of EGFP and EYFP. The decrease in pK_a in the fluorinated variants was reported to be in the following order: parent protein > (2-F)Tyr-protein > (3-F)Tyr-protein (see Fig. 4). This corresponded well with the pK_a values of the free amino acids, Tyr (10.0) > (2-F)Tyr (9.04) > (3-F)Tyr (8.5) [27]. Interestingly, the pK_a value of (3-F)Tyr-EYFP considerably decreased by about 1.2 units, while all other variants showed only a change of 0.1–0.5 units. This was at least partly attributed to the stacking interactions between (3-F)Tyr66 and (3-F) Tyr203.

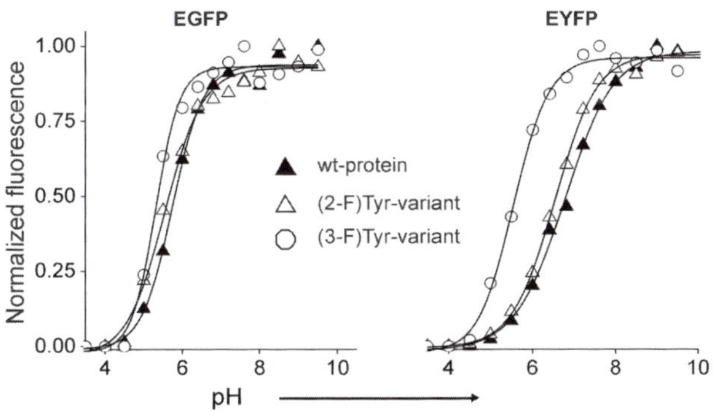

Fig. 4 Influence of chromophore fluorination on fluorescence pH titration profiles in parent and variant EGFPs (*left*) and EYFPs (*right*). The pKa value of EYFP[(3-F)Tyr] considerably decreased by about 1.2 units, while all other variants showed only a change of 0.1–0.5 units

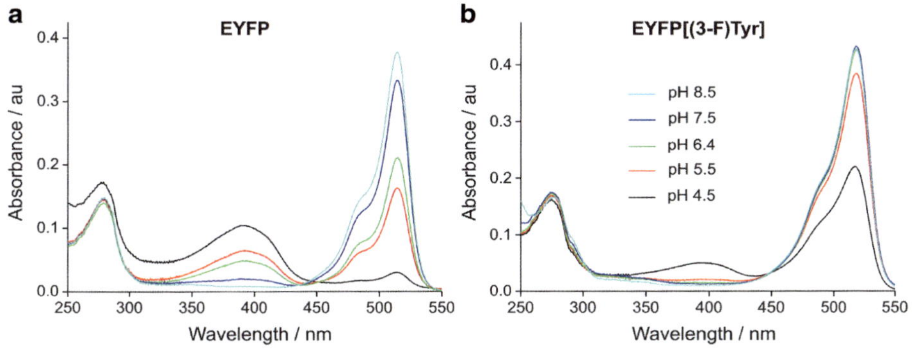

Fig. 5 pH dependency of absorbance and fluorescence of EYFP and its fluorinated variant EYFP [(3-FTyr]. While EYFP absorbance is highly dependent on the pH (**a**), EYFP[(3-F)Tyr] absorbance has a strongly reduced sensitivity to pH (**b**)

Although the absorption spectra of native EGFP and EYFP were strongly pH dependent, this tendency was less pronounced in the fluorinated variants (see Fig. 5) [17]. In EGFP[(2-F)Tyr], EGFP[(3-F)Tyr] and EYFP[(3-F)Tyr] significantly lower tendencies for the occurrence of the non-fluorescing protonated chromophore forms (absorbance at 395 nm) were detected. Among these variants, the decrease in pH to a value of 4.5 had the smallest effect on EYFP[(3-F)Tyr], which is reasonable since this variant also exhibited the lowest pK_a when compared to the respective parent protein EYFP. However, EYFP[(2-F)Tyr] did not follow the general tendency for stabilization of the anionic chromophore state at low pH and behaved exactly like the parent protein.

Additionally, the tendency of aggregation for all fluorinated EGFP and EYFP variants was tested and an interesting correlation was found. Namely, there is a direct relation between the appearance of the protonated chromophore species at lower pH and the occurrence of aggregation after prolonged storage at 4°C. Interestingly, the EYFP[(2-F)Tyr] exhibited a higher resistance to aggregation when compared to the parent protein. The other fluorinated variants, however, exhibited a higher tendency for aggregation. This behavior was assigned to favorable or unfavorable effects of the fluorine atoms on the folding of the chromophore surrounding protein matrix since EYFP[(2-F)Tyr] not only showed lower aggregation tendency but also significantly higher extinction coefficients and relative fluorescence. In contrast, EGFP[(2-F)Tyr] exhibits inferior spectral properties (~25% decreased absorbance and fluorescence intensities) and a higher tendency to aggregate.

2.4 Chromophores with (4-Am)Phe, (O-Me)Tyr and (3-Am)Tyr

2.4.1 GFPuv

Wang et al. reported site-directed substitution of Tyr66 of GFPuv with (4-Am)Phe and (4-Me)Tyr [14] (see Fig. 1). They found that the absorbance (range: 375–435 nm) and fluorescence (range: 428–498 nm) maxima and quantum yields correlate with the structural and electronic properties of the tyrosine substituents (see Tables 1 and 2). Upon (*O*-Me)Tyr incorporation, only one absorbance maximum at 394 nm was observed, which is plausible since (*O*-Me)Tyr66 does not provide a protonation site. The fluorescence is blue-shifted and the quantum yield considerably decreased (0.76 → 0.37). The fluorescence maximum can be detected at 460 nm, which fits well with the fluorescence maximum of the protonated form of the *av*GFP chromophore (disappearing in a pico second timescale because of ESPT [3]). Hence, this *av*GFP variant trapped the fluorescence of the protonated chromophore form and might be a good model to study the spectral properties of this chromophore species. Furthermore, (4-Am)Phe was introduced into the chromophore and also exhibited a decreased yield in fluorescence. The GFP[(4-Am)Phe], although blue-shifted (8 nm) when compared with the fluorescence emission

Table 2 Influence of phenyl ring substitution of residue 66 in the chromophore on excitation and emission wavelength. The maximum wavelengths of both absorbance and emission peaks increase with the electron-donating ability of the *para* substituents. The excitation maxima of Br and I substituted chromophores (marked by *asterisk*) do not fit exactly in this scheme. A possible explanation is the interference of electron withdrawing effect and spin-orbital coupling (see text).

Substituent	Excitation wavelength/nm	Emission wavelength/nm
–Br	375*	428
–I	381*	438
–H	360	442
$–O–CH_3$	394	460
–O–H	397	460
$–NH_2$	435	498
$–O^-$	475	506

maximum of GFPuv, had a Stokes shift of 63 nm, which makes it the second most red-shifted noncanonical *av*GFP-based autofluorescent protein (GdFP has a Stokes shift of 100 nm).

Comparing all generated noncanonical GFPuv variants from their study with the spectral properties of neutral and anionic *av*GFP, Wang et al. found that the maximum wavelengths of both absorbance and emission peaks increased with the electron-donating ability of the *para* substituents [14]. However, Br and I substituted chromophores do not fit exactly in this row since they show red-shifted excitation maxima in comparison to the nonsubstituted chromophore. This observation is difficult to explain since the electron withdrawing effect might interfere with the spin-orbital coupling effects of halogen atoms as already discussed above.

With respect to applicability, Wang et al. argue that one of their generated noncanonical mutants (GFP[(4-Me)Tyr]) could substitute BFP in the BFP-EGFP FRET pair. This pair is not commonly used due to relatively low quantum yield (0.24) and photostability of BFP. GFP[(4-Me)Tyr], however, has a higher quantum yield (0.37) and good photostability, and would therefore significantly enhance the applicability of the FRET pair with excitation possibility at ~400 nm and fluorescence detection at ~500 nm.

However, GFPuv[(4-Am)Phe] and GFPuv[(4-Me)Tyr], although already useful for application, had considerably lower quantum yields than the original GFPuv. To optimize the quantum yield of such noncanonical GFP variants, Sisido and coworkers [28] reported a combination of *in vitro* incorporation of NCAAs and random mutagenesis. Position S65 and Y145 were chosen for optimization since these positions were already effective in improving the intensity of GFP and BFP variants. They found that mutations Y145F, Y145L, and Y145M led to an increase in fluorescence intensity up to fourfold. In this way, sequence optimization was proposed to be a valuable method for optimization of unnatural autofluorescent proteins.

The generation of novel monomeric red fluorescent proteins is plausible as well. For example, the introduction of a donor–acceptor type fluorophore such as ortho-, or meta- amino-tyrosines should induce a large Stokes shift for the GFP chromophore. The augmentation of a chromophore with an electron-donating group (amine group) may result in even redder emission compared to GdFP (Prof. Hyundong Yun, personal communication).

2.4.2 *ds*Red-Monomer

Recently, Goulding et al. incorporated (3-Am)Tyr (see Fig. 1) into *ds*Red-monomer [16]. The incorporation caused a red-shift of 12 nm (603 → 615 nm) in fluorescence emission, which is in high agreement to the proposed influence of electron donating groups on the fluorescence wavelength maximum (see Table 1). Accordingly, incorporation of (3-F)Tyr led to a blue shift in fluorescence maximum because of the electron withdrawing effect of the Fluorine atom. In contrast to the experiments performed with *av*GFP derived FPs, the quantum yield was significantly elevated by incorporation of NCAAs. In *ds*Red, the incorporation of (3-Am)Tyr increased the QY from 0.04 to 0.16 and (3-F)Tyr introduction led to an increase from 0.04 to 0.25. The authors speculate that this increase may occur due to the higher bulkiness of the fluorinated or aminated chromophores and that this higher bulkiness shields the chromophore better from the surrounding of the molecule. Improper shielding was dedicated to be the main reason for the low quantum yield of the monomeric *ds*Red in comparison to its tetrameric form since in the monomer only one layer of β-strands is separating the chromophore from the solvent. However, it is difficult to believe that a single fluorine- or amino-group could protect the chromophore from the solvent influence in its cavity. To resolve these controversial issues, more research has to be performed on these noncanonical *ds*Red variants (e.g., combination of guided evolution with NCAA incorporation).

2.5 Chromophores with Chalcogen Containing Trp Analogs

Sulfur and selenium as heteroatoms are known to induce physicochemical changes, e.g., they are efficient static fluorescence quenchers. Unlike carbon, nitrogen, and oxygen, selenium and sulfur atoms possess vacant *d*-orbitals in the outer shell and can therefore act as electron acceptors. Thus, sulfur and selenium containing systems should be more permissive to interact with nearby charges, which might induce dipoles, additional dispersion forces, polarizabilities, exciplex formation, resonance energy transfer, or the formation of charge transfer complexes [29].

To study the effects of chalcogen atoms in the chromophore, the Trp analogs [2,3]Sep, [3,2]Sep, [2,3]Tpa, and [3,2]Tpa (see Fig. 1) were introduced into ECFP [15]. No fluorescence could be detected, although the proteins were reported to be properly folded [15]. Furthermore, absorbance around 400 nm was observed for all ECFP variants indicating the presence of a chromophore. However, mass analyses resulted in heterogeneous spectra without regular peak repetition, which was explained by uncontrolled chemical reactions that occured during or after chromophore formation.

In contrast, incorporation of [2,3]Sep, [3,2]Sep, [2,3]Tpa, and [3,2]Tpa (see Fig. 1) in EGFP (incorporation only takes place at position Trp57 outside of the chromophore) resulted in clear mass spectra proofing incorporation. Interestingly, their incorporation caused differences in fluorescence intensity when compared to EGFP (see Table 1). This was not seen for any other Trp analog incorporation in EGFP [11, 15]; however, Trp57 is coplanar to the chromophore and interacts with it by fluorescence resonance energy transfer [30]. 30% increase in fluorescence intensity could be achieved by replacement of Trp57 by [3,2]Tpa [15]. It was argued that the electronically rich sulfur/selenium species in a particularly favorable geometric arrangement of the chromophore neighborhood facilitated the increase in absorbance and fluorescence intensities. Indeed, upon introduction of Met and selenomethionine in the vicinity of the chromophore, similar findings were reported [31, 32].

With BT (see Fig. 1), another Sulfur-containing Trp analog was recently introduced site specifically into CFP6 (see Sect. 2.3.1) by Tirrell and co-workers. CFP6 [BT] showed blue shifts of 34 and 16 nm in the absorption and fluorescence emission maxima, respectively. This is congruent with the data obtained from incorporation of [2,3]Tpa and [3,2]Tpa. Furthermore, the chromophore's extinction coefficient was similarly lowered by BT incorporation (BT: 8,100 M^{-1} cm^{-1}, [2,3] Tpa: 6,840 M^{-1} cm^{-1} and [3,2]Tpa: 6,350 M^{-1} cm^{-1}). However, in contrast to [2,3]Tpa- and [3,2]Tpa-containing ECFP, CFP6[BT] exhibited fluorescence emission and showed a relatively high Stokes shift of 56 nm (CFP6: 37 nm) with only threefold reduction in quantum yield when compared to CFP6.

Tirrell and co-workers argued that CFP6[BT] could be used as a FRET partner for GFP since its fluorescence spectrum overlaps with the GFP's excitation spectrum and it can be efficiently excited with a violet diode laser.

3 NCAA Incorporation and Structural Integrity of FPs

3.1 Introduction and General Remarks

Besides the spectroscopic properties, structural integrity of a fluorescent protein may change upon incorporation of NCAAs. On the contrary, structural details of this protein family can be visualized and investigated from a different perspective with the help of NCAAs. Therefore, in the following the effects of NCAA incorporation into *av*GFP and its mutants will be discussed in the light of information gained mainly from high-resolution X-ray structures.

The canonical tertiary structure of *av*GFP and all its mutants is built from 11 β-strands and a central helix holding the chromophore in the context of a central cavity, which is solvent inaccessible. This basic structure, the so-called β-barrel fold, is maintained in all soluble mutants. Another feature of the GFP family is the chromophore, which is generated through a self-catalyzed intramolecular posttranslational modification (see Introduction). However, chromophore formation is not an intrinsic property of the tripeptide but rather of the proper folding of the surrounding protein matrix. This is further evidenced by several observations, e.g., denatured GFP or chemically synthesized model chromophores do not show the characteristic fluorescence at ambient temperatures. In contrast, high fluorescence of the isolated chromophore is detectable at 77 K [33, 34]. Thus, chromophore fluorescence at ambient temperature is possible only in the context of its solvent-free and rigid protein cavity, and its general optical properties are very sensitive to mutations of surrounding residues [35, 36]. From this point of view, it is easy to anticipate the GFP family as a potent model for NCAA incorporation experiments.

Surprisingly, all known fluorescent proteins with an *av*GFP-like β-barrel fold exhibit only four conserved residues. In the *av*GFP primary sequence, these residues are Tyr66,

Gly67, Arg96 and Glu222. Nevertheless, for Tyr66 mutants and their noncanonical variants, the fluorescence properties are preserved and even can be tremendously changed as discussed above. We observed recently that Trp57 and Met218 in *av*GFP play a crucial role for proper folding, chromophore maturation, and fluorescence [15]. For instance, Trp57 in *av*GFP is 15 Å apart from the chromophore and is not replaceable with Leu [37]. However, its direct fluorescence is not detectable due to resonance energy transfer to the chromophore [38]. In addition, Met218 cannot be replaced in *av*GFP by another canonical amino acid without protein misfolding and loss of fluorescence [15].

3.2 Aliphatic NCAAs

First attempts to introduce 5′,5′,5′-trifluoroleucine (Tfl) or 6,6,6-trifluoromethionine (Tfm) (see Fig. 1) into an EGFP mutant containing only two Met residues (M78L, M88L, M153T, M233K) failed or gave only marginal incorporation efficiency according to amino acid analysis [39]. In the EGFP-2M mutant, one Met residue (M1) is solvent exposed, whereas the second (M218) is buried. Although Met-containing species dominated over Tfm-containing variants, ^{19}F-NMR analysis was performed. Thereby, two signals were detectable; both could be assigned to the two different Tfm residues due to their different structural context. However, the comparatively low expression yields of EGFP-2M[Tfm] might be caused by the increased van der Waals radius of Tfm (84.4 Å) compared to Met (71.2 Å) [40], which disfavors its accommodation especially in the protein interior and/or causes unfavorable interactions. In addition, such increase in size may interfere with a proper recognition of the amino acid by the translational apparatus. This was confirmed in recent studies by Tirrell and co-workers, who reported that trifluoronorleucine could only be incorporated into proteins after evolution of the methionyl-tRNA synthetase, whereas norleucine is recognized at the same level as methionine by the unmodified methionyl-tRNA synthetase [41].

Although EGFP-2M contains 19 Leu residues, ESI-MS analysis after incorporation experiments in the presence of Tfl revealed only species with a maximum of 5 substituted residues. In addition, a tremendously decreased expression yield was observed [39]. Furthermore, amino acid analysis confirmed a Tfl content of only $< 7\%$. The increased size of the fluorinated amino acid might be the explanation for this behavior, since the trifluoromethyl moiety has the size of an isopropyl group [42]. Such increased bulkiness may be especially unfavorable in the context of the fact that Leu is mainly distributed in the tightly packed protein core [42]. Therefore, the accumulation of numerous local perturbations in the side chain packing probably induces significant disturbances in the protein core and subsequent unfolding of the related GFP variant.

Nevertheless, Tirrell and co-workers succeeded in incorporating Tfl into a mutant of GFPuv (Q80R, F99S, M153T and V163A) into which they included two further mutations (S65G and S72A) [43]. However, the fluorinated GFP variant was expressed only in insoluble (inclusion body) form. To overcome this problem, they applied random mutagenesis and after 11 rounds – each screened several times by FACS – they identified a mutant termed GFPm-11.3.3. This mutant tolerated Tfl in its sequence and, in addition, after Tfl incorporation exhibited even higher median fluorescence intensity than cells expressing GFPm with Leu. This protein mutant capable of tolerating these substitutions has six original Leu positions replaced by other canonical amino acids (L42I, L53I, L119I, L194R, L207I, L221H). Four of them are distributed in the protein interior and two (194 and 221) are surface exposed. In addition, two new Leu residues (F46L, S202L) were introduced into the protein's sequence; F46L is well known to increase chromophore maturation rate in YFP [44].

The fluorinated GFPm-11.3.3 variant exhibits higher fluorescence intensities as well as enhanced refolding rates [43]. This might be due to the fact that the CF_3 groups inside the protein enhance the hydrophobicity of the protein core. Thus, composition refinement, which selected for tolerance toward the presence of fluoroalkyl groups in the hydrophobic core, leads to higher quantum yields (better shielding of the chromophore), better expression yields, and solubility.

Another approach to improve the refolding properties of GFP is the incorporation of fluorinated proline derivatives (see Fig. 1), which was demonstrated in EGFP [45]. In general, the refolding rate of GFP and its mutants is known to be very slow (10–1,000 s) [37]. Such behavior is often associated with *cis/trans* isomerization of peptidyl-proline bonds in protein structures [46, 47]. Proline is the only canonical amino acid with a secondary α-amino group due to its five-membered pyrrolidine ring, which is directly incorporated into the protein's backbone. This ring can adopt two conformations, which differ in the position of Cγ, termed Cγ-*exo* pucker and Cγ-*endo* pucker (Fig. 6) [48, 49]. EGFP consists of 10

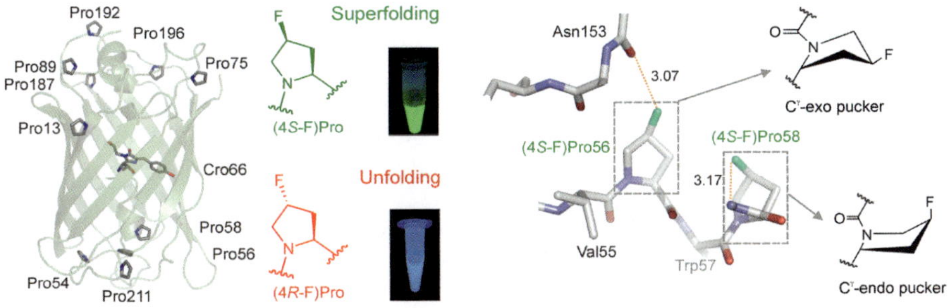

Fig. 6 Effects of 4-fluoroproline enantiomers on the stability of EGFP. The primary sequence of EGFP consists of ten Pro residues as indicated in the structure on the *left*. Substitution of all these Pro residues with (4*R*-F)Pro leads to the formation of unstructured inclusion bodies, whereas the introduction of 10 (4*S*-F)Pro residues yielded a protein with enhanced (re)folding properties. Inspection of the crystal structure revealed 12 new stabilizing interactions after (4*S*-F)Pro incorporation. Two examples are shown on the *right*. Furthermore, with EGFP as model protein the importance of Pro puckering for protein folding was identified for the first time in a complex protein structure [45]

Pro residues (see Fig. 6) of which nine form *trans* and only one (P89) forms a *cis* peptide bond. Since GFPs are in general rigid proteins, it is reasonable to assume that besides the *cis*/*trans* peptidyl-proline bond, Pro ring puckering plays a structural role [48, 49].

However, this was not obvious from studies on peptide and small proteins since they are usually very flexible in solution. For example, in these model systems (4*R*-F)Pro additionally stabilizes the *trans* conformation of peptidyl-proline bonds (since 4(*R*)-fluoroproline favors Cγ-*exo* puckering). Similarly, (4*S*-F)Pro stabilizes the *cis* conformation due to the favored Cγ-*endo* puckering. In addition, Renner et al. demonstrated that 4-fluorine containing Pro48 stabilizes/destabilizes barstar in dependence of the enantiomer used [50–53].

However, the role of Pro-puckering in folding of complex protein structures is for the first time identified in EGFP. In particular, EGFP[(4*R*-F)Pro] is a nonfluorescent insoluble protein, which could not be refolded. In contrast, EGFP[(4*S*-F)Pro] is a soluble fluorescent protein (Fig. 6). In solution, it is present predominantly in the monomeric state, with some features superior to the parent protein. For example, during the folding/refolding studies by fluorescence spectroscopy, it turned out that the recovery of fluorescence after denaturation/renaturation is much better for the variant (95%) compared to the parent protein (60%). Also, the refolding kinetics is two times enhanced for the fluorine variant (Table 3). In addition, the fluorinated variant of EGFP is less prone to aggregation and crystallizes faster.

High-resolution three-dimensional structures of EGFP and EGFP[(4*S*-F)Pro] (PDB: 1EMG and 2Q6P) provided an explanation for these observations. A closer inspection of the crystal structure of EGFP[(4*S*-F)Pro] revealed that nine out of ten (4*S*-F)Pro residues show Cγ-*endo* puckering and only (4*S*-F)Pro56 has a Cγ-*exo* pucker. This side chain conformational preference of (4*S*-F)Pro in the context of the rigid GFP structure might be one of the main reasons for the enhanced refolding properties. In other words, the proper stereochemical positioning of fluorine atoms in the GFP protein matrix caused preorganization of the majority of its Pro-side chains. Such subtle modifications, which confer extraordinary conformational stability on a protein without perturbing its structure endow EGFP[(4*S*-F)Pro] with extraordinary properties [54]. This is further confirmed by the mapping of 12 new stabilizing interactions in the interior of EGFP[(4*S*-F)Pro] up inspection of its three-dimensional structure (Fig. 6) [45].

Table 3 Comparison of the refolding kinetics and the fluorescence recovery of EGFP and the superfolding EGFP[(4*S*-F)Pro]. Upon incorporation of 4(*S*)-fluoroproline, EGFP exhibits enhanced refolding kinetics as *k* for the fast phase as well as for the slow face is approximately doubled. In addition, the fluorescence recovery is enhanced [45]

Protein	Refolding kinetics		Fluorescence
	Fast phase $k_1/(10^{-2}\ s^{-1})$	Slow phase $k_2/(10^{-2}\ s^{-1})$	Recovery after 24 h, RT
EGFP	1.41	0.15	60%
EGFP[(4*S*-F)Pro]	3.01	0.36	>95%

It is interesting to note the presence of a Pro-rich PVPWP motif in GFPs with unknown significance. It is speculated that the proline residues control the spatial orientation of V55 and W57 as they play important roles in the protection of the chromophore from collisional quenching by the solvent. Neither V55 nor W57 can be substituted by another canonical amino acid without the loss of fluorescence [37, 45].

3.3 Aromatic NCAAs

3.3.1 Nonchromophore Positions

The chromophore of *av*GFP and its mutants requires an aromatic amino acid in position 66 to exhibit fluorescence. Thus, the incorporation of noncanonical aromatic amino acids at this position opens the possibility to directly change the structural and optical properties of the chromophore. The supplementation-based incorporation (SPI) method [55] enables residue-specific replacement throughout the whole protein structure. Therefore, structural consequences for all nonchromophore position substitutions will be discussed first followed by a closer look to structural effects of NCAAs upon their direct incorporation into the chromophore structure.

GFP consists of 11 Tyr residues of which Tyr39, Tyr151, Tyr182, Tyr200 are surface exposed, Tyr143 is partially exposed and Tyr74, Tyr106, Tyr145, Tyr92 are buried in the protein core, whereas Tyr66 is part of the chromophore. Substitution of all Tyr residues with either (2-F)Tyr or (3-F)Tyr led to small changes in absorbance and fluorescence emission maxima as described above. Crystal structures of the two variants determined at 2.2 Å (EGFP[(2-F)Tyr]) and 1.6 Å (EGFP[(3-F)Tyr]) [17] revealed that the overall fold was not affected by NCAA incorporation. Careful inspection of these and other known structures of fluorinated proteins [56] revealed crystallographic distances between fluorine atoms and hydrogen donors allowing for weak interactions. However, fluorine cannot compete with stronger hydrogen-bond acceptors such as oxygen and nitrogen. Interestingly, in all positions, (2-F)Tyr exhibits only one conformation (see Fig. 7). None of the fluorine atoms in the (2-F)Tyr residues of EGFP[(2-F)Tyr] is involved in particular interactions [17]. In contrast, three (3-F)Tyr residues (92, 143, and 151) adopt two conformations in EGFP[(3-F)Tyr] (Fig. 7) [17, 57]. The majority of these observations correspond well with the assumption that there is a correlation between aromatic ring flipping and their burial in the protein's core. For example, (3-F)Tyr145 is involved in various contacts with atoms of neighboring residues (Pro58 and His169) as well as water molecules and only one conformer is detectable. Furthermore, the fluorine atom of (3-F)Tyr106 is in close contact (2.95 Å) with the ring carbons of Phe130. Surprisingly, buried (3-F)Tyr92 and partially exposed (3-F)Tyr143 adopt both possible conformational states. The fluorine of (3-F)Tyr92 is in interaction distance with the amide oxygen of Phe84 and most probably also with

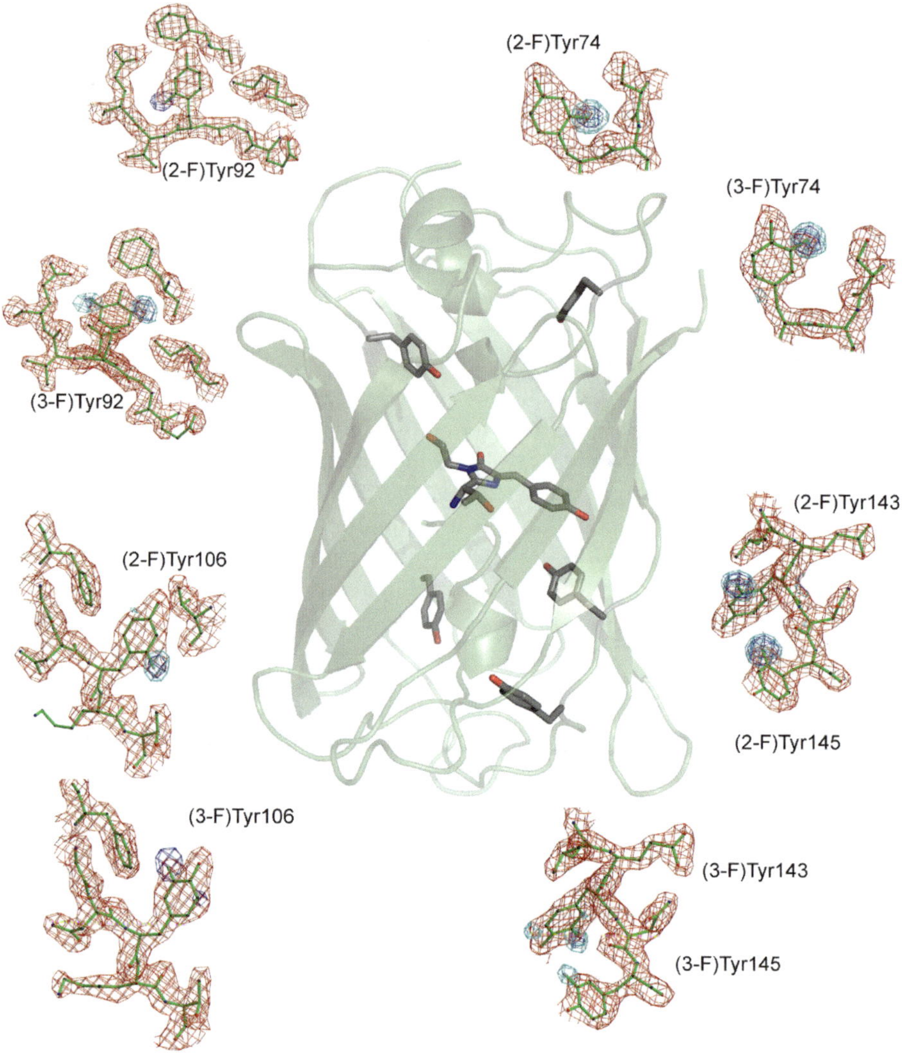

Fig. 7 Examples for the conformation of (2-F)Tyr and (3-F)Tyr residues in EGFP. Whereas in case of EGFP[(2-F)Tyr] all fluorinated Tyr residues adopt only one conformation, some of the (3-F)Tyr residues in EGFP[(3-F)Tyr] exhibit two conformations [17, 57]. This might be explained by different spatial demands of the two regioisomers

the ring carbons of Phe64, whereas (3-F)Tyr143 seems not to be involved in any interaction [17, 57].

The existence of two conformations of fluorinated amino acids in protein structures was also reported for Trp analogs such as (4-F)Trp, (5-F)Trp, and (6-F)Trp [58]. For instance, in the ^{19}F NMR spectra of EGFP (contains a single W57 residue) and ECFP (contains W57 and, in addition, W66 as part of the chromophore) two and four resonances appear, respectively. Thus, the fluorinated Trp residues exist in two states characterized by different broadening of the peaks, which indicates slow exchange processes between the two states. Although Seifert et al. did not provide detailed description for the Trp57 substitutes, the set of double signals cannot be explained by the phenomena of dimerization or aggregation. This was demonstrated by measurements with different labeled protein concentrations that resulted in identical spectra. In addition, the ^{19}F NMR spectrum of ECFP[(6-F)Trp] in the denatured state exhibits two resonances for (6-F)Trp57 as well. Most probably, proline *cis/trans* conformations in the P56-W57-P58 sequence are the main cause for the observed spectral behaviors. Interestingly, incorporation of (4-F)Trp and (6-F)Trp into GFPuv did not result in double peaks in ^{19}F NMR and led only to marginal differences in spectral shapes.

Most recently, it was shown that expression of ECFP in the presence of (4-Aza) Trp and (7-Aza)Trp (see Fig. 1) yielded in high amounts of insoluble nonfluorescent protein [59]. The small fraction of soluble protein has similar properties as the parent protein but the fluorescence signal is much broader. Control experiments were performed with EGFP, which confirmed that position 57 in the GFP molecule is a crucial determinant of the folding process. Both ECFP and EGFP containing either (4-Aza)Trp or (7-Aza)Trp were characterized by significantly higher amounts of insoluble protein compared to the parent proteins (up to 90% insoluble fraction). All attempts to perform proper refolding of these insoluble fractions failed, although their solubilization was possible. It is worth to note that proteins refolded in this way did not exhibit any defined secondary structure as revealed by NMR. In addition, ESI-MS as well as spectroscopic analyses of these samples indicated a complete Trp substitution by its aza-analogs but chromophore formation did not take place. However, soluble chromophore-free *av*GFP mutants are already reported. For instance, *cis/trans* isomerization of Xxx-Pro bonds can be affected by mutagenesis of Xxx [60], which leads to hindered chromophore formation.

Interestingly, in case of the (Aza)Trp variants of ECFP and EGFP, the small soluble protein fractions turned out to be mixtures of parent protein and variant. High-performance liquid chromatography (HPLC) separation of these mixtures revealed remarkable differences in retention times. Thus, the hydropathy of the whole protein is changed by introduction of either only one (EGFP) or two (ECFP) hydrophilic azatryptophans, although the fluorescence profiles of these species were not affected significantly. Again, this strongly indicates that the correct establishment of the protein's tertiary structure is much more important than the composition of the tripeptide sequence (Xxx65–Xxx66–Xxx67) for the chromophore formation [61, 62].

As ECFP and EGFP give similar amounts of insoluble protein upon azatryptophan incorporation, position 57 must be the crucial residue as it is the single Trp/(Aza)Trp position (aside from W66 in ECFP). Trp57 is part of the PVPWP motif and therefore involved in the water shielding protection of the chromophore from solvent water [45]. Corresponding to the literature, the Trp substitution may affect the *cis/trans* isomerization of the adjacent proline residues [60]. In addition, the substitution pattern of the Trp57's indole ring has also a great influence on the folding and stability of EGFP and ECFP. A closer inspection of the ECFP crystal structure (1OXD) revealed the accommodation of Trp57 in the minicore cavity of the protein with C4–C7 buried in the hydrophobic core, whereas N1 and C2 are solvent exposed.

However, the observation that both (7-Aza)Trp and (4-Aza)Trp gave significantly high amounts of insoluble protein cannot be explained solely by spatial requirements. Indeed, C7 of Trp57 in ECFP is in interaction distance to several atoms in the vicinity for instance with backbone nitrogen and oxygen atoms (Fig. 8). Introduction of nitrogen at C7 means an additional free electron pair at this site, which may result in repulsive interactions [59].

It is well known that various Trp derivatives substituted at position 4 could be incorporated into EGFP and ECFP including (4-Am)Trp and (4-Me)Trp [11]. These data as well as the crystal structure confirm that there is enough space in the vicinity of C4 of Trp57. But why does (4-Aza)Trp incorporation result in insoluble protein? It seems that the markedly increased hydrophilicity of (4-Aza)Trp compared to Trp causes perturbations in the hydrophobic minicore sufficient enough to interfere with proper folding of the protein. The same is most likely also true for (7-Aza)Trp. In this case, the combination of repulsive interactions of the free electron pair in ring position 7 and the increased hydrophilicity would explain the higher amount of insoluble (7-Aza)Trp-containing fluorescent proteins compared to (4-Aza)Trp-containing fluorescent proteins, which only suffer from the changed hydropathy.

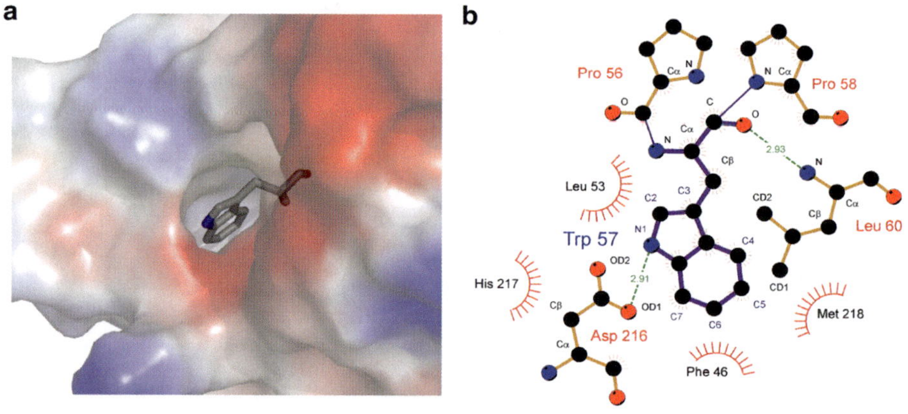

Fig. 8 Trp57's local structural environment in ECFP. Surface representation of Trp57 and its surrounding in ECFP (**a**). Trp57 has contacts to a number of atoms in its vicinity as depicted in (**b**). Ligand bonds are given in *purple*, nonligand bonds in *orange* and hydrogen bonds in *green* (including distances). Nonligand residues involved in hydrophobic contacts are depicted as *red semicircles* and atoms involved in hydrophobic contacts are depicted as *black circles*. It is worth to note that from these analyses it seems that there is more spatial freedom at position 4 than at position 7 for substituent accommodation of the indole side chain [59]

3.3.2 NCAA in the Chromophore (Position 66)

The structure of the chromophore of *av*GFP and its mutants was subject of several studies. For example, Ward et al. were the first documenting conformational flexibility of the chromophore [63]. Chen and co-workers could demonstrate that the chromophore *cis/trans* isomerization is not possible but the excited state could adopt a twisted conformation. However, the chromophore is held in a strained configuration by the surrounding residues, whereas its planarity (see Fig. 9) is a result of its delocalized π-electrons [64]. In this context, we discuss the effects of aromatic NCAAs in position 66 to the structure of *av*GFP and its mutants.

As discussed above, incorporation of (2-F)Tyr and (3-F)Tyr results only in slight changes of the spectral characteristics of EGFP and EYFP [17, 57]. Noteworthy, (2-F)Tyr and (3-F)Tyr have also different effects on the chromophore's structure upon incorporation (see Fig. 10). Whereas the chromophore adopts only one conformation in both variants of EYFP – mainly due to the Tyr203 which occupies more space than the Thr203 of EGFP – the same holds true only for (2-F)Tyr in EGFP since the chromophore including (3-F)Tyr shows two conformations [17, 57]. In general, structures of all crystals under study show a rather rigid architecture for the chromophore and its surrounding as indicated by the B factors [57]. Furthermore, the phenolate oxygen of the chromophore interacts with Thr203 and His148 and, in addition, the chromophore is in hydrophobic contact with Thr62, Val150, and Phe165 [17].

Even if there are only slight differences in the occupancies of the two states of EGFP[(3-F)Tyr] (distribution is approx. 3:2), one conformation is termed "major" and the other one "minor". Whereas in the "major" conformation the fluorine atom is involved in a rather unusual interaction with Cγ2 of Thr203 (maybe weak -CF ··· HC- hydrogen bond), the "minor" conformation is characterized by a kind of hydrogen bond between the fluorine atom and the hydroxyl group of Ser205 as well as a conserved water molecule [17, 57]. However, even with this information the observed higher occupancy of the major conformer remains hard to explain.

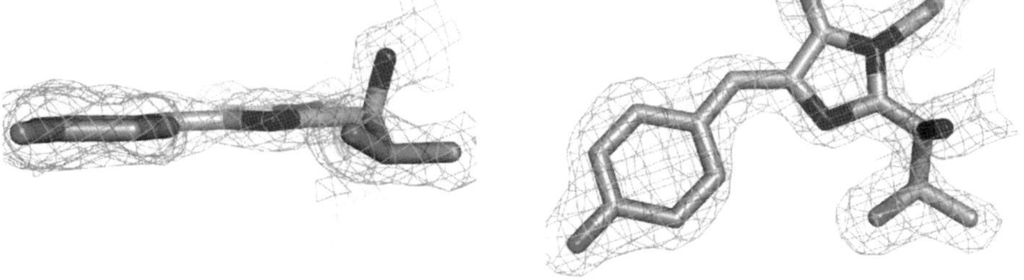

Fig. 9 Electron density of the *av*GFP chromophore in planar (*left*) and top view (*right*). The electron density reveals that the chromophore structure is nearly planar

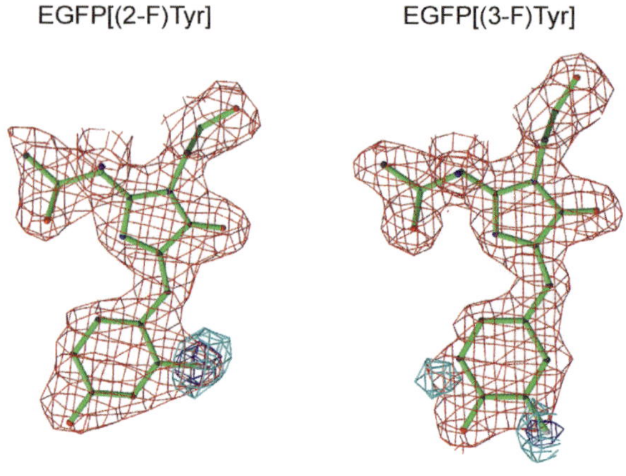

Fig. 10 Conformation of the EGFP chromophore upon incorporation of (2-F)Tyr or (3-F)Tyr. Interestingly, crystal structure analyses revealed only one chromophore conformation in case of EGFP[(2-F)Tyr]. In contrast, in EGFP[(3-F)Tyr], the fluorinated tyrosine moiety can adopt two conformations [17, 57]. The ratio of occupancy is 3:2. These results correspond well with the observations from nonchromophore Tyr positions (*vide supra*)

Thus, the question a rose whether the two conformers result from flipping of the Tyr moiety in position 66 like shown for Trp analogs (*vide infra*) [58]. But until now direct dynamic evidence could not be provided. However, it is more likely that the two conformational states are a result of Tyr flipping during chromophore formation and the residue is kind of frozen in the final chromophore in a statistical manner [17, 57]. In addition, even if by modeling and calculations some rotational freedom was attributed to the chromophore, *cis/trans* photoisomerization can be excluded as reason for the two states [63, 64].

However, the appearance of only one chromophore conformer in EGFP[(2-F) Tyr] must have sterical reasons since in this variant the chromophore has less hydrophobic contacts with its vicinity and no interaction of the fluorine atom can be assigned [17]. Indeed, crystal structure analysis revealed that after adoption of the alternative conformation the fluorine atom would induce a direct clash with the ring nitrogen of the chromophore.

For ECFP[(4-F)Trp], ECFP[(5-F)Trp] and ECFP[(6-F)Trp], ^{19}F NMR spectroscopy confirmed that fluorinated Trp66 as a part of the ECFP chromophore exists in two states (Fig. 11) and that the population of the states is temperature dependent [58]. The slow exchange process was determined by ^{19}F NMR to occur in a time frame of 1.2–1.4 ms and differences in enthalpy and entropy of the two states could be measured. Chromophore isomerization is an unlikely process in EGFP (*vide supra*) and can be discarded as reason for the two states as well as different protonation states or dimer formation. More likely, the different states originate from changes in the vicinity of the chromophore, i.e., in its electronic micro-environment leading to different shifts in the NMR spectrum (Fig. 11). As candidates, Tyr145 and His148 were identified. In GFP, His148 and Thr203 are in contact with the phenolic ring but the β-barrel is distorted in the region of residues 144–150, which is not hydrogen bonded to the adjacent backbone residues 165–170 [65]. By structure comparison, it seems likely that Tyr145 and His148 can adopt two conformations in ECFP [11, 58]. In the "major" conformation, Tyr145 points toward the protein interior, whereas

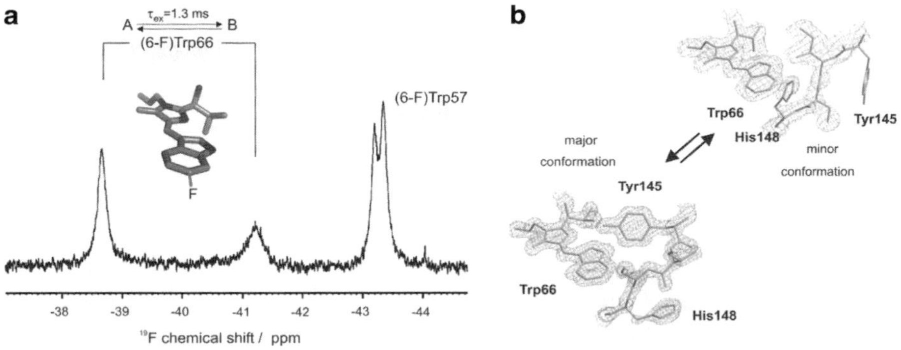

Fig. 11 ^{19}F NMR analysis of ECFP[(6-F)Trp] and modeling of the chromophore's vicinity. ^{19}F NMR upon fluorotryptophan incorporation into ECFP revealed a set of four signals for two residues [Trp57 and Trp66, the spectrum of EGFP[(6-F)Trp] is given as example (**a**)]. Therefore, it is reasonable to assume that the fluorinated Trp residues exist in two states. As flipping of the aromatic residues and phenomena such as aggregation can be excluded as reason for this observation (see text for a detailed discussion) modeling of the chromophore's vicinity was performed (**b**). Thereby, it turned out that Tyr145 and His148 might be able to adopt two conformations resulting in a different electronic environment of the Trp66 residue responsible for the splitting of the NMR signal. For Trp57, a similar explanation may holds true

His148 is directed toward the solvent. Both residues are involved in interactions with the chromophore. Conversely, in the "minor" conformation Tyr145 is turned toward the solvent without any interaction with the chromophore and His148 is directed toward the protein. At this point, it is worth to note that in EGFP both residues are directed to the chromophore [11]. Moreover, the two different conformational states could only be detected in solution (^{19}FNMR). In the crystal structure of ECFP[(4-F)Trp], it was not possible to detect two conformations [15].

However, pH titration does not affect the emission profile of the protein as would be expected by a changed His148 conformation. Most probably, hydrophobic residues in the neighborhood hinder the entry of water molecules or the imino nitrogen is involved in a hydrogen bond and therefore not accessible for titration [58]. In addition, the observed enthalpy–entropy compensation is in agreement with a flipping His148 [58]. At this point, it is worth to note that the two conformational states are not the reason for the complex photophysics of ECFP (*vide supra*).

Interestingly, the crystal structure of ECFP[(4-F)Trp] reveals an interaction between the fluorine atom and the sulfur atom of M218 [15]. Sulfur is highly polarizable and is known to interact with aromatic residues [67, 68]. The preferred distance between sulfur and the ring centroid is $r < 6$ Å and in the case of ECFP [(4-F)Trp] fluorine and sulfur posses a distance of 3.5 Å. According to the known interaction geometries, the sulfur points also toward one of the ring hydrogen of (4-F)Trp66 [15].

In contrast to the fluorine analogs, incorporation of 4-aminotryptophan ((4-Am) Trp) seems not to induce two conformational states of the chromophore in ECFP

[11]. However, the above described tremendous red-shift and extreme large Stokes shift of the ECFP[(4-Am)Trp] variant (called golden fluorescent protein GdFP) is only due to (4-Am)Trp at position 66, whereas expectedly (4-Am)Trp57 does not affect the spectral properties of the protein.

Furthermore, the chromophore seems to be tightly packed in GdFP and the crystalline state possesses a rather rigid architecture as there is no additional electron density indicating different conformational states. In addition, the region of residues Tyr145–Asn149 shows no flexibility as it is, for instance, the case in the "minor" conformation of ECPF. In contrast, whereas the ECFP chromophore exhibits eight mainly hydrophobic interactions with residues in its vicinity and one hydrogen bond (to Ser205), the chromophore of GdFP interacts only with five neighboring residues [11]. However, the amino group of (4-Am)Trp66 may interact with Phe165. This would enlarge the interacting network of aromatic residues Phe165-His148–Tyr145. These residues are rather rigid and involved in hydrophobic contacts with the chromophore. This enhanced network interaction would also explain the higher stability of GdFP compared to ECFP and its lower tendency for aggregation as well as its more co-operative unfolding process.

As already mentioned, azatryptophan inhibits proper folding of EGFP/ECFP and chromophore formation (*vide supra*) [59]. In contrast, chalcogen-containing Trp analogs ([2,3]Sep, [3,2]Sep, [2,3]Tpa and [3,2]Tpa) support proper folding of ECFP as the protein is mainly expressed in the soluble fraction and shows the same electrophoretic mobility as the parent protein in native two-dimensional protein gel analysis. But, on the contrary, the variants do not exhibit the characteristic fluorescence properties even if they are colored at daylight [15]. The fact that these ECFP variants are colored suggests chromophore formation but quenched fluorescence. Unfortunately, mass spectrometry only revealed a high heterogeneity of the samples.

Expectedly, these analogs cause no shifts of the fluorescence emission maximum in EGFP with single replacement at position 57. The proteins are expressed in high yields, which enabled their crystallization and X-ray structure elucidation. Expectedly, crystal structures of EGFP[3,2]Tpa, EGFP[2,3]Sep and EGFP[3,2]Sep revealed the same overall topology as EGFP, and no significant changes in the environment of residue 57 (Fig. 12) [15]. However, incorporation of [3,2]Tpa into EGFP increased the absorption and fluorescence emission intensity by approx. 25%. This is comparable with the effects of *in vitro* incorporation of selenomethionine into *av*GFP or the introduction of Met into position 69 into EYFP [32]. With chalcogen-containing Trp analogs, this effect occurs solely in case of [3,2]Tpa but not [2,3]Tpa or in the Sep-variants of EGFP. Therefore, this must be an electrostatic effect in combination with a particularly favorable geometric arrangement of the chromophore's vicinity [15]. This is reasonable as the distance between the sulfur of Tpa57 and sulfur of M218 is 4.7 Å for [3,2]Tpa but 6.6 Å for [2,3]Tpa. In addition, even if the sulfur–selenium distance for M218 and [3,2]Sep57 is also 4.7 Å, [3,2]Sep does not cause the same effect as [3,2]Tpa.

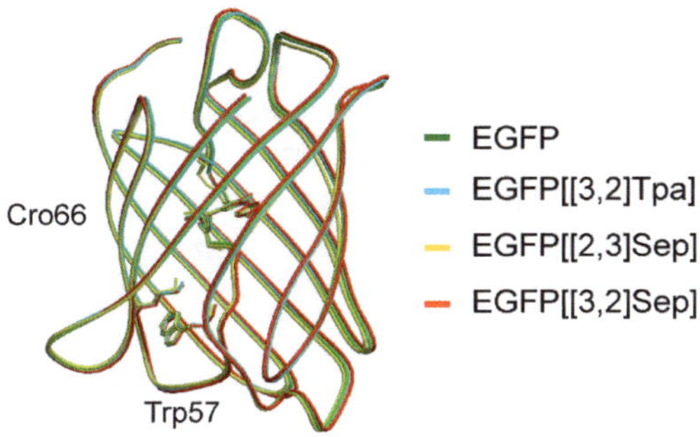

Fig. 12 Overlay of the amino acid backbone as revealed by crystal structure analysis of EGFP and its chalcogen-containing variants. Expression of EGFP in presence of [3,2]Tpa, [2,3] Sep and [3,2]Sep results in proteins with identical overall topology

4 Applications and Further Development of Autofluorescent Proteins Using NCAAs

Besides the use of NCAAs for structural analysis as discussed above, especially Sep derivatives are beneficial for X-ray crystallography as they can help to overcome problems in phase calculation [69]. But the highest potential of fluorescent protein variants with NCAA lies doubtlessly in the application of their changed fluorescent characteristics. A prominent example is the fusion protein of golden fluorescent protein and human annexin A5 (GdFP-anxA5) as novel apoptosis detection tool as recently reported by Kurschus and co-workers [70].

This approach made use of the extremly red-shifted fluorescence emission ($\lambda_{max,em}$ = 574 nm) and the 1:1 ratio of fluorophore (GdFP) and lipid binding protein (anxA5). Both are indeed improvements, as on the one hand fluorescence markers with orange/red emission are of special interest because the autofluorescence background of the cell is low in this spectral region. On the other hand, the defined label ratio is superior to commercial fluorescence labeled anxA5 as this is often a mixture of heterologous products because of the chemical reaction employed for labeling. Very often Lys residues are labeled which, unfortunately, are partially in the vicinity of the binding domain of anxA5. Thus binding may interfere with fluorescence [71, 72]. These limitations can be circumvented by using a fusion protein; however, GFP and its classical derivatives have their emission maximum at <530 nm and the more red-shifted *ds*Red and its family is not suitable for general applications as they still tend to oligomerize and aggregate.

In contrast, GdFP is almost as red-shifted as *ds*Red and exhibits an extraordinary thermostability and keeps its monomeric state even after long-term storage at 4°C [70]. Moreover, its approximately twofold higher stability to pH in the range of 5.5–7.0 when compared to anxA5-FITC provides the possibility for applications at lower pH values. Furthermore, the 4-amino group of (4-Am)Trp, which is part of the GdFP chromophore and replaces the single Trp residue in anxA5 upon SPI expression, does not affect lipid binding of the fusion protein. Therefore, GdFP-anxA5 is suitable for the quantification of apoptotic and necrotic cells and, in addition, its use in combination with other dyes in flow cytometry or fluorescence microscopy. The only disadvantage of this system is the relatively broad emission peak, whereas the quantum yield is similar to ECFP [70].

It should be mentioned again that most commonly used FPs usually have poor brightness when illuminated with UV light <366 nm. In order to address this general limitation, Skerra and co-workers [73] recently combined in a FRET system the fluorescence activity of a FP with the spectrally matching fluorescence of a NCAA translated into the protein during its expression. In particular, they succeeded in incorporating site-specifically (~20 Å away from the protein's fluorophore) the fluorescent (7-hydroxycoumarin-4-yl-)ethylglycine in ECFP. They chose ECFP as the FRET acceptor for this NCAA because its absorption maximum in the range of 430–450 nm overlaps with the 7-hydroxycoumarin emission maximum (λ_{max} approx. 450 nm). Expectedly, the ECFP modified in this way exhibited efficient FRET between its two fluorophores, having 'cyan' emission at 476 nm upon excitation in the near-UV at 365 nm.

Finally, new classes of autofluorescent proteins could be generated using NCAA. For instance, we could recently demonstrate that a single Trp → (4-Aza) Trp substitution turned colorless anxA5 into a blue fluorescent protein [74]. The same effects are known for (6-Aza)Trp and (7-Aza)Trp but these chromophores suffer either from lower biocompatibility or fluorescence intensity compared to (4-Aza)Trp [75]. In the future, we will explore the utilities of designing autofluorescent proteins by using fluorescent amino acids as building blocks for protein biosynthesis.

References

1. Sample V, Newman RH, Zhang J (2009) The structure and function of fluorescent proteins. Chem Soc Rev 38(10):2852–2864
2. Stepanenko OV, Verkhusha VV, Kuznetsova IM, Uversky VN, Turoverov KK (2008) Fluorescent proteins as biomarkers and biosensors: throwing color lights on molecular and cellular processes. Curr Protein Pept Sci 9(4):338–369
3. Chattoraj M, King BA, Bublitz GU, Boxer SG (1996) Ultra-fast excited state dynamics in green fluorescent protein: multiple states and proton transfer. Proc Natl Acad Sci USA 93 (16):8362–8367
4. Tsien RY (1998) The green fluorescent protein. Annu Rev Biochem 67:509–544
5. Crameri A, Whitehorn EA, Tate E, Stemmer WPC (1996) Improved green fluorescent protein by molecular evolution using DNA shuffling. Nat Biotechnol 14(3):315–319
6. Cormack BP, Valdivia RH, Falkow S (1996) Facs-optimized mutants of the green fluorescent protein (gfp). Gene 173(1):33–38
7. Heim R, Tsien RY (1996) Engineering green fluorescent protein for improved brightness, longer wavelengths and fluorescence resonance energy transfer. Curr Biol 6(2):178–182
8. Ormö M, Cubitt AB, Kallio K, Gross LA, Tsien RY, Remington SJ (1996) Crystal structure of the *Aequorea victoria* green fluorescent protein. Science 273(5280):1392–1395
9. Heim R, Prasher DC, Tsien RY (1994) Wavelength mutations and posttranslational autoxidation of green fluorescent protein. Proc Natl Acad Sci USA 91(26):12501–12504
10. Matz MV, Fradkov AF, Labas YA, Savitsky AP, Zaraisky AG, Markelov ML, Lukyanov SA (1999) Fluorescent proteins from nonbioluminescent anthozoa species. Nat Biotechnol 17(10): 969–973

11. Bae JH, Rubini M, Jung G, Wiegand G, Seifert MHJ, Azim MK, Kim JS, Zumbusch A, Holak TA, Moroder L, Huber R, Budisa N (2003) Expansion of the genetic code enables design of a novel "gold"' class of green fluorescent proteins. J Mol Biol 328(5):1071–1081
12. Shaner NC, Steinbach PA, Tsien RY (2005) A guide to choosing fluorescent proteins. Nat Methods 2(12):905–909
13. Kwon I, Tirrell DA (2007) Site-specific incorporation of tryptophan analogues into recombinant proteins in bacterial cells. J Am Chem Soc 129(34):10431–10437
14. Wang L, Xie JM, Deniz AA, Schultz PG (2003) Unnatural amino acid mutagenesis of green fluorescent protein. J Org Chem 68(1):174–176
15. Budisa N, Pal PP, Alefelder S, Birle P, Krywcun T, Rubini M, Wenger W, Bae JH, Steiner T (2004) Probing the role of tryptophans in *Aequorea victoria* green fluorescent proteins with an expanded genetic code. Biol Chem 385(2):191–202
16. Goulding A, Shrestha S, Dria K, Hunt E, Deo SK (2008) Red fluorescent protein variants with incorporated non-natural amino acid analogues. Protein Eng Des Sel 21(2):101–106
17. Pal PP, Bae JH, Azim MK, Hess P, Friedrich R, Huber R, Moroder L, Budisa N (2005) Structural and spectral response of *Aequorea victoria* green fluorescent proteins to chromophore fluorination. Biochemistry 44(10):3663–3672
18. Yarbrough D, Wachter RM, Kallio K, Matz MV, Remington SJ (2001) Refined crystal structure of dsred, a red fluorescent protein from coral, at 2.0-angstrom resolution. Proc Natl Acad Sci USA 98(2):462–467
19. Lossau H, Kummer A, Heinecke R, PollingerDammer F, Kompa C, Bieser G, Jonsson T, Silva CM, Yang MM, Youvan DC, MichelBeyerle ME (1996) Time-resolved spectroscopy of wild-type and mutant green fluorescent proteins reveals excited state deprotonation consistent with fluorophore-protein interactions. Chem Phys 213(1–3):1–16
20. Malo GD, Pouwels LJ, Wang MT, Weichsel A, Montfort WR, Rizzo MA, Piston DW, Wachter RM (2007) X-ray structure of cerulean gfp: A tryptophan-based chromophore useful for fluorescence lifetime imaging. Biochemistry 46(35):9865–9873
21. Demachy I, Ridard J, Laguitton-Pasquier H, Durnerin E, Vallverdu G, Archirel P, Levy B (2005) Cyan fluorescent protein: molecular dynamics, simulations, and electronic absorption spectrum. J Phys Chem B 109(50):24121–24133
22. Sinha HK, Dogra SK, Krishnamurthy M (1987) Excited-state and ground-state proton-transfer reactions in 5-aminoindole. Bull Chem Soc Jpn 60(12):4401–4407
23. Craggs TD (2009) Green fluorescent protein: structure, folding and chromophore maturation. Chem Soc Rev 38(10):2865–2875
24. Kasha M (1952) Collisional perturbation of spin orbital coupling and the mechanism of fluorescence quenching. A visual demonstration of the perturbation. J Chem Phys 20(1): 71–74
25. Bonnett R, Harriman A, Kozyrev AN (1992) Photophysics of halogenated porphyrins. J Chem Soc Faraday Trans 88(6):763–769
26. Bondi A (1964) Van der waals volumes and radii. J Phys Chem 68(3):441–451
27. Brooks B, Phillips RS, Benisek WF (1998) High-efficiency incorporation in vivo of tyrosine analogues with altered hydroxyl acidity in place of the catalytic tyrosine-14 of delta(5)-3-ketosteroid isomerase of *Comamonas (Pseudomonas) testosteroni*: effects of the modifications on isomerase kinetics. Biochemistry 37(27):9738–9742
28. Kajihara D, Hohsaka T, Sisido M (2005) Synthesis and sequence optimization of gfp mutants containing aromatic non-natural amino acids at the tyr66 position. Protein Eng Des Sel 18 (6):273–278
29. Budisa N, Alefelder S, Bae JH, Golbik R, Minks C, Huber R, Moroder L (2001) Proteins with beta-(thienopyrrolyl)alanines as alternative chromophores and pharmaceutically active amino acids. Protein Sci 10(7):1281–1292
30. Visser NV, Borst JW, Hink MA, van Hoek A, Visser AJWG (2005) Direct observation of resonance tryptophan-to-chromophore energy transfer in visible fluorescent proteins. Biophys Chem 116(3):207–212

31. Griesbeck O, Baird GS, Campbell RE, Zacharias DA, Tsien RY (2001) Reducing the environmental sensitivity of yellow fluorescent protein – mechanism and applications. J Biol Chem 276(31):29188–29194
32. Budisa N (2003) Expression of "tailor-made" proteins via incorporation of synthetic amino acids by using cell-free protein synthesis. Cell-free protein expression. Springer-Verlag, Heidelberg
33. Niwa H, Inouye S, Hirano T, Matsuno T, Kojima S, Kubota M, Ohashi M, Tsuji FI (1996) Chemical nature of the light emitter of the aequorea green fluorescent protein. Proc Natl Acad Sci USA 93(24):13617–13622
34. Bell AF, He X, Wachter RM, Tonge PJ (2000) Probing the ground state structure of the green fluorescent protein chromophore using Raman spectroscopy. Biochemistry 39(15): 4423–4431
35. Kummer AD, Kompa C, Lossau H, Pollinger-Dammer F, Michel-Beyerle ME, Silva CM, Bylina EJ, Coleman WJ, Yang MM, Youvan DC (1998) Dramatic reduction in fluorescence quantum yield in mutants of green fluorescent protein due to fast internal conversion. Chem Phys 237(1–2):183–193
36. Kummer AD, Wiehler J, Rehaber H, Kompa C, Steipe B, Michel-Beyerle ME (2000) Effects of threonine 203 replacements on excited-state dynamics and fluorescence properties of the green fluorescent protein (gfp). J Phys Chem B 104(19):4791–4798
37. Palm GJ, Wlodawer A (1999) Spectral variants of green fluorescent protein. In: Green fluorescent protein, vol 302 (Ed. Conn P M). Methods in enzymology. Academic, San Diego, pp 378–394
38. Prendergast FG (1999) Biophysics of the green fluorescent protein. Methods Cell Biol 58: 1–18
39. Budisa N, Pipitone O, Siwanowicz I, Rubini M, Pal PP, Holak TA, Gelmi ML (2004) Efforts towards the design of 'teflon' proteins: in vivo translation with trifluorinated leucine and methionine analogues. Chem Biodivers 1(10):1465–1475
40. Houston ME, Harvath L, Honek JF (1997) Synthesis of and chemotactic responses elicited by fmet-leu-phe analogs containing difluoro- and trifluoromethionine. Bioorg Med Chem Lett 7(23):3007–3012
41. Yoo TH, Tirrell DA (2007) High-throughput screening for methionyl-trna synthetases that enable residue-specific incorporation of noncanonical amino acids into recombinant proteins in bacterial cells. Angew Chem Int Ed 46(28):5340–5343
42. O'Hagan D, Rzepa HS (1997) Some influences of fluorine in bioorganic chemistry. Chem Commun 7:645–652
43. Yoo TH, Link AJ, Tirrell DA (2007) Evolution of a fluorinated green fluorescent protein. Proc Natl Acad Sci USA 104(35):13887–13890
44. Nagai T, Ibata K, Park ES, Kubota M, Mikoshiba K, Miyawaki A (2002) A variant of yellow fluorescent protein with fast and efficient maturation for cell-biological applications. Nat Biotechnol 20(1):87–90
45. Steiner T, Hess P, Bae JH, Wiltschi B, Moroder L, Budisa N (2008) Synthetic biology of proteins: tuning gfps folding and stability with fluoroproline. PLoS ONE 3(2):e1680
46. Wedemeyer WJ, Welker E, Scheraga HA (2002) Proline cis-trans isomerization and protein folding. Biochemistry 41(50):14637–14644
47. Dugave C, Demange L (2003) Cis-trans isomerization of organic molecules and biomolecules: implications and applications. Chem Rev 103(7):2475–2532
48. Pal D, Chakrabarti P (1999) Cis peptide bonds in proteins: residues involved, their conformations, interactions and locations. J Mol Biol 294(1):271–288
49. Milner-White EJ, Bell LH, Maccallum PH (1992) Pyrrolidine ring puckering in cis and trans-proline residues in proteins and polypeptides. Different puckers are favoured in certain situations. J Mol Biol 228(3):725–734
50. Renner C, Alefelder S, Bae JH, Budisa N, Huber R, Moroder L (2001) Fluoroprolines as tools for protein design and engineering. Angew Chem Int Ed 40(5):923–925

51. Holmgren SK, Taylor KM, Bretscher LE, Raines RT (1998) Code for collagen's stability deciphered. Nature 392(6677):666–667
52. Holmgren SK, Bretscher LE, Taylor KM, Raines RT (1999) A hyperstable collagen mimic. Chem Biol 6(2):63–70
53. Eberhardt ES, Panasik N, Raines RT (1996) Inductive effects on the energetics of prolyl peptide bond isomerization: implications for collagen folding and stability. J Am Chem Soc 118(49):12261–12266
54. Moroder L, Budisa N (2010) Synthetic biology of protein folding. Chem Phys Chem 11(6): 1181–1187
55. Budisa N (2004) Prolegomena to future experimental efforts on genetic code engineering by expanding its amino acid repertoire. Angew Chem Int Ed 43(47):6426–6463
56. Xiao GY, Parsons JF, Tesh K, Armstrong RN, Gilliland GL (1998) Conformational changes in the crystal structure of rat glutathione transferase m1-1 with global substitution of 3-fluorotyrosine for tyrosine. J Mol Biol 281(2):323–339
57. Bae JH, Pal PP, Moroder L, Huber R, Budisa N (2004) Crystallographic evidence for isomeric chromophores in 3-fluorotyrosyl-green fluorescent protein. Chem Bio Chem 5(5):720–722
58. Seifert MH, Ksiazek D, Azim MK, Smialowski P, Budisa N, Holak TA (2002) Slow exchange in the chromophore of a green fluorescent protein variant. J Am Chem Soc 124(27): 7932–7942
59. Hoesl MG, Larregola M, Cui H, Budisa N (2010) Azatryptophans as tools to study polarity requirements for folding of green fluorescent protein. J Pept Sci 16(10):589–595
60. Andrews BT, Schoenfish AR, Roy M, Waldo G, Jennings PA (2007) The rough energy landscape of superfolder gfp is linked to the chromophore. J Mol Biol 373:476–490
61. Barondeau DP, Putnam CD, Kassmann CJ, Tainer JA, Getzoff ED (2003) Mechanism and energetics of green fluorescent protein chromophore synthesis revealed by trapped intermediate structures. Proc Natl Acad Sci USA 100(21):12111–12116
62. Reid BG, Flynn GC (1997) Chromophore formation in green fluorescent protein. Biochemistry 36(22):6786–6791
63. Ward WW, Prentice HJ, Roth AF, Cody CW, Reeves SC (1982) Spectral perturbations of mutants of recombinant *Aequorea victoria* green-fluorescent protein. Photochem Photobiol 35(6):803–808
64. Chen MC, Lambert CR, Urgitis JD, Zimmer M (2001) Photoisomerization of green fluorescent protein and the dimensions of the chromophore cavity. Chem Phys 270(1):157–164
65. Wachter RM, Elsliger MA, Kallio K, Hanson GT, Remington SJ (1998) Structural basis of spectral shifts in the yellow-emission variants of green fluorescent protein. Structure 6 (10):1267–1277
66. Villoing A, Ridhoir M, Cinquin B, Erard M, Alvarez L, Vallverdu G, Pernot P, Grailhe R, Merola F, Pasquier H (2008) Complex fluorescence of the cyan fluorescent protein: comparisons with the h148d variant and consequences for quantitative cell imaging. Biochemistry 47 (47):12483–12492
67. Reid KSC, Lindley PF, Thornton JM (1985) Sulfur-aromatic interactions in proteins. FEBS Lett 190(2):209–213
68. Viguera AR, Serrano L (1995) Side-chain interactions between sulfur-containing amino acids and phenyalanine in alpha-helices. Biochemistry 34(27):8771–8779
69. Bae JH, Alefelder S, Kaiser JT, Friedrich R, Moroder L, Huber R, Budisa N (2001) Incorporation of beta-selenolo 3, 2-b pyrrolyl-alanine into proteins for phase determination in protein x-ray crystallography. J Mol Biol 309(4):925–936
70. Kurschus FC, Pal PP, Baumler P, Jenne DE, Wiltschi B, Budisa N (2009) Gold fluorescent annexin a5 as a novel apoptosis detection tool. Cytom A 75A(7):626–633
71. Ernst JD, Mall A, Chew G (1994) Annexins possess functionally distinguishable ca2+ and phospholipid binding domains. Biochem Biophys Res Commun 200(2):867–876

72. Ernst JD, Yang L, Rosales JL, Broaddus VC (1998) Preparation and characterization of an endogenously fluorescent annexin for detection of apoptotic cells. Anal Biochem 260(1): 18–23
73. Kuhn SM, Rubini M, Mueller MA, Skerra A (2011) Biosynthesis of a Fluorescent Protein with Extreme Pseudo-Stokes Shift by Introducing a Genetically Encoded Non-Natural Amino Acid outside the Fluorophore. J Am Chem Soc 133:3708–3711
74. Lepthien S, Hoesl MG, Merkel L, Budisa N (2008) Azatryptophans endow proteins with intrinsic blue fluorescence. Proc Natl Acad Sci USA 105(42):16095–16100
75. Merkel L, Hoesl MG, Albrecht M, Schmidt A, Budisa N (2010) Blue fluorescent amino acids as in vivo building blocks for proteins. Chem Bio Chem 11(3):305–314

Part II
Switching on the Molecular Level

Vibrational Spectroscopy of Fluorescent Proteins: A Tool to Investigate the Structure of the Chromophore and Its Environment

Valentina Tozzini and Stefano Luin

Abstract The design of fluorescent protein (FP) mutants with tailored properties benefits from the comprehension of chromophore structure, interactions, energy landscapes, and dynamics. Vibrational spectroscopy can often provide detailed information on these characteristics for proteins in their natural aqueous environment, during their (photo)dynamics and without the need of crystallization. Here we will review the experimental and theoretical techniques that have been used to analyze the relations between vibrational spectra and different structural, photophysical, and chemical properties of FPs, in particular the ones able to selectively address the chromophore and its close environment, like (pre)resonance Raman, difference-IR absorption measurements, and their computational simulations. Starting from the preliminary results aimed at identifying vibrational modes in the neutral and anionic GFP chromophore, we will discuss selected experiments that unraveled, often thanks to comparison with theoretical studies, the structure of the chromophore in some FP mutants, the impact of *cis–trans* isomerization and different protonation states in reversibly photoswitchable proteins, and the structural changes and proton-transfer pathway in the photoexcited state.

Keywords Density Functional Theory · Fluorescent protein chromophores · GFP · IR absorption · Raman · Time-resolved spectroscopy · Vibrational spectroscopy

Contents

V. Tozzini and S. Luin (✉)
NEST, Scuola Normale Superiore and Istituto Nanoscienze-CNR, Piazza San Silvestro 12, 56127 Pisa, Italy
e-mail: s.luin@sns.it

G. Jung (ed.), *Fluorescent Proteins I*, Springer Ser Fluoresc (2012) 11: 133–170, DOI 10.1007/4243_2011_31,
Published online: 6 September 2011

1 Introduction

Vibrational information is specific to the chemical bonds and symmetry of a molecule; therefore, every molecule has its distinctive vibrational spectrum, a fingerprint by which the molecule can be identified. For this reason, vibrational spectroscopy techniques can in principle be used both for studying the presence and localization of molecular species and for following structural changes within chemical, physical, or biological processes. Indeed, the interest of vibrational spectroscopy of the green fluorescent protein (GFP) and its mutants was driven by the interest on the structural changes accompanying its mechanism of fluorescence and its complex photophysics [1–4], and on the role of the protein matrix in modulating the spectral properties of the chromophore. This knowledge was and still is necessary to help the design of GFP mutants with novel and engineered spectroscopic properties.

The structure of a molecule cannot be directly determined by its vibrational spectrum, as is the case, e.g., for X-ray crystallography. However, the comparison of experimental results on GFP mutants with theoretical calculations and with the vibrational spectra of isolated chromophores homologues (like the HBDI shown in Fig. 1a) give to vibrational spectroscopy the sensitivity to detect important structural details such as the protonation state and the effect of light absorption on the structure of bound chromophores, e.g., the eventually induced isomerization. Moreover, vibrational spectroscopies are nondestructive techniques that often do not require any lengthy sample preparation (like crystallization) and allow monitoring the results of ongoing (photo)reactions of molecules in their natural environment, i.e., aqueous solutions for proteins.

In this chapter, we will review the results of vibrational spectroscopy in GFP mutants, with an emphasis on the results on their chromophore in different environments. In Sect. 2, after a short general introduction on vibrational properties of molecules, we will present the experimental techniques that have been most used to analyze vibrational spectra of fluorescent proteins. The theoretical methods for calculating the vibrational spectra of fluorescent protein chromophores will be reviewed in Sect. 3, with particular attention on methods based on the time-dependent density functional theory (TDDFT). Section 4 is dedicated to the

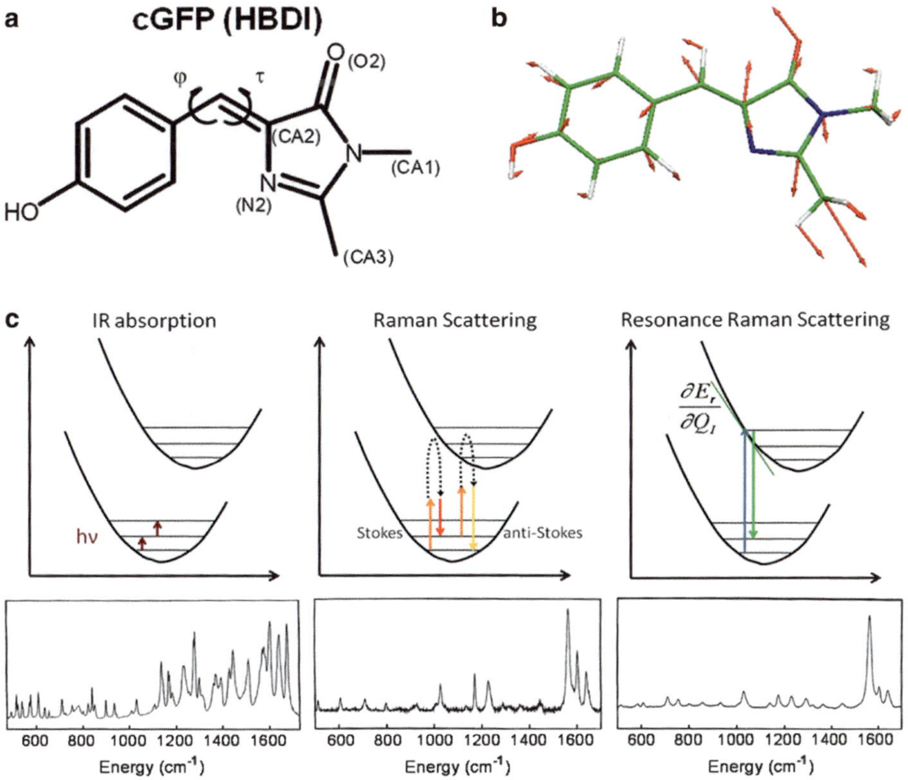

Fig. 1 (**a**) Molecular structures of an analogues of the chromophore of GFP in its *cis* form (cGFP) [HBDI: (Z)-4-(4-hydroxybenzylidene)-1,2-dimethyl-1 H-imidazol-5(4 H)-one]. In the figure are also labeled the angles τ (involved in the "*cis–trans*" isomerization) and φ, and the PDB nomenclature for some of the atoms. (**b**) Example of a delocalized mode on cGFP (calculated mode at 612 cm^{-1}, corresponding to the experimental peak at 606 cm^{-1}); the chromophore is represented in *stick* mode, with different colors for different atoms; the *arrows* indicate the relative displacement (or velocity) of the atoms within the vibrational mode. (**c**) *Top*: schematic representation of the quantum transitions involved in IR absorption spectroscopy, Raman, and resonance Raman scattering. *Up-(down-)arrows* represent the energy of the photons of impinging (scattered) radiation; in the central panel, *dotted lines* schematize the fact that all transitions are summed in the calculation of the dynamic electronic polarizability responsible for the Raman scattering. *Bottom*: comparison of IR-absorption, Raman, and resonance Raman spectra of the model chromophore HDBI

presentation of some of the results obtained by vibrational spectroscopy on GFP mutants: in the first part, the techniques used to interpret the vibrational spectrum of the chromophore and the influence of the protein environment on its vibrational modes will be discussed; in the second part, some examples will help to explain in which ways vibrational spectroscopy can help in understanding structural changes and dynamics of the chromophore and how these affect the spectroscopy properties and in general the physical behavior of GFP mutants.

2 Experimental Techniques

A molecular vibration occurs when atoms in a molecule are in periodic motion while the molecule as a whole has constant translational and rotational motion. A molecule can vibrate in many ways, or better, a complicated vibrational motion can be considered, in a first approximation, as a superposition of simple harmonic motions; each of these is called a vibrational mode. In a mode, all the atoms move harmonically in (or out-of) phase with a single frequency, in a concerted motion with directions and amplitudes typical of the given mode; one example of these directions and amplitudes for a delocalized mode of the GFP chromophore is shown in Fig. 1b. Molecules with N atoms have 3N-6 vibrational modes (3N-5 if the molecule is linear), each characterized by its vibrational frequency, ν. A molecular vibration is excited when the molecule absorbs a quantum of energy $E = h\nu$, where h is the Planck constant. The energies of vibrations are usually measured in inverse-wavenumbers $\bar{k}$, with units cm^{-1}, linked to the usual units of energy by the relation $E = hc\bar{k}$, where c is the velocity of light. The fingerprint-region energies for organic molecules are mostly within the infrared (IR) portion of the electromagnetic spectrum, in particular in the mid IR, ranging from ~300 cm^{-1} to ~4,000 cm^{-1} (corresponding to wavelength λ ~ 30–2.5 μm).

The vibrational states of a molecule can be probed in a variety of ways. Within biophysics and biology, the most often used techniques are infrared absorption (IR) and inelastic light scattering (Raman), which will be described in Sect. 2.1 and 2.2; a scheme of the quantum processes at the basis of these techniques is shown in the top panels of Fig. 1c, while examples of the results on HBDI are shown in its bottom panels. In Sect. 2.3 multiphoton and time-resolved variations of these techniques will be presented.

2.1 IR Absorption

The infrared absorption spectrum of a sample is recorded by transmitting a beam of infrared light through the sample, analogously to the UV/Vis absorption technique. Examination of the transmitted light reveals how much energy is absorbed at each wavelength. This can be done with a monochromatic beam, which changes in wavelength over time (dispersive configuration), or by using a multichannel dispersive spectrograph or a Fourier transform instrument (FTIR configuration) to measure all wavelengths at once. From this, a transmittance or absorbance spectrum can be produced, as schematized in Fig. 2a.

In the dispersive configuration, the setup is usually very similar to the ones used for UV/Vis spectrometers. Most often, a beam of infrared light is split into two separate beams passing through the sample and through a reference (often the same substance where the sample is dissolved). In the lock-in configuration, the splitter actually alternates the light path between sample and reference (usually 10–100 times a second); the beams are both directed toward a detector, and the resulting

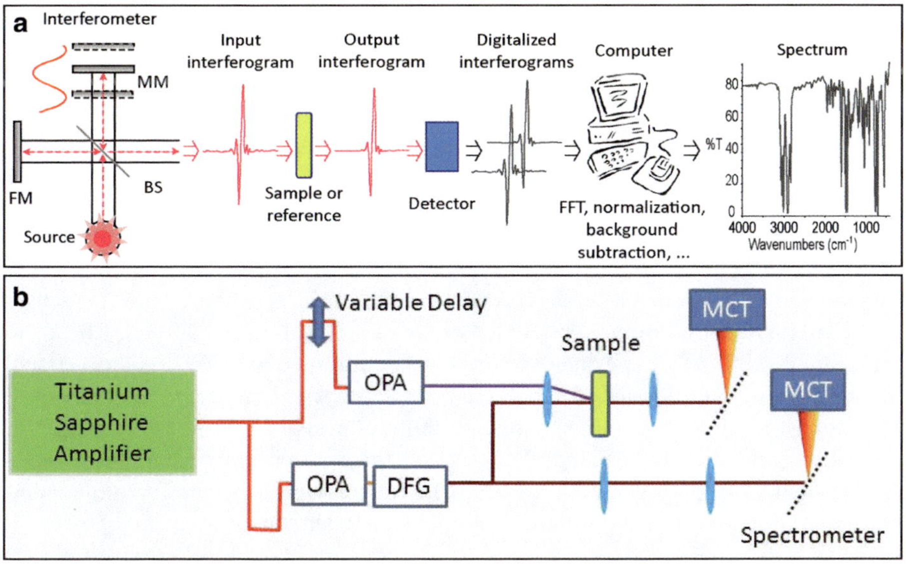

Fig. 2 Setups for IR absorption spectroscopy. (**a**) Scheme for a Fourier transform infrared spectroscopy (FTIR) experiment. The schematized interferometer is a Michelson–Morley one, the *red curve* schematizes the intensity on the detector of a monochromatic light (e.g., the *red line* of a HeNe laser) as a function of the mobile mirror (MM) position. *Purple arrows* represent propagation of IR light. *FM* fixed mirror, *BS* beamsplitter, *FFT* fast Fourier transform. (**b**) Possible scheme for a setup for picoseconds time-resolved infrared spectroscopy (TRIR) reproduced from [5], © 2011 Science and Technology Facilities Council

signal is then demodulated to measure directly the difference between the two light intensities. The comparison of the signals with and without the sample allows to obtain the transmission (T) spectrum, which can be converted to an absorbance spectrum ($A = -\log_{10}(T)$). The main differences with a UV/Vis spectrometer consist in the different light sources, detectors, and optic materials, which must be IR transparent. It should also be noted that typical cross sections and molar extinction coefficients in the IR range are usually much lower compared to those of the excitation of electronic levels, responsible for the UV/Vis absorption and fluorescence spectra of molecules.

In the FTIR configuration, the "dispersion" is obtained by using a Michelson–Morley interferometer with a mobile mirror; in this way, the information on wavelengths is mapped in the interferogram (i.e., the intensity of the signal as a function of the mirror position), from which the spectrum can be reobtained with a Fourier transform. This technique allows obtaining a better signal-to-noise (S/N) ratio in the same time when compared to the dispersive configuration. This advantage stems from the facts that information for all wavelengths are collected for each point of the interferogram, that Fourier transformation can filter out some part of the noise, and that there is no necessity to change filters or gratings during the scan and the detectors can be used always in the same "optimized" configuration. Moreover, the wavelength scale is usually calibrated using a He–Ne laser, and is therefore very accurate. Some advantages arise also from the facts that the impact of stray light is negligible and that the resolution is the same at all wavelengths.

A disadvantage of IR absorption arises from the fact that many materials and substances, and in particular water, have rich IR spectra that might obscure the ones of the substance under investigation. This requires a careful preparation of the sample, which can be contained in solvents with low IR absorption at the wavelengths of interest (e.g., Cl_4, $CHCl_3$), can be dispersed in solid pellets or films dried between polished salt (KBr) disks, could be in gaseous phase, or, if liquid, must be sandwiched in a thin layer between two salt disks. A different configuration is used in the attenuated total reflection (ATR) spectroscopy technique: in this case, an IR-transparent, high refractive index internal reflection element (IRE) is "immersed" into or pressed onto the sample, or the sample is deposited on it. A beam of infrared light is passed through the IRE so that it totally reflects at least once off the internal surface in contact with the sample; in practice, the light partially enters the solution (evanescent wave) and can be absorbed. The reflected light is therefore attenuated, causing a measurable reduction in the output signal dependent upon the absorbance of the sample. ATR has been used, e.g., to study changes in the IR spectrum during bacteriorhodopsin photoreaction on a thin film of the protein covering the IRE [6].

However, most of the preparations mentioned above are not the best options for protein samples. Moreover, the application of infrared spectroscopy on fluorescent proteins is challenging, because special means have to be used to resolve the vibrational bands of the chromophore in the presence of the strong IR absorption by the protein backbone or by other aminoacidic residues. To overcome these difficulties, it is possible to analyze differential spectra, which highlight changes between different configurations of a given fluorescent protein; starting from these kinds of experiments, it is possible to relate the changes in the IR spectrum with specific changes in the molecular structure of the sample. In the case of fast reversible changes, it is even possible to collect only the differential spectrum, e.g., with a lock-in technique. A variation of this technique, the time-resolved infrared absorption (TRIR), will be described in Sects. 2.3 and 4.4.

2.2 Raman and SERS

Raman spectroscopy is a technique used to study vibrational, rotational, and other low-frequency modes in a system. It relies on inelastic scattering of monochromatic light. The light interacts with excitations in the system, resulting in the energy of the scattered photons being shifted down (up) by an amount corresponding to the energy of the excited (depleted) mode, for Stokes (anti-Stokes) Raman scattering, as schematized in the second panel of Fig. 1c. For molecular spectroscopy, the derivative of the molecular polarization potential (i.e., the amount of deformation of the electron cloud) with respect to the vibrational coordinate will determine the Raman scattering intensity (see Sect. 3). The pattern of shifted frequencies is determined by the vibrational (and rotational) states of the sample. Also, the polarization of the incoming and scattered light can be used to infer properties of the sample (polarized or optically active Raman).

Typically, a sample is illuminated with a laser beam in the visible, near infrared, or near ultraviolet range; since water and glass are transparent in these ranges, the sample can be dissolved in water and cuvettes for fluorescence experiments can be used. Light from the illuminated spot is collected with a lens system and sent through a monochromator. Wavelengths close to the laser line (stray light), due to elastic (Rayleigh) scattering, are filtered out, while the rest of the collected light is dispersed onto a detector.

Spontaneous Raman scattering is typically very weak, and as a result the main difficulty of Raman spectroscopy is separating the weak unelastically scattered light from the intense Rayleigh scattered laser light. Historically, Raman spectrometers used holographic gratings and multiple dispersion stages to achieve a high degree of laser rejection. In the past, photomultipliers were the detectors of choice for dispersive Raman setups, which resulted in long acquisition times. However, modern instrumentation almost universally employs notch or edge filters for laser rejection, FT (Fourier-transform based) or multichannel spectrographs, and CCD detectors.

One of the main advantages of the Raman spectroscopy stems from the resonance phenomenon: there is a huge increase of Raman intensity when the energy of the incoming or of the scattered photons is close to a fundamental excitation of the system (third panel in Fig. 1c). Most importantly, the increase is specific for the modes that interact with the in-resonance excitation, in particular for vibrational modes localized on the part of the molecule responsible for the considered electronic transition, which is, in fluorescent protein, the chromophore. Moreover, the use of preresonance conditions (i.e., photons with energy slightly lower than necessary for the excitation of a higher electronic energy level, so that it cannot excite the fluorescence of the sample) makes it possible to selectively enhance chromophore signal with negligible impact of the residual fluorescence, and the vibrational spectrum of the chromophore can be investigated also while inside fully folded proteins. For example, this technique helped to determine the protonation of the chromophore before and after photoactivation in wild-type GFP (*wt*GFP) [7]. Raman spectroscopy has also been used together with differential IR absorption to collect evidence for isomerization and decarboxylation in the photoconversion of the Red Fluorescent Protein DsRed [8].

In addition to yielding information on a chromophore's vibrational frequencies, resonance Raman spectra can provide insight into the shape of the potential energy surface of the excited state. The vibrational modes that are enhanced most strongly are those for which the excited-state surface is shifted significantly with respect to the ground state [9]. As an example, Schellenberg et al. observed that the resonance Raman spectra of GFP and of the chromophore in solution were different in transition intensities, suggesting that the protein modifies the excited-state potential energy surface of the chromophore [9].

There are a number of advanced types of Raman spectroscopy aimed at overcoming the difficulty given by the usually low cross-section for Raman scattering. Among these, we mention surface-enhanced Raman scattering (SERS), tip-enhanced Raman (TERS), stimulated Raman (analogous to stimulated emission), and hyper Raman.

SERS spectroscopy is a surface-sensitive technique that results in the enhancement of Raman scattering by molecules in proximity of (adsorbed on or linked to) metal surfaces [10, 11]. This enhancement is particularly strong when the metallic surfaces possess nanometric-size roughness or features, and it can reach values up to 10^{14} when appropriate nanoparticles are used and when Raman resonance is exploited as well (surface enhanced resonant Raman scattering – SERRS) [12]. Two effects are operative in SERS: the electromagnetic (EM) enhancement mechanism is related to resonances between the surface plasmons of the metal nanostructures and the impinging and scattered electromagnetic radiation during the Raman process, giving rise to enhanced local electric fields; the EM enhancement is more effective with certain configurations of the metallic surface, which could happen in the so called "hot spots." The second enhancement mechanism is attributed to the so-called chemical or electronic effect, and is linked with the interaction between the electronic states of the molecule and of the metals, which produce more states through which the Raman scattering could occur and that give rise to more opportunities for resonances.

SERS of EGFP, E^2GFP and of their common chromophore allowed to address their low-energy vibrational modes [13], which usually have a too low signal-to-noise ratio to be detected with normal Raman scattering (see Sects. 4.2 and 4.3). Moreover, the high enhancement of the SE(R)RS sensitivity offers the possibility of single-molecule (SM) detection [14]. Indeed, SM-SERRS has been used to study EGFP, highlighting changes in the spectra with time scales ranging from the ms to several seconds [15]. The origin of the fluctuations at the fastest time scales was not clear; the spectral jumps in a time scale of seconds were interpreted as produced by dynamic conversions between the protonated and the deprotonated forms of the chromophore in individual EGFPs [15]. However, the strong local fields responsible for the enhancement in SERS can hinder the interpretation of the collected time-resolved spectra, especially in mutants with many possible structural changes such as reversible switchable fluorescent proteins (RSFPs).

2.3 Nonlinear Techniques: CARS and Time-Resolved Pump-and-Probe-Based Techniques

While Raman is formally a second-order process, the intensity of the signal is only proportional to the incoming light intensity, because the Raman process is usually based on a "spontaneous emission." The Raman signal can be increased using the stimulated Raman scattering (SRS) technique [16]. SRS is analogous to the well-known phenomenon of stimulated emission. In SRS, two laser beams at ω_p and ω_S (pump and Stokes beam, respectively) coincide on the sample (Fig. 3a). Only when the difference frequency, $\Delta\omega = \omega_P - \omega_S$ (Raman shift), matches a particular molecular vibrational frequency Ω, amplification of the Raman signal is achieved by virtue of a stimulated process. Consequently, the intensity of the Stokes beam experiences a gain, and the intensity of the pump beam experiences a loss, as shown in Fig. 3b. If the Stokes beam is not monochromatic, it is also possible to collect directly a SRS spectrum (Fig. 3c, d).

The intensity of the pump beam can be lost also due to higher order phenomena, like the coherent anti-Stokes Raman scattering (CARS) schematized in Fig. 3a, b. In this process, two photons of the pump beam are adsorbed, a photon is emitted by stimulation of the Stokes beam, and as a result a photon is emitted coherently with frequency $\omega_{AS} = \omega_P + \Omega$. Being a coherent process, CARS signal interferes with other coherent 3-photon processes that can produce photons of frequency ω_{AS}, causing a more complicated peak shape than the simple lorentzian one expected for homogeneously broadened Raman peaks. Nevertheless, CARS spectra have been successfully used, e.g., to study the structure of the immature and mature form of the red fluorescent protein DsRed; the Raman peaks, obtained after a fitting procedure of the CARS spectra, have been compared to the known ones of S65T-GFP [18]. The advantages of the CARS technique over simple Raman stem from the nonlinearity of the process. This allows to increment the Raman signal over the fluorescence (linear) background, e.g., even when exploiting the double resonance for both the incoming and the scattered light for selective probing of different DsRed species in resonance with their visible absorption [18].

The SRS can be exploited also in time-dependent, femtosecond-stimulated Raman spectroscopy (FSRS). In this technique, femtosecond-pulsed lasers are used, and the narrow pump beam and the broad Stokes probe are delayed by a known amount of time from an actinic pump, which excites the system in a higher-energy electronic state (Figs. 3e, f). Therefore, low-frequency vibrational motions can be sampled with time steps shorter than the vibrational period. In this aspect, this technique is similar to other femtosecond-resolved techniques: as examples, a pump-probe technique, with both pump and probe in resonance with the absorption of EGFP, was used to test the coherent dynamics of single-electron vibronic wave packets following ultrafast excitation [19], highlighting two modes at 497 cm^{-1} (period 67 fs) and 593 cm^{-1} (period 59 fs); later, diffractive-optic ultrafast transient grating spectroscopy was used to observe EGFP modes below 100 cm^{-1}, one of which (at 95 cm^{-1}) was assigned to an intramolecular torsional motion in the excited state. However, FSRS has the advantage of collecting information on the structure of the molecule within its dynamics in the excited state, through its Raman spectra, as will be shown in Sect. 4.4.

A similar technique, albeit simplified and with poorer time resolution, has been exploited by using an IR absorption probe delayed from the actinic probe; this technique, called time resolved infrared (TRIR), has been used, e.g., to determine the excited-state structure of the GFP chromophore [20]. A scheme of the set-up used to study the proton pathways in the excited state proton transfer (ESPT) of

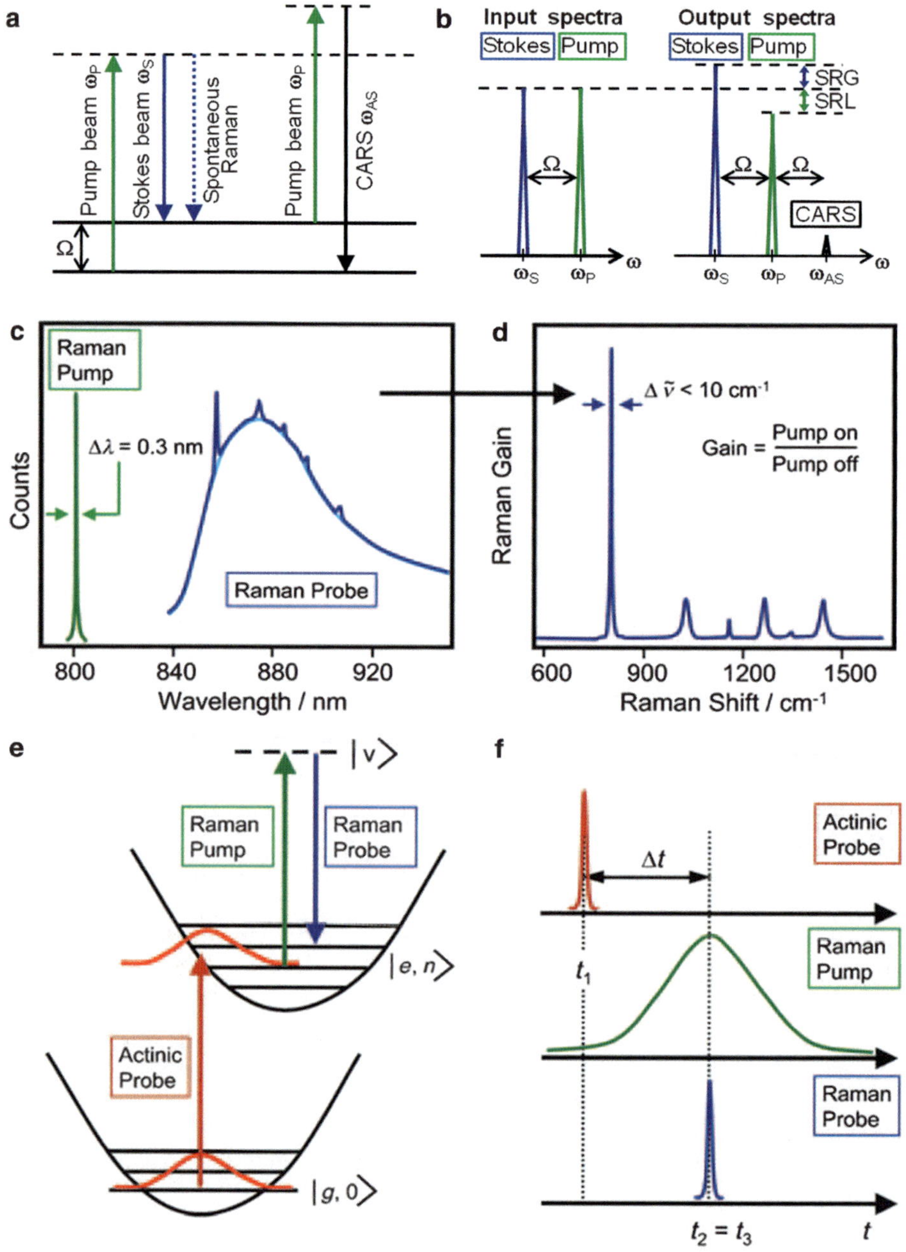

Fig. 3 (**a**) Energy diagram for stimulated Raman scattering (SRS) and coherent anti-Stokes Raman scattering (CARS). (**b**) Scheme for the input and output spectra of SRS. SRS leads to an intensity gain in the Stokes beam (SRG) and to a scattering Raman loss (SRL) in the pump beam. Also shown (not to scale) is the CARS signal generated at the anti-Stokes frequency ω_{AS}. (**c, d**) Diagrams depicting the SRS spectroscopy: in panel **c** are shown the Raman pump line (*green line*) and the probe continuum without (*cyan line*) and with the peaks due to stimulated Raman

various GFP mutants is shown in Fig. 2b (see Sect. 4.4). By comparing the time-resolved spectra collected with and without the actinic pump, it is possible to analyze the vibrational differences of the molecule in the excited state and therefore to explore its induced structural changes.

3 Theory: The Density Functional Theory Framework

The accurate evaluation of vibrational frequencies and spectra requires the use of quantum chemistry (QM) modeling techniques, treating explicitly the electronic system. One in principle should solve the complete electrons-nuclei Schrödinger equation

$$\hat{H}_{\text{mol}}\Psi_I = \left[\hat{T}_{\text{K}} + \hat{T}_e + \hat{V}_{\text{nuc,nuc}} + \hat{V}_{\text{nuc},e} + \hat{V}_{e,e}\right]\Psi_I = E_I\Psi_I,$$

where Ψ is the total wavefunction depending on the complete set of nuclear and electronic coordinates $\{\mathbf{R},\mathbf{r}\}$ and the operators in [] are the kinetic energy of nuclei, the kinetic energy of electrons, and the electrostatic potential energy of nuclei, of electrons–nuclei and of electrons interactions, respectively. In order to simplify the problem, basically all the solving techniques rely on the Born-Oppenheimer (BO) approximation, which consists in separating the total wavefunction in the product of nuclear (χ_a) and electronic ($\Psi_{e,i}$) wave functions (the last one having the coordinates of the nuclei $\tilde{\mathbf{R}}$ as parameters)

$$\Psi_I\left(\mathbf{R},\,\mathbf{r}\right) = \chi_a\left(\mathbf{R}\right)\cdot\Psi_{e,i}\left(\mathbf{r},\,\tilde{\mathbf{R}}\right)$$

under the assumption that the electron dynamics are much faster than the nuclear ones. Using this form of the wavefunction, the total Hamiltonian separates, producing two Schrödinger equations

$$\left[\hat{T}_K + E_{e,i}(\mathbf{R})\right]\chi_a(\mathbf{R}) = E_{\text{tot}}\,\chi_a(\mathbf{R})$$

$$\left[\hat{T}_e + \hat{V}_{\text{nuc,nuc}} + \hat{V}_{\text{nuc},e} + \hat{V}_{e,e}\right]\Psi_{e,i}\left(\mathbf{r},\tilde{\mathbf{R}}\right) = E_{e,\text{i}}\left(\mathbf{R}\right)\Psi_{e,i}\left(\mathbf{r},\,\tilde{\mathbf{R}}\right)$$

for nuclei and electrons, respectively. The second line is the Schrödinger equation of an electron system at fixed nuclear configuration $\tilde{\mathbf{R}}$ and its eigenvalues depend

Fig. 3 (continued) transitions (*blue line*); in (**d**) the gain spectrum, i.e., the ratio of the probe continua with Raman pump pulse on and off, is depicted as a *blue line*. (**e**, **f**): Diagrams depicting the femtosecond-stimulated Raman scattering (FSRS) technique: (**e**) energy level diagram showing the electronic and vibrational resonances; (**f**) timing diagram for the actinic (*red*), Raman pump (*green*), and Raman-continuum probe (*blue*) pulses. Panels (**c**–**f**) reproduced with permission from [17], coyright 2010 John Wiley and Sons parametrically on $\tilde{\mathbf{R}}$. The nuclear equation is generally treated at the classical level or quantized within the harmonic approximation, while the electron equation can be solved with a variety of different QM methods. A widely used one is the Density Functional Theory (DFT; for a review on DFT, see [21]), which is a good compromise between the size of the system (up to ~100 atoms when considering the chromophore and the other atoms in its pocket in the protein) and accuracy.

The DFT is based on the one-to-one correspondence between the multivariable ground state electron wave-function and two single variable functions, i.e., the external field (e.g., that of the nuclei) and the ground state density $\rho(\mathbf{r})$. Consequently, the uniqueness of a density functional describing the ground state energy and a minimal principle for it can be derived, which would in principle allow to find the exact ground state electron density (and the wavefunction). In practice, however, the DFT is commonly used within the Kohn–Sham (KS) scheme, which exploits the expansion of the electron wavefunctions in single-electron orbitals ϕ_i, leading to the very simple form for the electron density

$$\rho(\mathbf{r}) = \sum_i^N |\phi_i(\mathbf{r})|^2$$

that, combined with the minimal principle, allows rewriting the electron Schrödinger equation in a set of separated equations for electrons

$$\left\{ -\frac{1}{2}\nabla^2 + v(\mathbf{r}) + V_H(\mathbf{r}) + \frac{\delta E_{\mathrm{xc}}[\rho]}{\delta n(\mathbf{r})} \right\} \phi_i(\mathbf{r}) = \sum_j \Lambda_{ij}\phi_j(\mathbf{r})$$

i.e., the single electron Schrödinger equation in an effective potential, called the Kohn–Sham potential, which is the functional derivative of the KS energy (total electron energy) with respect to the density. In this way, the N electron problem is remapped in a set of N equivalent single electron equations in an effective external potential depending on the electron density. Even if this set of equations has to be solved self-consistently, there is a great save in computational cost. Λ_{ij} are Lagrange multipliers that account for the orbitals orthonormality and the Kohn– Sham energy is

$$E^{\mathrm{KS}}[\rho] = T_{\mathrm{s}}[\rho] + \int d\mathbf{r} v(\mathbf{r})\rho(\mathbf{r}) + \frac{1}{2}\int d\mathbf{r} V_{\mathrm{H}}(\mathbf{r})\rho(\mathbf{r}) + E_{\mathrm{xc}}[\rho],$$

where $v(\mathbf{r})$ is the external potential (implicitly dependent on the nuclear coordinates), $V_{\mathrm{H}}(\mathbf{r})$ is the Hartree potential (the coulomb part), T_{s} the kinetic part

$$T_{\mathrm{s}}[\rho] = \sum_i^N \left\langle \phi_i \middle| -\frac{1}{2}\nabla^2 \middle| \phi_i \right\rangle$$

and $E_{xc}[\rho]$ is the exchange and correlation energy functional. This is the only unknown part of the system, where most of the many-body effects are hidden. Several different approximations are commonly used for it, which are variants and evolutions of the local density approximation; this assumes that a system with a general nonuniform electron density $\rho(\mathbf{r})$ is locally coincident with a uniform electron gas at the same density

$$\mathrm{E}_{\mathrm{xc}}[\rho] = \int \rho(\mathbf{r}) e_{\mathrm{xc}}(\rho(\mathbf{r}))\mathrm{d}\mathbf{r},$$

where $e_{xc}(\rho)$ is the exchange and correlation energy density (single variable) function of a uniform electron gas evaluated for instance through exact sum rules and Monte Carlo simulations [22–24]. In the last 20 years, tens of different energy functionals treating the exchange and correlation at different levels were proposed and their performances evaluated on different systems [25]. Some functionals of the last generation include a part of explicit exchange through a Fock-like approach. This improves the accuracy, especially for the calculations of properties involving excited states, or of the ground state properties at the very fine level, but at the expenses of a much larger computational cost.

A further simplification involves the type of basis set – localized orbitals or plane waves – used for the expansion of the wavefunctions and the strictly related question of the number of explicit electrons considered. In most cases, the core electrons are basically inert during the system dynamics: the frozen core approximation allows treating much larger systems considering only the valence electrons. This is quite easily implementable within the localized bases schemes because the core orbitals are expanded using only single or few bases functions whose coefficients are simply kept fixed during dynamics. The expansion in plane waves, conversely, brings great save in computational cost due to the possibility of using the fast Fourier transforms, but implies the use of pseudopotentials for the nuclei to mimic the effect of their core frozen electrons. Again, in the latest years a large number of different pseudopotentials with different levels of accuracy and transferability were optimized and tested on different systems [26].

3.1 *Calculation of the Vibrational Frequencies*

The KS energy returns the ground state energy when the KS equation system is self-consistently resolved. Considering its implicit dependence on the nuclear coordinates **R**, this means that $E_{e,0}(\mathbf{R}) = E^{\mathrm{KS}}[\boldsymbol{\rho}](\mathbf{R})$, i.e., the KS energy takes the part of an effective potential energy for the nuclei. The vibrational frequencies and vibrational modes are by definition the eigenvalues and eigenvectors of the dynamical matrix, that is the matrix of second derivatives of $E_{e,0}(\mathbf{R})$ with respect to **R**, appearing in the second order expansion of the energy with respect to the equilibrium configuration of the nuclei:

$$E_{e,i}(\mathbf{R}) = E_{e,i}(0) + \sum_I \left(\frac{\partial E_{e,i}}{\partial R_I}\right)_0 \Delta R_I + \sum_{I,J} \left(\frac{\partial^2 E_{e,i}}{\partial R_I \partial R_J}\right)_0 \Delta R_I \Delta R_J + \ldots$$

This expansion could be evaluated in principle on every potential energy surface $E_{e,i}$ obtaining the vibrational properties of the excited states. However, care must be taken in using DFT for excited states since it is rigorously a ground state theory.

The dynamical matrix can be evaluated either by numerical difference applying small finite displacements, or directly within the perturbative approach [27].

The evaluation of the vibrational frequencies of the system is usually the preliminary step for the calculation of the vibrational spectra. As will be discussed through explicit examples in the next section, the DFT approach leads to systematic errors, especially in the high-frequency range, which are up to the order of 5–10%, stemming from the approximation introduced by the energy functionals, by the frozen core approximation, and by the finite basis set expansion. However, the trends of variation of frequencies with respect to external environment variations are generally accurately reproduced.

3.2 Calculation of the Vibrational Spectra

The production of a vibrational spectrum involves the interaction of an external field (typically the electromagnetic field) with the vibrational modes of the system. Once this interaction occurs, one might simply look at the variation of the probe field after the interaction, as in the IR absorption experiments, or analyze the result of the scattering process, as in the Raman experiments.

The absorption rate $W_{I \to F}$ of a radiation of frequency ω excting a material system from a state I to a state F is

$$W_{I \to F} \propto E^2_{\text{inc}}(\omega) |\mu_{FI}|^2,$$

where E_{inc} is the electric field of the incident light and μ the electric dipole. After the BO decoupling, one can treat the nuclear part quantum mechanically within the harmonic approximation [28].

$$\mu_{FI} = \mu^0_{fi} \langle v_f | v_i \rangle + \sum_I \langle v_f | Q_I | v_i \rangle \frac{\partial \mu_{fi}}{\partial Q_I} + \ldots,$$

where v_i are the nuclear vibrational states of the mode at frequency ω_I and for the ith electron state, and Q_I the corresponding vibrational coordinate. In the case of the IR spectroscopy, however, the energy of the incident radiation is not sufficient to electronically excite the system, thus v_f and v_i are both vibrational states of the ground electronic state. Consequently, the first term of the sum gives a negligible contribution due to orthonormality of v_f and v_i. In addition, given the selection rules for eigenstates of the harmonic oscillator, the second term is nonvanishing only when v_f and v_i differ by one, implying that the system has adsorbed an amount of energy corresponding to $\hbar\omega_I$. The intensity of the spectral lines is proportional to the square of the derivative of the dipole with respect to the vibrational coordinate.

The Raman process is the inelastic scattering of light by an interaction with the vibrations of the system. Consequently, the Raman amplitudes are proportional to the response function of the system to an external time dependent electric field, i.e., the dynamical electron polarizability:

$$\alpha^{\alpha\beta}_{\text{Stokes}} = \sum_{I,r} \sqrt{\frac{n_I+1}{2\hbar\omega_I}} \frac{\partial}{\partial Q_I} \left[\frac{\mu^{\alpha}_{0r}\mu^{\beta}_{r0}}{E_r - E_0 - \hbar\omega - 1/2i\Gamma} + \frac{\mu^{\beta}_{0r}\mu^{\alpha}_{r0}}{E_r - E_0 + \hbar\omega - \hbar\omega_I + 1/2i\Gamma} \right],$$

where ω is the frequency of the incident radiation, ω_I is the frequency of the Ith vibrational mode and Q_I its eigenvector. μ_{0r} is the electric dipole evaluated between the ground and the excited state r over which the sum is made and α, β are Cartesian coordinates;

$$\langle n_I + 1 \rangle = \frac{1}{1 - e^{\hbar\omega_I/KT}},$$

where T is temperature and K is the Boltzmann constant. If the Raman signal is collected at 90° with respect to the direction of the incoming light, the spectral intensity is:

$$I(\pi/2) = \frac{(\omega - \omega_I)^4}{32\pi^2 c^3 \epsilon_0} E^2_{\text{inc}} (\alpha^2_{yy} + \alpha^2_{zy}).$$

The calculation of IR and Raman spectra involves the evaluation of the derivative of some operators (the dipole moment and the electronic polarizability, respectively) with respect to a normal coordinate. Again, these are evaluated using the finite difference method and the linear response theory [29]. We observe that the DFT – a ground state theory – gives results in good quantitative agreement with experimental intensities, in spite of the fact that the calculation of the polarizability involves a sum over the electronic excited states.

The above formula for the electronic polarizability involves an explicit dependence of the sum terms on the incident radiation frequency, ω. When the Raman experiment is performed in far-from-resonance conditions, i.e., with an incident frequency far from any electronic excitation of the system, this dependence can be neglected with a great save of computational cost. Conversely, the full calculation is necessary in the so called preresonance conditions. However, in the very near resonance conditions a different approximation can be considered. If the photon energy or the incident light is very close to an electronic excitation energy, one single term of the sum over r is dominant. Additionally, one assumes that in these conditions the variation of the polarizability with respect to the normal coordinates is mainly due to the denominator, and neglects the variation of the dipoles. In these conditions, one can write [30]:

$$\frac{\partial}{\partial Q_I}\frac{\mu_{0r}^{\alpha}\mu_{r0}^{\beta}}{E_r - E_0 - \hbar\omega - 1/2i\Gamma} \simeq -\frac{\mu_{0r}^{\alpha}\mu_{r0}^{\beta}}{(E_r - E_0 - \hbar\omega - 1/2i\Gamma)^2}\frac{\partial E_r}{\partial Q_I}$$

(the derivative of E_0 vanishes if the system is in the equilibrium geometry). The appearance of the excited state energy can be explained in the following way: in resonance conditions, the system is actually excited and stays for some time on the excited state, where the atoms acquire velocities according to the forces, which coincide with the derivative $\partial E_r/\partial Q_I$. Then it goes back to the ground state and moves according to the acquired velocities. Thus, the modes whose displacement has a large projection onto the excited state force are selected. Although the very on-resonance conditions are usually avoided due to the dominance of other phenomena, such as fluorescence, this picture is still a good representation also for the near-resonance conditions.

As in the case of the polarizability and dipole moment derivatives, one could calculate the derivative of the E_r with respect to Q directly or numerically. However in this case there is an alternative approach that consists in simulating the whole process of excitation and subsequent dynamical evolution of the system. This approach is general and can be applied to any kind of external perturbation and has the advantage of going beyond the linear response approximation. The algorithm is the following: starting from the relaxed system, (a) induce changes due to the external perturbation, (b) simulate the dynamics of the system for a sufficiently long time after perturbation, and (c) analyze the trajectory to extract the vibrational spectrum.

Step (a) depends on the kind of process one is considering and generally produces a set of nuclear displacements or equivalently a set of atomic velocities. In the case of resonance Raman, these are obtained by performing a few steps of excited state dynamics [31]. This and the subsequent point (b) require an algorithm to simulate the dynamics of the system. Within the DFT frame one can use the Born-Oppehneimer (BO) dynamics or the Car-Parrinello dynamics (CP) [32]. Within the BO scheme, one simply evolves the nuclei with Newtonian dynamics using the forces evaluated within the DFT approach as derivatives of the $E_r(R)$ with respect to R. The classical dynamics of the nuclear system is integrated numerically, and for each step of the nuclear system the KS equation for the electron system must be solved, involving a full electronic wavefunction optimization. In this way, a line on the exact BO surface $E_r(R)$ is followed.

The CP approach is an approximate way to solve the BO dynamics. A lagrangian system including both the nuclear coordinates and the electronic coordinates is defined. The electronic quantum coordinates are transformed in classical functional coordinates, the KS orbitals φ_i, and a fictitious classical mass μ is associated with them:

$$M_I\ddot{R}_I(t) = -\frac{\partial E^{\mathrm{KS}}}{\partial R_I} + \sum_{ij} \Lambda_{ij} \frac{\partial}{\partial R_I} \langle \phi_i | \phi_j \rangle$$

$$\mu\ddot{\phi}_i(t) = -\frac{\delta E^{\mathrm{KS}}}{\delta \langle \phi_i |} + \sum_{j} \Lambda_{ij} | \phi_j \rangle .$$

This allows evolving the electronic and nuclear system at the same time: after the full electronic optimization at the starting nuclear step, there is no need to re-optimize the electronic system, which is simply evolved with a single electronic calculation for each nuclear step. Since the electronic part is the bottleneck of the calculation, this generally allows a save in computational cost. The electron system dynamics is driven by the value of the fictitious electron mass. The exact BO approximation is recovered if $\mu = 0$, but this would need an infinitesimally small timestep for the evolution of the whole system. Exact BO or CP can be used for the step (a) and to generate a long trajectory for step (b).

Once a dynamical trajectory of the system is available, the vibrational spectrum can be obtained by the Fourier transform (FT) of the velocity self-correlation function [point (c)] [33]:

$$S(\omega) = \mathrm{FT}\,[c(\tau)] = \mathrm{FT}\,[\langle v(t) v(t+\tau) \rangle].$$

$S(\omega)$ will have resonances (peaks) at the frequencies corresponding to the vibrational modes of the system, whose height will be larger for the modes that have larger superposition with the starting displacement generated by the perturbation, i.e., the spectrum will reflect the external perturbation. The width of the peaks includes several effects. One is the necessarily finite length of the simulation, which implies a numerical enlargement of the peaks in the Fourier space. This adds up to the physical broadening due to anharmocity that implies also interactions between the modes and displacement of the frequencies from those evaluated within the harmonic approach. In addition if the dynamics is evaluated within the CP scheme an additional softening of the frequencies of about 1–3% is due to the fictitious electron mass μ that has the effect of slowing down the dynamics [34].

A schematic representation of the IR, Raman, and Resonance Raman processes is reported in Fig. 1c, together with the corresponding spectra of the same molecule, i.e., the GFP model chromophore HDBI. The difference between IR and Raman spectra is apparent: the selection rules for the two processes are very different, thus the Raman active modes in this molecule are much less than the IR active modes. The difference between Raman and Resonance Raman spectra are less evident in this kind of molecule. They are more evident, e.g., when the chromophore is in the protein, where the Resonance Raman technique is able to select the modes of the chromophore, related to the electronic excitation, from those of the protein.

4 Results on Fluorescent Proteins and Their Chromophores

Since each frequency identifies a specific vibrational mode, which corresponds to a specific pattern of displacement, vibrational spectroscopy can give structural information. The modes localized on bonds have characteristic frequencies that depend on the order of the bond (double, single) and on the ligands. For instance, the C=O stretching mode has a characteristic frequency of ~1,700 cm^{-1}, the C=C bond of ~1,650 cm^{-1}. However, the fine tuning of the frequency depends on the chemical environment of the bond. The situation for the delocalized modes is even more complex, because their frequency depends on the specific structure of the molecule. Comparison of the experiments with the theory, or with the results in simplified models or isotope-substituted samples, allows on one side to link each peak in a spectrum with a given vibrational mode, and on the other side to understand the correlations between changes in the structure and in the vibrational spectrum of a molecule.

In the following Sect. 4.1, we report the state-of-the-art knowledge about the vibrational properties of the GFP chromophores. In Sect. 4.2, we will discuss how the protein environment can affect the structure and the vibrational modes of the chromophores in fully folded proteins. In Sect. 4.3, the results on photochromic fluorescent proteins epitomize the methods through which one can study the relations between the function of the FP and structure of its chromophore using vibrational spectroscopy; more results, based on time-dependent vibrational spectroscopies, will be reviewed in Sect. 4.4.

4.1 Identification of Vibrational Modes of Model Chromophores

The mode-identification process in the FP chromophores started over 10 years ago and involved a number of theoretical, experimental, and mixed theoretical experimental papers [4, 9, 13, 35–41]. For the mode assignment, the comparison between experiment and theory is fundamental. In fact, the experiment can identify only the vibrational mode frequencies (i.e., resonances in IR or Raman spectra). Some indications on the atoms involved in each mode can be extracted by the frequency displacement in isotopically substituted molecules; however, the precise pattern of displacement of a mode can be obtained only through the comparison with theoretical studies. In particular, Raman experiments on synthetic analogues of the GFP chromophore highlighted the differences in the vibrational spectra linked to the different protonation states [36].

Tables 1 (a) and (b) report a summary of the assignment of the modes updated at the current knowledge. There is an overall agreement on the assignment of the first five modes in the range 1,700–1,530 cm^{-1}, also considered the fingerprint high-frequency modes. In the neutral chromophore, they are assigned to bond-stretching vibrations of C=O (only IR active), exocyclic C=C, phenolic ring (two different symmetries, in principle only one Raman active) and C=N. The studies showed that

Table 1 Assignment of the modes of the neutral (a) and anionic (b) chromophore modes. The average frequency mode in the isolated chromophore and in the protein (in italic) is reported in the first column. A description of the mode and the theoretical and experimental methods used are reported, together with other information (measurement/calculation in the isolated/solvated chromophore or in the protein) and the references [4, 9, 13, 35–40] for each mode

Mode frequency (cm^{-1}) chromophore – *protein*	Mode description	Activity	Measurement in isolated chromophore	Measurement in protein	Assignment (theory)
(a) Neutral chromophore					
~1,690	C=O stretch	IR	[4, 9, 35, 37, 39]	[4]	[4, 35, 37–40]
1,643 – *~1,635*	C=C stretch	IR, R, RR	[4, 9, 35–37]	[4, 36]	[4, 35, 37–40]
1,603 – *~1,595*	Phenol	IR, R, RR	[4, 9, 35–37]	[4, 36]	[4, 35, 37, 40]
1,580 – *1,570*	Phenol + C–O	IR	[4, 35, 37]	[4, 36]	[4, 35, 37, 40]
1,563 – *1,564*	C=N (+C=C)	IR, R, RR	[4, 9, 35–37, 39]	[4, 36]	[4, 35, 37, 39, 40]
1,515	Phenol	IR, R	[9, 35–37]	[36]	[35, 37]
1,446	C–N + C–H def	IR, R, RR	[9, 35, 37]		[35, 37]
1430 – *1,430*	Phenol	IR	[35]		[35]
1,393		IR	[35]		
1,371		IR	[35]		
1,362		IR, R, RR	[4, 9, 35]	[4]	[4]
1,317	Phenol	R, RR	[4, 9, 35, 37]	[4]	[4, 37]
1,301		IR	[35]		
1,290		R, RR	[9, 35]		
1,282		IR	[35]		
1,277		IR	[35]		
1,235		IR	[35]		
1,230	Phenol	IR, R, RR	[4, 9, 35, 37]	[4]	[4, 37]
1,184		IR	[35]		
1,175	Phenol + C–H def	IR, R, RR	[4, 9, 35, 37]	[4]	[4, 37]
1165 – *1,169*	Imid C–C C–N	IR,RR, SERS	[9, 13, 35]		[13]
1,138		IR, R, RR, SERS	[9, 13, 35]		[13]
1,102 – *1,098*	Imid C–C C–N	RR, SERS	[4, 9, 13]	[4]	[4]
1,030	Imid C–C C–N	IR, R, RR, SERS	[9, 13, 35, 37]	[13]	[13, 37]

(continued)

Table 1 (continued)

Mode frequency (cm^{-1}) chromophore – *protein*	Mode description	Activity	Measurement in isolated chromophore	Measurement in protein	Assignment (theory)
1,009		IR, R	[35]		
930		IR, R, RR, SERS	[9, 13, 35]		[13]
910 – *910*		IR, SERS	[13, 35]	[13]	[13]
850 – *875*	Rings breathing	IR, R, RR, SERS	[9, 13, 35]	[13]	[13]
840		IR	[35]		
823		IR	[35]		
805 – *830*		R, RR, SERS	[9, 13, 35]	[13]	[13]
781		IR, SERS	[13, 35]	[13]	[13]
750		IR, R, RR, SERS	[9, 13, 35]	[13]	
715 – *720*	Pl rings + bridge	IR, R, RR, SERS	[4, 9, 13, 35]	[4, 13]	[4, 13]
657 – *650*		IR, SERS	[13, 35]	[13]	[13]
636		IR	[35]		
606	Pl imid + methil	IR, R, RR, SERS	[4, 9, 13, 35]	[13]	[4]
580		IR, R, RR	[9, 35]		
536		IR	[35]		
520	Phenol + OH	SERS	[13]		[13]
512		IR, R, RR	[9, 35]		
489 – *500*	Phenol + OH	SERS	[13]	[13]	[13]
415 – *400*	Phenol + OH	SERS	[13]	[13]	[13]
364		SERS	[13]	[13]	[13]
(b) Anionic chromophore					
1,665 – ~*1,663*	C=O stretch	IR	[9, 37]		[9, 37–39]
1,630 – ~*1,620*	C=C stretch	IR, RR	[9, 36, 37]	[36]	[37–39]
1,580 – *1,603*	Phenol (+C–O)	IR, RR	[9, 36, 37]	[36]	[37]
1,560 – *1,565*	C=N (+C=C)	IR, RR	[9, 36, 37]	[36]	[37, 39]
1,530	Phenol	RR	[9, 36, 37]	[36]	[37]

1,500	Phenol	IR, RR	[9, 36, 37]	[36]	[37]
1,445		IR	[9, 36]		
1,435	Phenol + C–H def	RR	[9, 37]		[37]
1,365		R, RR	[4, 9, 36, 37]	[4]	[4, 37]
1,310	phenol	R	[4, 37]	[4]	[4, 37]
1,295		IR, RR	[9]		
1,262		IR	[9]		
1,246		R	[37]		[37]
1,238		RR	[4, 9]	[4]	[4]
1,181 – *1,180*		R, RR,SERS	[4, 9, 36]	[4, 13]	[4, 13]
1,170 – *1,070*	Phenol + CH def	IR, RR	[9, 37]	[13]	[13, 37]
1,140	Imid ring	IR, R	[9, 37]		[37]
1,035 – *1,030*	Imid + C=O bend	IR, RR, SERS	[9, 37]	[13]	[13, 37]
1,000		IR	[9]		
930		IR, RR	[9]		
906 – *910*		IR, SERS	[9]	[13]	[13]
850 – *875*	Rings breathing	IR, RR, SERS	[4, 9]	[4, 13]	[4, 13]
800 – *790*		IR, RR. SERS	[9]	[13]	[13]
764		IR	[9]		
755		RR	[9]		
711 – *720*	Pl rings + bridge	RR, SERS	[9]	[4, 13]	[4, 13]
608	Pl chromophore	IR, RR	[9]	[4]	[4]
583		RR	[9]		
518		IR	[9]		
–500		SERS		[13]	[13]
–400		SERS		[13]	[13]

R Raman, *RR* resonant Raman, *IR* infrared absorption, *SERS* surface-enhanced Raman scattering; *Pl* planar, imid imidazolinone, *def* deformation none of these modes is perfectly localized, and in particular the C=N and one of the phenol ring modes show delocalization over the bridge and the phenol C-O bond, respectively. In the anionic chromophore, the modes are the same, although the frequencies are generally smaller, especially that of one of the two phenol modes.

4.2 Dependence of Vibrational Modes on Chromophore Structure and Its Interactions Inside Various Fluorescent Proteins

Not only the frequency but also the Raman and IR activities depend on the protonation state of the chromophore. In addition, both the frequency and the activity are modified by the protein matrix and depend on mutations in the chromophore or in the chromophore environment. This makes the situation rather complex. However, the studies of the last years suggest a key for the interpretation of the behavior of these modes, which is summarized in Fig. 4.

Considering for instance the C=C mode (which can be easily identified as the first high-frequency Raman active mode), its frequency is strictly related to the C=C bond strength (and length). It is observed to decrease in frequency passing from the neutral chromophore to the anionic one. This is related to the fact that the

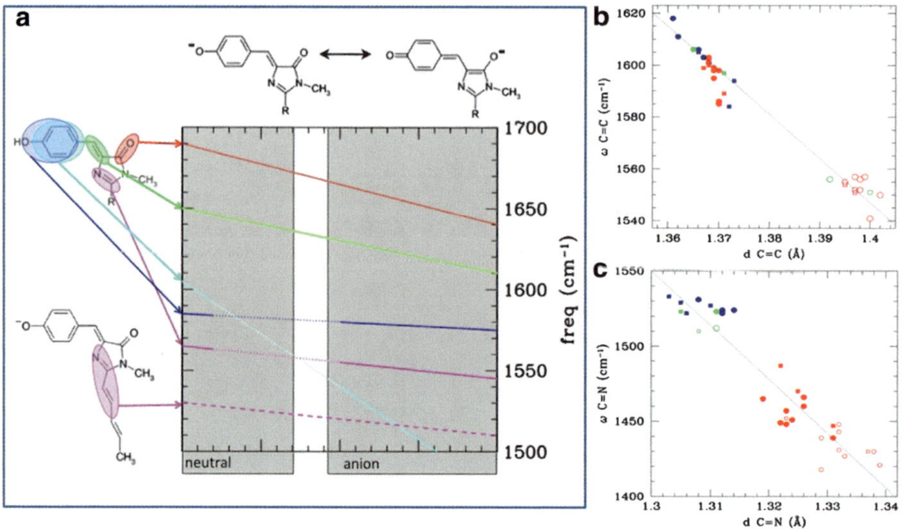

Fig. 4 (**a**) Relationship between structure and mode frequencies for the first five high-frequency fingerprint modes. The mode localization areas are described by the colored regions in the structures on the left of the graph (for the mode description see also Table 1). The *colored lines* in the graphs report qualitatively the mode frequency as a function of the resonance structure of the molecule, modulated by the chromophore environment. (**b** and **c**) Mode frequency of the C=C and C=N stretching modes (*green* and *magenta* in panel **a**) as a function of the length of the corresponding bond. *Red*, *green*, and *blue dots* correspond to RedFP, GFP, and BFP chromophores, respectively. *Filled dots* correspond to neutral chromophores, *empty dots* to anionic chromophores

bond length scheme changes passing from a benzenoid one, typical of the neutral chromophore, to a mixed benzenoid–quinonoid, typical of the anionic structure [38], where the exocyclic C=C bond has a mixed character (single-double). This structural transition is also driven by the interaction of the chromophore with the protein matrix. In fact, the hydrogen bonds, especially those on the phenolic and imidazolidinone oxygens, influence the bond order of the chromophore. This explains the range of frequencies that this mode assumes in different mutants of GFP. In turn, this frequency mode could be used to give indications on the level of bond length alternation present in the chromophore, and in particular to discriminate neutral from anionic forms of the chromophore. A similar behavior is found also in the chromophores of different homologues of GFP (e.g., in orange and red proteins [39]). This picture is confirmed by the observation that this mode frequency is linearly related to the absorption frequency [36]: in fact, also this property is related to the bond length alternation [38]. A very similar behavior is found in the C=O mode [38], although this is more difficult to be used as a fingerprint mode because it is not Raman active.

The three subsequent modes are more difficult to recognize and interpret because their frequencies are very close and their character superimpose, especially in the range of intermediate bond length alternation, in the anionic chromophore. However, they are also likely to decrease their frequency going toward the quinonoid form, although with different velocities. In particular, the C=N mode is less sensitive to the change in the bond length alternation because it preserves its double bond character in both resonance forms. For this reason, since its frequency is almost constant and its Raman activity quite high, it can be considered a reference mode. Conversely, it is very sensitive to the presence or absence of the chromophore tail stemming from the second post-translational oxidation in orange/red proteins: its frequency is ~30 cm^{-1} smaller in the red proteins [41]. This could be used to identify the occurrence of the second oxidation (Fig. 4).

Other modes in the range 1,000–1,500 cm^{-1} were considered, but in this region the calculations are less accurate for the presence of many delocalized modes at similar frequencies, which increases the difficulty in their description.

The lower frequency modes ($<1{,}000$ cm^{-1}) were only more recently considered due to the fact that they are less easily measured, and especially for the protein, they need resonance Raman techniques. They correspond to vibrations delocalized over the whole chromophore, less sensitive to the environment, but more sensitive to the specific chromophore structure.

As an example, we analyze here two low-frequency modes that are clearly visible in the Raman spectra of the chromophores and in the SERS spectra of the proteins. All the other modes have small intensity, thus these two can be considered the low frequency fingerprint modes of the chromophores. One is located at ~705 cm^{-1} in the neutral *cis* conformer of the chromophore [4] in water and organic solvents. Its frequency is quite insensitive to the solvent and appears only slightly shifted in different mutants of GFP (~710 cm^{-1} [13]) and in the blue and cyan chromophores ([4] and Fig. 5b). The comparison with theoretical DFT calculations indicates that it is a largely delocalized planar vibration that involves the

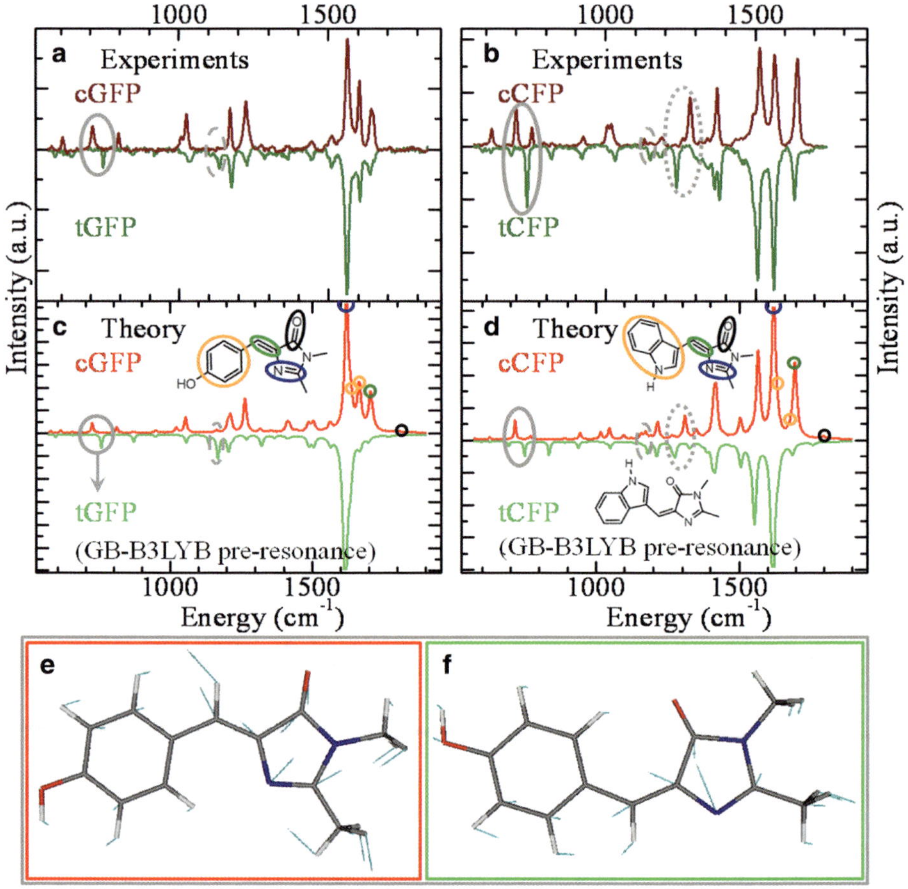

Fig. 5 Experimental (**a**, **b**) and calculated *in vacuo* (**c**, **d**) preresonance Raman spectra of cGFP and tGFP (**a**, **c**) and cCFP, tCFP (**b**, **d**); the spectrum of the *trans* form is shown inverted around the x-axis. Calculations are based on the B3LYP functional with Gaussian bases (GB), considering an excitation at 514.5 nm. As known, GB-B3LYP approach tends to overestimate the mode energies [42]; this is reflected in the slightly different scale between theoretical and experimental results. *Gray ellipses* highlight corresponding fingerprint modes for the *cis–trans* transition. *Small circles* emphasize stretching modes mostly localized on double bonds in the chromophore region, highlighted in the insets of panels (**c**) and (**d**) with corresponding colors: *black*, C=O; *green*, C=C on the bridge; *orange*, C=C on the phenolic (panel **c**) and indolic (panel **d**) rings; *blue*, C=N. (**e**, **f**) Schematic representations of atomic displacements in the modes responsible for the peaks inside the *solid gray ellipse* in panel (**c**) for cGFP, and tGFP, respectively. Panels (**a**) and (**b**) reprinted with permission from [4]. Copyright 2009 American Chemical Society

deformation of both rings and of the bridge (see Fig. 5e). Very interestingly, this mode appears blue shifted of about 30 cm^{-1} after irradiation of the system in the UV. An identical shift is observed in the calculation on the *trans* configuration of the chromophore, clearly indicating this mode as the fingerprint one for the *cis–trans* transition, as will be discussed in more details in Sect. 4.3.

Another interesting mode is located at ~600 cm^{-1} and it is weak in the neutral *cis* chromophore, absent in the neutral *trans* chromophore and very strong in the *cis* anionic chromophore ([4], Fig. 5a, and black ellipses in Fig. 6a, b). This mode is observed in the mutant EGFP while it is very weak in the E^{2}GFP [13], in agreement with the fact that in these mutants the dominant forms of the chromophore are anionic and neutral, respectively. The comparison with theory indicate that this mode is a delocalized planar vibration with a deformation of both the rings and of the C–O phenolic bond (it is shown in Fig. 1b), which explains the strong variation of the Raman intensity between neutral and anionic forms.

4.3 Raman Study of the Chromophore States in Photoswitchable Mutants

Reversibly switchable fluorescent proteins (RSFPs) were developed, which can undergo repeated transitions between different states, e.g., bright and dark forms. This property makes RSFPs particularly attractive as active labels in biological systems for selective photolabeling applications or subdiffraction imaging.

The photophysical basis for this peculiar behavior, whose knowledge is needed to guide the detailed tailoring of mutant properties, will be discussed in details elsewhere in of this book. Here, we will discuss mostly the vibrational spectroscopy studies that shed light on the process. It is useful to remember that, in some fluorescent proteins [43–45], X-ray crystallographic studies showed that the brightness of these proteins is closely related to the conformation of the chromophore: the chromophore in the dark state is in a *trans* configuration, while in the bright state it shows a *cis* configuration. A similar mechanism was proposed as a general feature for RSFPs in the family of GFP mutants [46]. X-ray studies of protein crystals, however, are not applicable to proteins in solution or undergoing small conformational changes upon irradiation, and cannot discern the protonation states of chromophores and of other residues [47]. On the contrary, spectroscopic and in particular Raman studies of chromophores while embedded in folded proteins can provide detailed information on the chromophore structure, on its interactions with the environment, and in general on the energy landscapes and the dynamics of the processes involved [19].

The first vibrational study of the photoswitching in a GFP mutant (GFP$_{uv}$, with spectral and photoconversion characteristics very similar to *wt*GFP) has been carried out by van Thor et al. using difference FTIR spectroscopy [2]. This work demonstrated the impact of different protonation states of the chromophore in the original and the photoconverted forms, and that there are only small changes in the structural backbone of the protein, since only small changes in the amide I and II spectral markers for protein conformational changes are detected; in this work, the authors could detect no changes in any glutamic acid residue since they could not observe any difference in the 1,710–1,760 cm^{-1} region (characteristic of a –COOH mode) nor in the ~1,560 cm^{-1} region (characteristic of a –COO$^-$ mode). The authors

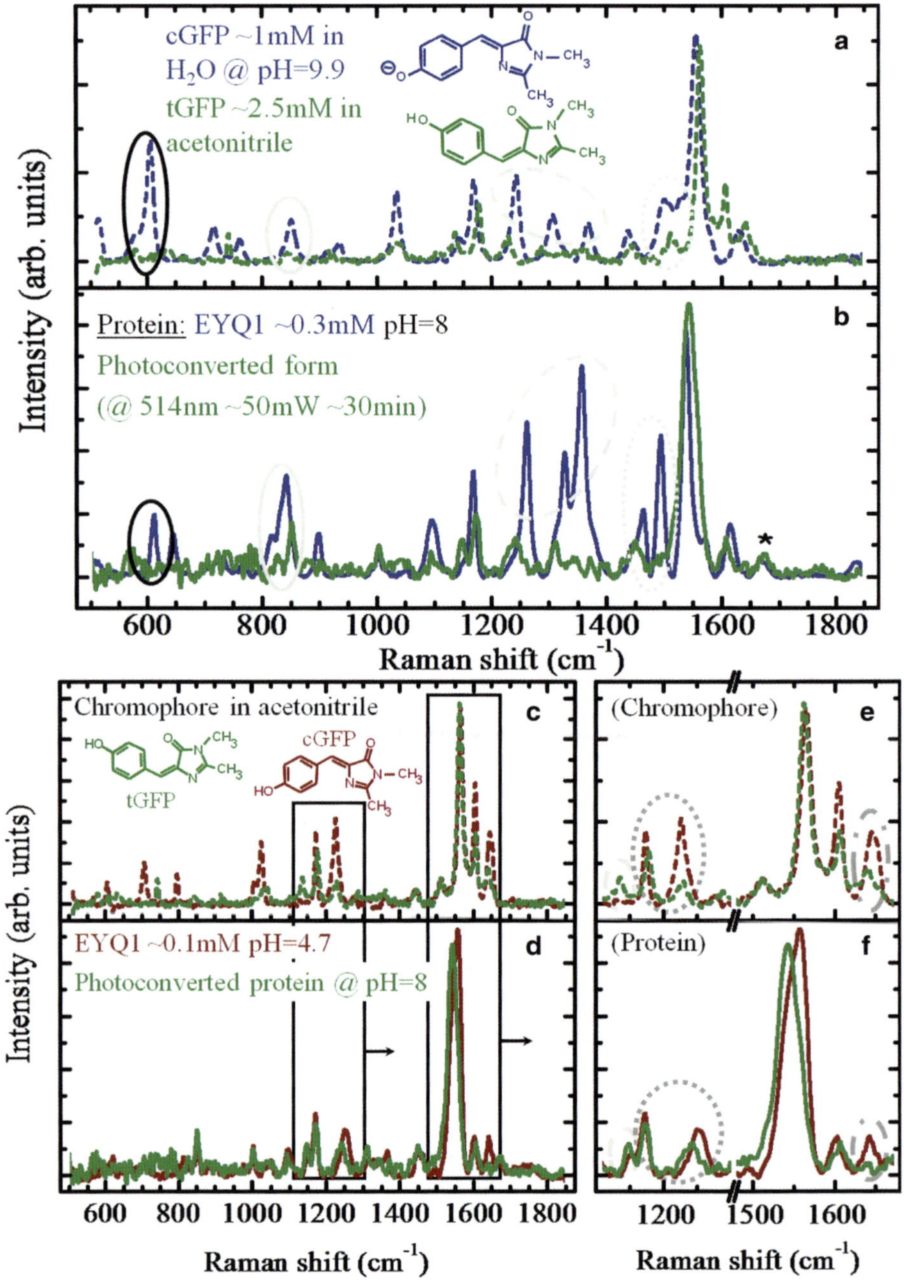

Fig. 6 Raman-driven analysis of the different isomerization and protonation states of the chromophore of EYQ1. (**a**) Raman spectra of anionic cGFP (in aqueous solution of NaOH at pH = 9.9, *dashed blue line*) and of tGFP (in acetonitrile, same data of *dark green* curve in Fig. 5a). (**b**) Raman spectra of native EYQ1 (*blue solid line*) and of its photoconverted form (*dark green*

excluded, therefore, Glu222 as the final receptor for the proton after photoswitching; however, on a later paper, van Thor et al. found evidences of decarboxylation of Glu222 upon photoconversion, on the basis of temperature-dependent photoconversion and spectroscopy, X-ray crystallography, mass spectroscopy, and electron spin resonance spectroscopy experiments [48]. Different from other photoswitchable fluorescent proteins, the authors could not observe any *cis–trans* photoisomerization signature. Possible explanations are that this does not happen in the specific mutant, or that the changes in the vibrational spectrum given by a *cis–trans* isomerization are much smaller in intensity than the ones caused by the deprotonation of the chromophore and therefore could not be observed.

Raman spectra were also studied to address the differences in the two metastable states of a red RSFP, revealing a possible involvement of *cis–trans* isomerization in the photoswitching process [47]. In order to assess the possibility that *cis–trans* isomerization be a general feature of RSFPs, we investigated systematically various chromophores, identified the fingerprints of their structural change, and then searched these features in the Raman spectrum of the corresponding complete folded proteins; in this way, a *cis–trans* isomerization has been demonstrated as implicated in the photoconversion of at least two photoswitchable mutants of the GFP (a blue one, BFPF, and a green–yellow one, EYQ1) [4].

As a first step, the photochromic behavior of three synthetic chromophores in their neutral form, namely the analogues of (a) GFP, (b) Y66F GFP (BFPF), (c) Y66W GFP (CFP) was studied [49]. Combined NMR, absorption and Raman spectroscopy, completed by accurate theoretical calculations, demonstrated that all the studied chromophores undergo the same *cis–trans* isomerization upon reversible photoconversion [4, 49].

In Fig. 5, we show the Raman spectra of both *cis* and *trans* forms of the chromophores of GFP (panel a, cGFP and tGFP, respectively), and, for the first time, of CFP (cCFP for the *cis* form and tCFP for the *trans* form, panel c). Calculations based on density functional theory for preresonant Raman spectra accurately reproduce the experiments, except for a slight linear rescaling of the energy scale [4, 42] (Figs. 5c, d); similar results were obtained for c/tBFPF [4]. These outcomes allowed finding the spectral fingerprints of after *cis–trans* isomerization of the chromophores. Based on this knowledge, we analyzed chromophore states in the complete, folded protein in BFPF and EYQ1, by comparing the Raman spectra of the photoconversion product with the ones of the chromophores. Our approach

Fig. 6 (continued) solid line), at pH = 8, after the subtraction of the baseline; the *asterisk* indicates an amide I mode. Ellipses in panels **a** and **b** with matching styles and colors emphasize some fingerprint Raman peaks for the anionic chromophore that are present in the spectrum of both the protein and the chromophore. (**c–f**) The *dark red solid curves* represent the Raman spectrum of native EYQ1 at pH = 4.7 after the subtraction of the baseline. For comparison are reported the Raman spectra of the photoconverted form at pH = 8 (*dark green solid curve*), and of neutral cGFP and tGFP (*dark red* and *dark green dashed curves*, same data of Fig. 5a). Panels **e** and **f** are magnifications of the highlighted regions in panels (**c**) and (**d**); the ellipses emphasize corresponding fingerprint spectral regions for the *cis–trans* transition. Adapted with permission from [4].

allowed us to identify the spectroscopic signatures of *cis–trans* isomerization and separate them from those linked to different protonation states of the chromophore [4].

With the help of Fig. 6, we will review here, as a paradigm, the rationale of the conclusion that, at pH = 8, EYQ1 illuminated at 514 nm undergoes a photoconversion from a form with a *cis* anionic chromophore to one with a *trans* neutral chromophore. More information about the materials and methods for this experiment can be found in [4]; the absorption and fluorescence spectra of the various forms of EYQ1 can be found in [50]. In Fig. 6b, the comparison between the Raman spectra of the native and the photoconverted EYQ1 (blue and green solid lines, respectively) demonstrates that the chromophore in the native form is deprotonated, while the chromophore in the photoconverted form is protonated. This results from the fingerprint-modes of the two chromophore forms discussed in the previous section or shown in Tables 1a, b (some of those are highlighted by ellipses in Fig. 6b) and is confirmed by the comparison between the spectra of the protein and of the chromophore with different protonation (Fig. 6a, blue dashed line for the anionic chromophore, green dashed line for a neutral chromophore, in particular in its *trans* form). Being in preresonance, indeed, the spectra in the protein are characterized by the chromophore modes; as discussed in Sect. 4.2, the small differences observed in the spectra of the synthetic chromophores in solution with respect to those inside the proteins stem from two factors: the protein backbone (substituted in the chromophore by methyl groups where the backbone should continue) and the different environment, i.e., a different network of hydrogen bonds, electrostatic interactions and van-der-Waals interactions. As an example, both contributions cause the weakening of some low-energy detected modes (solid and dashed gray ellipses in Fig. 5a), less (or not) visible in the noisier low-energy part of the protein spectra; indeed, these changes were reproduced upon adding elements of the protein environment in on-resonance TDDFT-based calculations, after the identification of the corresponding modes in pre-resonance calculations.

Despite these differences, the assignment of the chromophore protonation is clear from the data in Fig. 6a–b. The assignment of the stereoisomerization for the photoconverted form, however, required a direct comparison with the Raman spectrum of the protein in its native *cis* neutral A form, collected at pH = 4.7 (Fig. 6d, f, solid dark-red curves). The last spectrum is indeed very similar to the one of neutral cGFP (Fig. 6c, e, dashed dark-red curve), and the differences from the spectrum of the photoconverted form are small and consistent with the ones between neutral cGFP and tGFP. The enhanced red-shift after photoconversion of the C=N mode (responsible for the highest intensity peak in Fig. 6f) was present also in the spectrum of photoconverted BFPF [4], and is most probably due to the different environment in the protein pocket, since in the *trans* form the N atom is more exposed to the protein environment, as shown for E^2GFP by Nifosì et al. [51].

In conclusion of this section, Raman investigations allowed to demonstrate that the *cis–trans* isomerization (Z-E diastereomerization) of the chromophore is responsible for the photochromic properties of RSFPs in the GFP family, and to discern the contribution of different protonation states. These investigations highlighted the relevance of Raman spectroscopy for the study of ground and metastable states of optically active portions of proteins in solution, since Raman is a nondestructive technique that allows monitoring on-the-flow the products of photoconversion. Finally, these results support the hypothesis that *cis–trans* isomerization is a general feature in RSFPs.

4.4 Time Resolved Vibrational Spectroscopy for Analyzing ESPT

Time-resolved vibrational spectra have been used to determine the proton pathways in ESPT in various mutants of the GFP. The ESPT is explained in details in elsewhere in this book; here it is only necessary to recall that, upon photon-absorption-driven excitation of the neutral A state of the chromophore into A^*, this state is completely converted by deprotonation into the state I^* in less than 200 ps; this state decays by fluorescence emission into the state I (which differs from the anionic ground state B of GFP mutants for the position of some residues close to the chromophore), which, mostly, rapidly reconverts to state A.

An example of differential time-resolved infrared (TRIR) spectra collected from wild-type GFP by the group of Tongue [52, 53] is shown in Fig. 7a. Negative peaks correspond to modes of the protein that are present in the ground state, but not in the excited state. Among these, some do not change within 200 ps, and are assigned to modes localized on the neutral chromophore in its ground state (state A) by comparison with the vibrational IR spectrum of a chemically synthesized GFP chromophore analogue (HBDI) [52]. Some relatively small changes in the shown spectrum were later [53] assigned to vibrational modes of A^*, while no modes of I^* were observed, probably because their energies are shifted or their absorption cross sections are low. However, in the shown data, there is a bleaching peak around 1,560 cm^{-1} (frequency typical for a mode of a glutamate $–COO^-$), and another peak appearing at 1,712 cm^{-1} (frequency typical for a mode of a glutamic acid –COOH), within the same observed time scale. This behavior was not observed in mutants that do not experience ESPT. Moreover, the time scale is exactly the same for the fluorescence dynamics of the state A^*. This fact demonstrates the connection of the protonation of a glutamate residue with the transition from A^* to I^*. The initial claim by Stoner-Ma et al. [52] that the involved glutamate is the E222 was based on the known geometry of the chromophore environment and on comparison with previous difference-IR spectra of the product of irreversible photoconversion, where decarboxylation of E222 is involved. After this claim in 2005, several studies have been carried out to provide finer details on the dynamics of GFP. In the same year, van Thor et al. analyzed the similarities among the normal photocycle of GFP and the irreversible photoconversion by combining X-ray crystallography, static IR for intermediate photoreaction steps frozen at cryogenic temperatures, TRIR, and smart mutagenesis [3]. Stoner-Ma and collaborators extended then their TRIR measurements of wtGFP by recording polarization-resolved transient vibrational spectra and applying the method to study proteins in which the chromophore

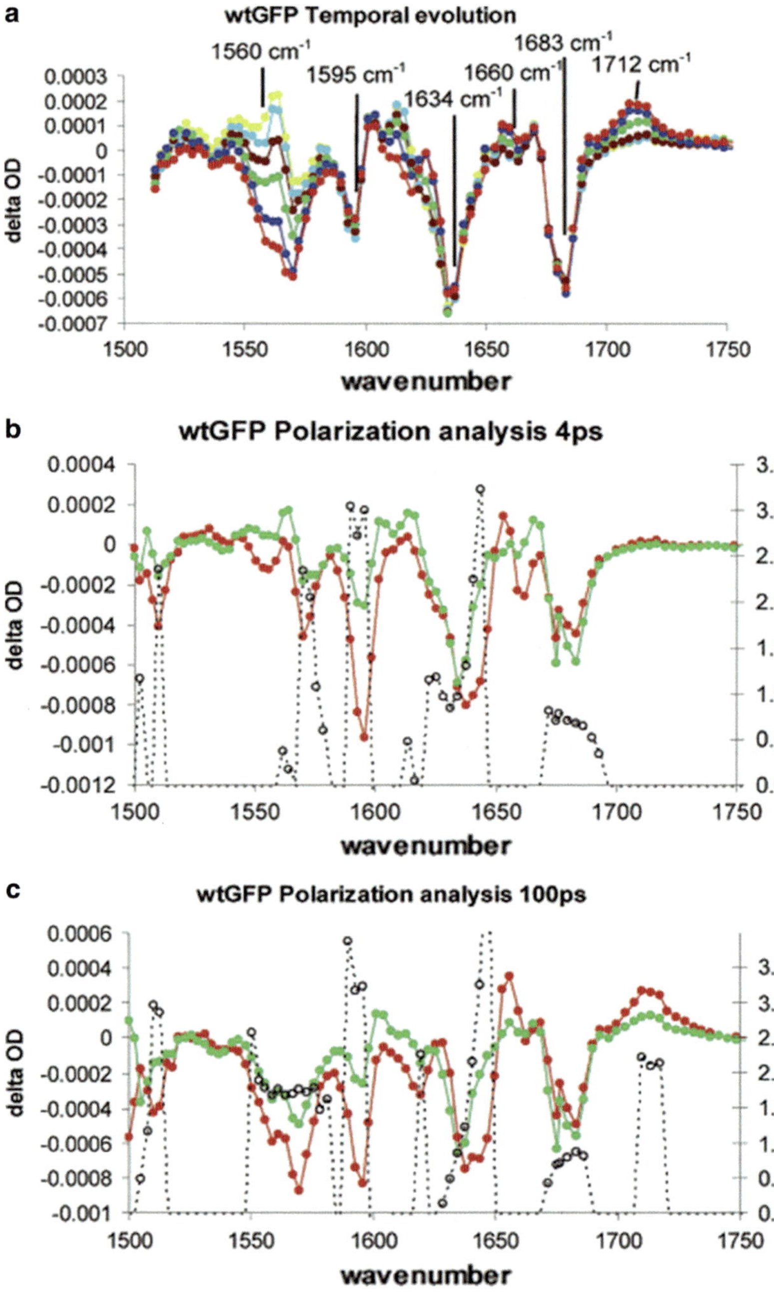

Fig. 7 (**a**) Time-resolved infrared (TRIR) spectra of *wt*GFP. Spectra were recorded at 2 ps (*yellow*), 4 ps (*turquoise*), 10 ps (*brown*), 30 ps (*green*), 100 ps (*blue*), and 200 ps (*red*) after excitation. Spectra were recorded under magic angle conditions to avoid the influence of polarization. The main transitions are labeled with their energy. (**b**, **c**) Polarization-resolved TRIR for

has been isotopically labeled [53]. The results were compared both with the polarization-resolved TRIR studies of the HBDI chromophore and its isotopes and with quantum chemical calculations of the vibrational spectrum of HBDI. The combination of these methods allowed them to assign most of the modes in the complex transient IR spectrum of wtGFP. In particular, polarization-resolved TRIR is able to measure the angle between the dipoles of the vibrational transitions and of the electronic transition involved in the excitation of the system by the actinic pump. Some of these data are shown in Fig. 7b, c, with the probe polarization oriented parallel ($\Delta A_{//}$) and perpendicular ($\Delta A_{\perp}$) to the pump polarization. The shown parameter D is obtained from the experimental data as:

$$D = \frac{\Delta A_{//}}{\Delta A_{\perp}}.$$

From this, the angle between the two dipoles can be obtained as follows:

$$\cos(\theta) = \left(\frac{2D - 1}{D + 2}\right)^{1/2}.$$

Knowing this angle and the direction of the electronic transition dipole, there are additional restraints to assign each peak to a given vibrational mode, whose dipole direction can be estimated either because it is localized on a given bond or from theoretical calculations. Already the observation of a complex behavior of the anisotropy parameter within what could seem a single peak, gives the possibility to discern among different contributions in the same spectral region. As an example, the composite negative peak around 1,630 cm^{-1} has contributes from the bleach of the ground state C=C mode, the A^* C=O mode, and possibly from a protein mode, presumably a carbonyl or amide I vibration which is sufficiently close to the chromophore to be influenced by the H-bond rearrangement which accompanies proton transfer [53]. The final conclusion was the confirmation that the final step on the time scale of 200 ps is the protonation of E222, without any evidence for any resolvable kinetic, consistent with a concerted proton motion along a wire involving water molecules and the S205 residue.

The same authors studied with similar techniques some blue-shifted mutants that do not support ESPT [54], and some mutants of these where a much faster ESPT with a different acceptor is recovered [55]. In all the studied cases, it was possible to gain information on how electronic excitation can be coupled to specific structural reorganization not only of the chromophore but also of the protein matrix. In the end, this will help the rational design of new mutants with improved qualities, being those a more stable A^* state for a more efficient blue fluorescent protein, a faster

Fig. 7 (continued) wtGFP, with parallel polarization in *red* and perpendicular polarization in *green*. The anisotropy parameter (*D*, see text) is shown in *black* with the right-hand axis: (**b**) 4 ps after excitation; (**c**) 100 ps after excitation. All panels are reproduced with permission from [53]; Copyright 2006 American Chemical Society

ESPT, or even better light activated proteins. Moreover, these experiments set the scene for using GFP mutants as a model system for understanding the molecular basis for proton transfer in biological systems.

While the previously cited experiments revealed the final proton acceptor for the ESPT in *wt*GFP, they could not determine what was the trigger for the reaction, which was considered the rate-limiting step. The only speculation was that the observed strong coupling of the electronic excitation to the protein structure may act as the trigger for the proton-transfer reaction.

Some more information about this arose from a different technique, the FSRS. In their published FSRS study [56], Mathies and coworkers could determine a mode at 280 fs (120 cm^{-1}) in the excited state of *wt*GFP just before the ESPT. An example of the collected data is shown in Fig. 8a, together with the assignment of the chromophore modes that mostly contribute in the given spectral region. The A^* spectrum appears with a 120-fs rise time and decays to I^* on the picoseconds timescale. Superimposed to this decay, there is an oscillation of the frequencies of some of the modes, as highlighted in Fig. 8b. In particular, both the frequencies and intensities (data not shown, see [56]) of two marker bands, the C–O and C=N stretching modes at opposite ends of the conjugated chromophore, oscillate out of phase with a period of 280 fs, indicating an out-of-phase oscillating behavior in the length and bond order of these two bonds (similarly to what discussed in Sect. 4.2). All these observations can be explained by the impulsively excited low frequency phenoxyl-ring wagging motions shown in Fig. 8c; this assignment is confirmed by the fact that in DFT calculations in the electronic ground state, a two-ring out-of-plane wagging mode of the chromophore was identified with a similar frequency of 110 cm^{-1}. This motion was identified as the trigger for the occurrence of ESPT by optimizing the geometry of the chromophore for the latter.

5 Conclusions and Perspectives

Vibrational spectroscopy can provide detailed information on molecular structure, orientation or interaction with neighboring species; in this way, it is similar to structural nuclear magnetic resonance (NMR). While the connections between the structure in a molecule and its vibrational spectrum are less straightforward than in the case of NMR signatures, IR and Raman spectroscopy have the advantage that they can be applied to much bigger specimens; moreover, carefully exploiting the (pre-)resonance conditions, Raman spectroscopy can address the structure of parts of a macromolecule.

In the previous sections, we have reviewed how structural changes in GFP and its mutants have been analyzed using vibrational spectroscopy. This process required first a careful assignment of the vibrational modes of the chromophore in these fluorescent proteins to each peak in their IR or Raman spectra. This assignment has been carried out mostly by comparison with the theoretical simulations described in Sect. 3 or by studying the spectral similarities and differences with

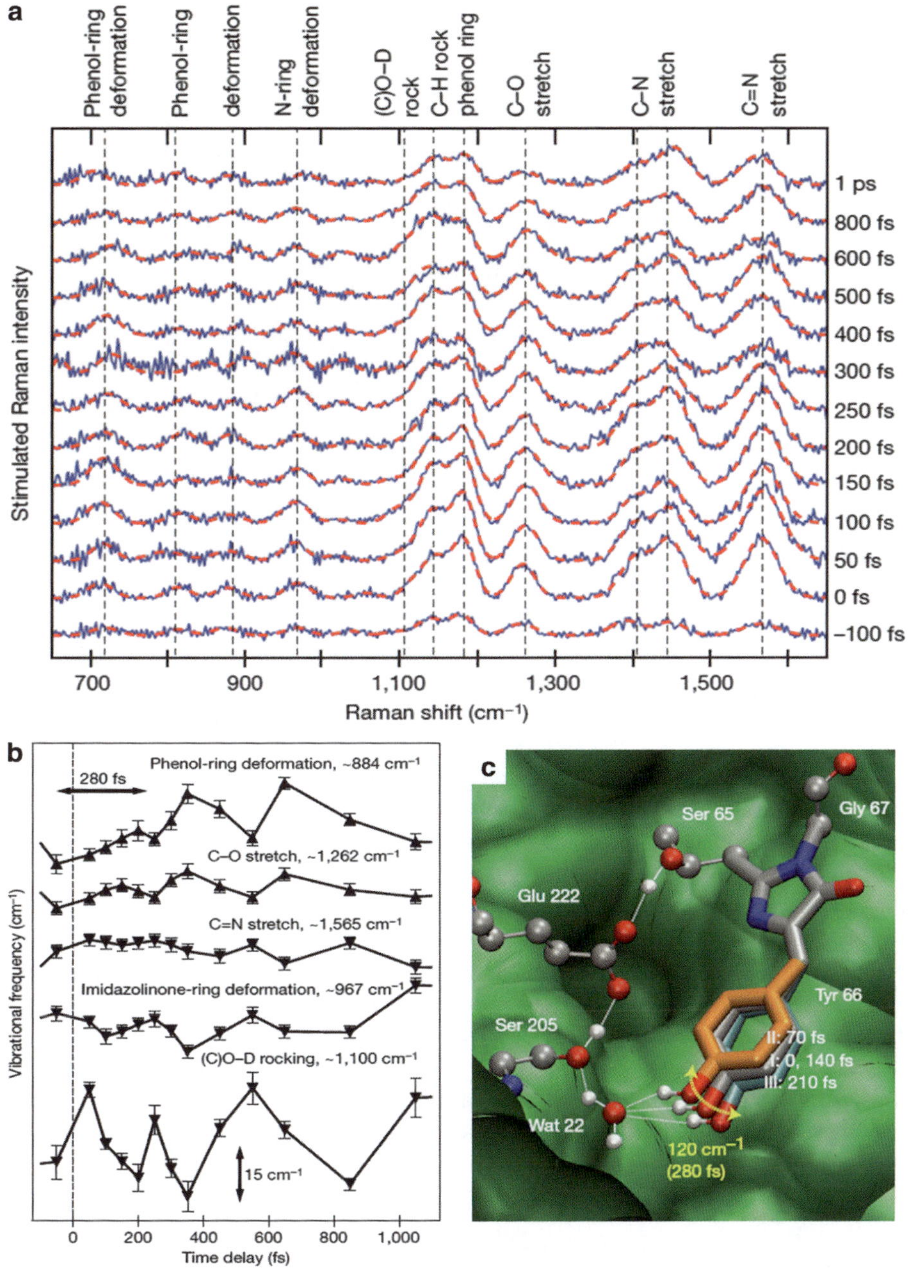

Fig. 8 (**a**) *Blue lines*: femtosecond-stimulated Raman scattering (FSRS) spectra of *wt*GFP in D_2O, from −100 fs to 1 ps following 396-nm excitation; Gaussian fits are shown as *dashed red lines* on each spectrum. (**b**) Time dependence of vibrational structural features frequencies; they oscillate with a period of ~280 fs. The C=N stretching, imidazolinone-ring deformation, and (C)O–D rocking modes are in phase, and this group of modes oscillates out of phase with the C–O

molecules similar to the chromophore, but with some changes in the structure itself, in its environment, or in the isotopic composition of some of the atoms. Based on this knowledge, it was possible to determine structural or protonation changes responsible for the peculiar photophysics of some mutants of the GFP, as described in Sects. 4.3 and 4.4, or even in mutants of other families of autofluorescent proteins; as a recent example, Raman spectroscopy allowed to link a reverse pH-dependence of chromophore protonation with the large Stokes shift observed in the red fluorescent protein mKeima [57].

Even more information can be achieved by single-molecule studies, which are allowed also for vibrational spectroscopy by the SM-SE(R)RS techniques (Sect. 2.2). Recently, Schleifenbaum et al. detected the same individual protein and its photoproducts via fluorescence and SERS imaging [58]. By studying single molecules, it was possible to assign distinct Raman bands to green and red fluorescence forms of the studied bichromophoric autofluorescent protein DsRed_N42H, revealing new insight into the photodegeneration processes of its two chromophores [58].

As shown in Sect. 4.4, also the dynamics of the chromophore can be studied by time-resolved infrared and Raman spectroscopy; with still higher time resolution (10 fs), achievable in future experiments, FSRS may allow even more detailed views into structural changes in the dynamics along the chemical reaction coordinate on the multidimensional potential energy surfaces of polyatomic molecules [17]. Another time-resolved nonlinear technique, which could be applied in the future for the study of the dynamics of FP chromophores, could be the 2-dimensional infrared (2D-IR) spectroscopy [59]. This can be considered an extension of the 2D-NMR, but addressing the vibrational modes of a molecule. The 2D-IR method allows to expose structural dynamics through the molecular vibrations, similarly to the FSRS experiment discussed at the end of Sect. 4.4, but by considering the anharmonic coupling between different modes. The essential advantages of 2D-IR are its ability to identify the different dynamic contributions to the spectral shapes and their intrinsic time resolution. Furthermore, the 2D-IR spectrum exposes directly not only the frequencies but also the anharmonicities of vibrational modes, requiring a more quantitative interpretation of the spectrum. It is also possible to study the time dependence of the coupling between different vibrational modes. 2D-IR has already been applied to study peptides, proteins, and hydrogen-bond dynamics [60], and is therefore a good candidate for vibrational-based structural and dynamical studies of proteins of the GFP family as well.

Fig. 8 (continued) stretching and phenol-ring deformation modes. (**c**) Perspective view of the phenol ring-wagging motion considered responsible for the oscillations mentioned above. Three positions in the phenol-ring motion are illustrated to show how the phenol ring swings from the native structure (I) toward (structure II) and possibly away (structure III) from the optimal ESPT geometry; green background, protein pocket. All panels are adapted from [56], with permission; copyright 2009 Nature Publishing Group

References

1. Bell AF et al (2001) Structure of green fluorescent protein chromophores probed by Raman spectroscopy. Biochemistry 40(29):8619–8619
2. van Thor JJ et al (1998) Characterization of the photoconversion of green fluorescent protein with FTIR spectroscopy. Biochemistry 37(48):16915–16921
3. van Thor JJ et al (2005) Structural events in the photocycle of green fluorescent protein. J Phys Chem B 109(33):16099–16108
4. Luin S et al (2009) Raman study of chromophore states in photochromic fluorescent proteins. J Am Chem Soc 131(1):96–103
5. Time-Resolved IR (TRIR) Absorption Spectroscopy http://www.clf.stfc.ac.uk/Facilities/Lasers+for+Science+Facility/Molecular+Structure+and+Dynamics/Techniques/14374.aspx, Accessed on August 16, 2011
6. Zscherp C, Heberle J (1997) Infrared difference spectra of the intermediates L, M, N, and O of the bacteriorhodopsin photoreaction obtained by time-resolved attenuated total reflection spectroscopy. J Phys Chem B 101(49):10542–10547
7. Bell AF et al (2003) Light-driven decarboxylation of wild-type green fluorescent protein. J Am Chem Soc 125(23):6919–26
8. Habuchi S et al (2005) Evidence for the isomerization and decarboxylation in the photoconversion of the red fluorescent protein DsRed. J Am Chem Soc 127(25):8977–8984
9. Schellenberg P et al (2001) Resonance Raman scattering by the green fluorescent protein and an analogue of its chromophore. J Phys Chem B 105(22):5316–5322
10. Talley CE et al (2005) Surface-enhanced Raman scattering from individual au nanoparticles and nanoparticle dimer substrates. Nano Lett 5(8):1569–74
11. Kneipp J et al (2009) Optical probing and imaging of live cells using SERS labels. J Raman Spectrosc 40(1):1–5
12. Kneipp J et al (2006) In vivo molecular probing of cellular compartments with gold nanoparticles and nanoaggregates. Nano Lett 6(10):2225–2231
13. Tozzini V et al (2003) The low frequency vibrational modes of green fluorescent proteins. ChemPhys 287:33–42
14. Nie S, Emory SR (1997) Probing single molecules and single nanoparticles by surface-enhanced Raman scattering. Science 275(5303):1102–1106
15. Habuchi S et al (2003) Single-molecule surface enhanced resonance Raman spectroscopy of the enhanced green fluorescent protein. J Am Chem Soc 125(28):8446–8447
16. Freudiger CW et al (2008) Label-free biomedical imaging with high sensitivity by stimulated Raman scattering microscopy. Science 322(5909):1857–1861
17. Nienhaus GU (2010) The "Wiggling and Jiggling of Atoms" leading to excited-state proton transfer in green fluorescent protein. Chemphyschem 11(5):971–974
18. Kruglik SG et al (2002) Resonance CARS study of the structure of "Green" and "Red" chromophores within the red fluorescent protein DsRed. J Am Chem Soc 124(37):10992–10993
19. Cinelli RAG et al (2001) Coherent dynamics of photoexcited green fluorescent proteins. Phys Rev Lett 86(15):3439–3442
20. Usman A et al (2005) Excited-state structure determination of the green fluorescent protein chromophore. J Am Chem Soc 127(32):11214–5
21. Thar J, Reckien W, Kirchner B (2007) Car–Parrinello molecular dynamics simulations and biological systems. In: Reiher M (ed) Atomistic approaches in modern biology. Springer, Berlin, pp 133–171
22. Ceperley D (1978) Ground state of the fermion one-component plasma: a Monte Carlo study in two and three dimensions. Phys Rev B 18(7):3126
23. Ceperley DM, Alder BJ (1980) Ground state of the electron gas by a stochastic method. Phys Rev Lett 45(7):566
24. Tanatar B, Ceperley DM (1989) Ground state of the two-dimensional electron gas. Phys Rev B 39(8):5005

25. Sousa SF, Fernandes PA, Ramos MJ (2007) General performance of density functionals. J Phy Chem A 111(42):10439–10452
26. Troullier N, Martins JL (1991) Phys Rev B 43:1993–2006
27. Filippone F, Parrinello M (2001) Vibrational analysis from linear response theory. Chem Phys Lett 345(1–2):179–182
28. Herrmann C (2007) First-principles approach to vibrational spectroscopy of biomolecules. In: Reiher M (ed) Atomistic approaches in modern biology. Springer, Berlin, pp 85–132
29. Putrino A, Sebastiani D, Parrinello M (2000) Generalized variational density functional perturbation theory. J Chem Phys 113(17):7102–7109
30. Jensen L et al (2005) Theory and method for calculating resonance Raman scattering from resonance polarizability derivatives. J Chem Phys 123(17):174110–11
31. Tozzini V, Giannozzi P (2005) Vibrational properties of DsRed model chromophores. Chemphyschem 6:1786–8
32. Marx D, Hutter J (2000) In: Modern methods and algorithms of quantum chemistry, J. Grotendorst (Ed.), John von Neumann Institute for Computing, Jülich, NIC Series, Vol. 1, pp. 301-449
33. Kohanoff J (1994) Phonon spectra from short non-thermally equilibrated molecular dynamics simulations. Comput Mater Sci 2(2):221–232
34. Tangney P, Scandolo S (2002) How well do Car-Parrinello simulations reproduce the Born-Oppenheimer surface? Theory and examples. J Chem Phys 116(1):14–24
35. Esposito AP et al (2001) Vibrational spectroscopy and mode assignments for an analog of the green fluorescent protein cromophore. J Mol Struct 569:25–41
36. Bell AF et al (2000) Probing the ground state structure of the green fluorescent protein chromophore using Raman spectroscopy. Biochemistry 39(15):4423–31
37. He X, Bell AF, Tonge PJ (2002) Isotopic labeling and normal-mode analysis of a model green fluorescent protein chromophore. J Phys Chem B 106(23):6056–6066
38. Laino T, Nifosì R, Tozzini V (2004) Relationship between structure and optical properties in green fluorescent proteins: a quantum mechanical study of the chromophore environment. ChemPhys 298:17–28
39. Nifosì R, Amat P, Tozzini V (2007) Variation of spectral, structural and vibrational properties within the intrinsically fluorescent proteins family: a density functional study. J Comput Chem 28(14):2366–2377
40. Andruniów T (2007) Vibrational analysis of a solvated green fluorescent protein chromophore. J Mol Model 13(6):775–783
41. He X, Bell AF, Tonge PJ (2002) Synthesis and spectroscopic studies of model red fluorescent protein chromophores. Org Lett 4(9):1523–1526
42. Scott AP, Radom L (1996) Harmonic vibrational frequencies: an evaluation of Hartree-Fock, Møller-Plesset, quadratic configuration interaction, density functional theory, and semiempirical scale factors. J Phys Chem 100(41):16502–16513
43. Henderson JN et al (2007) Structural basis for reversible photobleaching of a green fluorescent protein homologue. Proc Natl Acad Sci USA 104(16):6672–6677
44. Andresen M et al (2005) Structure and mechanism of the reversible photoswitch of a fluorescent protein. Proc Natl Acad Sci USA 102(37):13070–13074
45. Andresen M et al (2007) Structural basis for reversible photoswitching in Dronpa. Proc Natl Acad Sci USA 104(32):13005–13009
46. Nifosì R et al (2003) Photoreversible dark state in a tristable green fluorescent protein variant. J Phys Chem B 107(7):1679–1684
47. Loos DC et al (2006) Photoconversion in the red fluorescent protein from the Sea Anemone Entacmaea quadricolor: is cis-trans isomerization involved? J Am Chem Soc 128(19): 6270–6271
48. van Thor JJ et al (2002) Phototransformation of green fluorescent protein with UV and visible light leads to decarboxylation of glutamate 222. Nat Struct Biol 9(1):37–41

49. Voliani V et al (2008) Cis-trans photoisomerization of fluorescent-protein chromophores. J Phys Chem B 112:10714–10722
50. Bizzarri R et al (2010) Single amino acid replacement makes aequorea victoria fluorescent proteins reversibly photoswitchable. J Am Chem Soc 132(1):85–95
51. Nifosì R, Tozzini V (2006) Cis-trans photoisomerization of the chromophore in the green fluorescent protein variant E2GFP: a molecular dynamics study. ChemPhys 323(2–3):358–368
52. Stoner-Ma D et al (2005) Observation of excited-state proton transfer in green fluorescent protein using ultrafast vibrational spectroscopy. J Am Chem Soc 127(9):2864–2865
53. Stoner-Ma D et al (2006) Proton relay reaction in green fluorescent protein (GFP): polarization-resolved ultrafast vibrational spectroscopy of isotopically edited GFP. J Phys Chem B 110(43): 22009–22018
54. Stoner-Ma D et al (2008) Ultrafast electronic and vibrational dynamics of stabilized A state mutants of the green fluorescent protein (GFP): snipping the proton wire. ChemPhys 350(1–3): 193–200
55. Stoner-Ma D et al (2008) An alternate proton acceptor for excited-state proton transfer in green fluorescent protein: rewiring GFP. J Am Chem Soc 130(4):1227–1235
56. Fang C et al (2009) Mapping GFP structure evolution during proton transfer with femtosecond Raman spectroscopy. Nature 462(7270):200–204
57. Violot S et al (2009) Reverse pH-dependence of chromophore protonation explains the large stokes shift of the red fluorescent protein mKeima. J Am Chem Soc 131(30):10356–10357
58. Schleifenbaum F, Peter S, Meixner AJ (2009) Detecting the same individual protein and its photoproducts via fluorescence and surface-enhanced Raman spectroscopic imaging. J Phys Chem A 114(1):143–150
59. Hochstrasser RM (2007) Two-dimensional spectroscopy at infrared and optical frequencies. Proc Natl Acad Sci USA 104(36):14190–14196
60. Kim YS, Hochstrasser RM (2009) Applications of 2D IR spectroscopy to peptides proteins, and hydrogen-bond dynamics. J Phys Chem B 113(24):8231–8251

Proton Travel in Green Fluorescent Protein

Volkhard Helms and Wei Gu

Abstract Green fluorescence protein (GFP) wild type and some of its mutants undergo excited state proton transfer between the chromophore and the nearby Glu222 residue. This process has been covered in detail in the chapter written by Stephen Meech. Apart from this ultrafast photochemical reaction, multiple other proton-transfer processes take place in the GFP protein matrix, and these will be covered in this chapter. For example, proton exchange between the chromophore and the nearby bulk solvent may occur via His148 that is located in hydrogen-bonding distance from the chromophore and provides direct access to the bulk solvent. Moreover, two extended proton-transfer wires including titratable residues as well as a number of buried water molecules connect the chromophore to the protein surface. Based on a recent high-resolution X-ray structure of GFP, all titratable groups of the protein could be placed in one of these two large hydrogen-bonding clusters, suggesting that a multitude of proton-transfer processes can occur in the GFP matrix at any moment in time. While it is quite likely that similar proton pathways also exist in other soluble and membrane proteins, they are much harder to study. GFP is an exciting model system for monitoring those processes as they often directly affect the chromophore photophysics. The dynamics of proton exchange inside the GFP barrel and with bulk solvent has thus been characterized by fluorescence correlation spectroscopy (FCS) of the chromophore fluorescence and by pH-jump experiments. These studies showed that the autocorrelation of the chromophore fluorescence is affected either by pH-independent processes on microsecond to millisecond time scales or by pH-dependent processes on similar time scales. The former ones are likely proton equilibria occurring within the GFP barrel, and the latter ones are likely exchange processes with the solvent. Biomolecular simulation methods are now being developed, which will soon allow accessing such time scales by computational means. Then, we will hopefully be able to connect the spectroscopic findings with dynamic atomistic simulations of proton-transfer dynamics.

V. Helms (✉) and W. Gu
Center for Bioinformatics, Saarland University, Campus C7.1, 66123 Saarbruecken, Germany
e-mail: volkhard.helms@bioinformatik.uni-saarland.de

G. Jung (ed.), *Fluorescent Proteins I*, Springer Ser Fluoresc (2012) 11: 171–182,
DOI 10.1007/4243_2011_13, © Springer-Verlag Berlin Heidelberg 2011,
Published online: 2 April 2011

Keywords Fluorescence correlation spectroscopy · Hydrogen bond network · pH jump · Proton antenna · QHOP molecular dynamics

Contents

1 Background on Biological Proton Transfer

Proton-transfer (PT) reactions are essential parts of many biochemical and bioenergetic processes [1, 2], e.g., in enzymatic reactions [3] or along hydrogen-bonding networks through entire membrane proteins such as bacteriorhodopsin or cytochrome *c* oxidase. PT in hydrogen-bonded networks can either proceed via structural (Grotthuss-) diffusion or vehicle (Stokes-) diffusion [1]. In the latter case, the proton remains bound to one particular diffusing (water) molecule, whereas in the former case it changes continuously its partner by breaking and reforming covalent bonds [4]. In proteins, the thermodynamic and kinetic properties of PT processes are strongly influenced by electrostatic interactions with the environment of the proton-donating and -accepting chemical groups [5, 6]. By affecting both the relative energetic difference between the reactant and product states and the barrier height, environmental effects are of crucial importance for reaction rates, equilibria, and local pK_a values of titratable residues. Things may be complicated further because (1) proton- and electron-transfer processes are often coupled as, for example, in cytochrome *c* oxidase, (2) proton transport across entire proteins involves a large number of subsequent PT events, and (3) proton migration often involves concerted transfers and quantum-mechanical tunneling.

For many biological systems involved in proton pumping, such as the bacterial photosynthetic reaction center and cytochrome *c* oxidase, the biomolecular rate constant for proton uptake from bulk solution was found to significantly exceed the rate for proton diffusion through water to a single surface-bound titratable group [7]. Gutman and Nachliel gave a possible explanation for this behavior by proposing the so-called antenna effect [8]. There, they suggested that such rapid proton-transfer reactions indicate the involvement of multiple negatively charged Asp and Glu residues on the protein surface that are titratable and can capture protons from bulk solution. The closely located charged residues might thus extend the proton capture area of the surface and provide a local, two-dimensional buffer composed of rapidly proton-exchanging titratable sites.

Providing definite experimental evidence for proton-transfer phenomena in biomolecules is quite hard. One basic technique is to perform time-resolved experiments in H_2O and in D_2O. If the rate constants of photo-physical processes are slowed down considerably, this is commonly taken as clear evidence for the involvement of protons and is characterized by the kinetic isotope effect (KIE = k_D/k_H). In the case of green fluorescence protein (GFP), this is exactly what was done in the classical paper by Chattoraj et al., which demonstrated the existence of ESPT upon photo-excitation [9]. There, the decay of the photo-excited species A* and the rise of the I* are slowed down by a factor of about 5 when performing the experiments in D_2O relative to H_2O.

Spatial mapping of proton-transfer pathways in proteins is the realm of Fourier-transformed infrared spectroscopy (FTIR) and has been applied to map out PT pathways in bacteriorhodopsin, the bacterial photosynthetic reaction center, and in ras p21 [10]. For this, one typically takes time-resolved difference spectra between wild-type and mutant protein. GFP is an ideal system for this technique because the chromophore is naturally built into the proton barrel so that one can simultaneously apply optical and infrared spectroscopy. So far, FTIR analysis of GFP has focused on mapping the immediate chromophore surrounding [11].

2 Fluorescence Autocorrelation of GFP

Fluorescence correlation spectroscopy (FCS) experiments have been used to probe the fluorescence fluctuations of GFP proteins over time [12]. For example, one may probe whether a GFP protein that predominantly contains a neutral chromophore at time t_0 still contains a neutral chromophore after time $t_0 + t$. When fluorescence can arise only from the neutral chromophore state, then the reversible loss of a proton leads to a reversible loss of fluorescence thus resulting in fluorescence fluctuations. The fluctuations are subsequently analyzed by its autocorrelation $G(t)$. Changes of the protonation state of the chromophore may occur either due to proton exchange with other titratable groups of the protein or via proton exchange with bulk solvent [13]. Interestingly, the decay of the autocorrelation of the EGFP chromophore on the sub-millisecond time scale showed a clear pH dependence [12] (Fig. 1). This pH dependence of the decay time is a clear evidence that the protein interior is in dynamic exchange with the surrounding proton reservoir on a micro- to millisecond time scale. This on–off fluorescence flickering of GFP mutants is influenced by several environmental factors. The pH effect, attributed to reversible external protonation of the chromophore, was discussed for EGFP [12]. In addition, a fraction ($\approx$13%) of EGFP molecules showed a pH-insensitive flicker that seemed dependent on illumination intensity. Schwille and coworkers presented thorough intensity- and pH-dependent FCS studies of flicker dynamics in the yellow-shifted mutants T203Y and T203F [14] that allowed an effective separation of these two effects, pH vs. intensity. So far, it is unclear whether this pH dependence is of functional relevance. Yet, the coupling of internal proton transfer to the chromophore fluorescence makes GFP a unique probe to characterize and understand such proton-transfer processes.

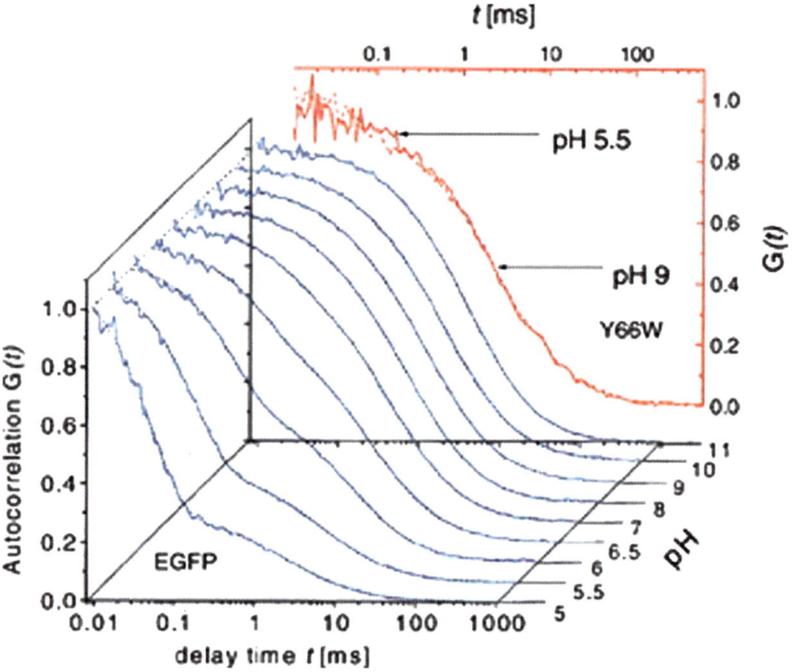

Fig. 1 The autocorrelation of the chromophore fluorescence of the GFP mutant Y66W and of EGFP. For Y66W, the spectra at pH 5.5 and at pH 9 are almost identical. The autocorrelation decays with a time constant of about 10 ms. For EGFP, the autocorrelation decays much faster at lower pH indicating a proton equilibrium with bulk solution. Figure reprinted from [12] with permission

3 pH-Jump Experiments

The picture obtained from FCS measurements is complemented by experimental data that analyzed changes of the chromophore photophysics following an external pH jump of the medium [15]. When exposed to laser light, small chemicals known as "caged protons" undergo ESPT and release protons into the medium. In this way, a pH jump of the medium from high to low pH can be generated within a laser flash of 10-ns duration. Following the pH jump, the following elementary processes will occur: (1) free diffusion of protons in the bulk solvent, (2) binding of protons from bulk solvent at the protein/water interface, (3) proton flow through a hydrogen-bonded network from the interface to neighboring titratable groups or water molecules around the buried chromophore, and (4) rearrangement of the side chains around the GFP chromophore, as well as of the chomophore itself, to enable PT to the deprotonated chromophore. The combination of all these factors will then lead to the pH-jump-initiated fluorescence changes at the chromophore site that can be detected spectroscopically.

NMR experiments and computer simulations have determined the diffusion time of excess protons in water as $D = 9.3 \times 10^{-5}\ \mathrm{cm}^2\ \mathrm{s}^{-1}$. In a simulation study of an acetic acid molecule immersed into a water box of 2.4-nm dimensions [16], we showed that protons traverse the entire box volume within a few nanoseconds. Thus, process (1) is unlikely to be the rate-determining step for proton entry. Process (2) is expected to occur on the sub-microsecond timescale as well since proton uptake at interfaces was shown to occur without any major kinetic barrier [17]. For example, protonation of fluorescein covalently attached to the surfaces of the proteins barstar and bovine serum albumin was found to occur within ~10 μs. Thus, processes (3) or (4) are the most likely candidates for the rate-determining process.

In the pH-jump experiments on EGFP, the observed time constant for protonation of the EGFP chromophore was ~300 μs upon changing the pH from 8 to 5, while that for the S65T mutant form of GFP was ~87 μs [15]. EGFP differs from S65T-GFP by the single additional mutation F64L. Being close to the site of the chromophore and to the putative hydrogen-bonding networks involved in proton pathway, the F64L mutation might indeed affect the efficiency of proton transfer from bulk solvent, thereby slowing down the overall process. Interestingly, the rate constants deduced from the pH-jump experiments are slower, by a factor of 3–5, than those obtained in the FCS experiments [12, 14]. This likely reflects that different processes are being probed by the pH-jump relaxation method and by the near-equilibrium FCS method [15]. In the pH-jump relaxation method, the concentration of protons in the bulk solvent is jumped to higher values, and the subsequent proton transfer to the chromophore buried inside the protein is monitored. This entire process occurs in several steps and, possibly, one of the steps, namely proton transfer through the protein matrix, is slower than the rest. In contrast, the FCS method monitors fluctuations of the protonation state of the chromophore, and the observed process most likely represents the shuttling of protons between the chromophore and one or more nearby side chain or water molecules.

Surprisingly, almost no dependence on temperature was found for the proton-transfer kinetics indicating a barrierless process [15]. This argues against a simple diffusion-limited protonation process that is expected to have an activation barrier. On the other hand, the process can be expected to be barrierless if the overall protonation process is rate limited by proton hopping through a H-bonded network, analogous to the Grotthuss mechanism [18] for the tunneling of protons through "proton wires" [19]. Since such a concerted proton relay would require a stable H-bonded network connecting the chromophore with the protein surface, the rate of proton flow would be expected to decrease with an increase in temperature due to weakening of the network. Thus, the observed very low temperature dependence might also be the result of a fortuitous combination of activated movement of protons and the temperature-dependent stability of the H-bonded network.

4 Hydrogen-Bonded Clusters in GFP X-Ray Structures

We will now turn to the data generated by X-ray crystallography. The Protein Data Bank makes available a very impressive amount of data with more than 100 crystal structures of GFP and mutant FPs. Recently, an ultra-high-resolution structure of a triple mutant F64L I167T K238N of GFP was determined at 0.90 Å resolution (Protein Data Bank entry 2WUR) that allowed for detection of 320 water molecules, compared to 105 waters at 2.30 Å resolution (1 EML) and 74 at 2.50 Å resolution (1 EME) [20].

At first, there seems no necessity to involve further residues in proton transfer apart from those involved in the multistep proton shuttle between the buried chromophore and Glutamate Glu222 that is part of the fluorescence cycle of wild-type GFP. The only time when proton transfer is functionally required is during chromophore biosynthesis when protons need to be transferred to the protein surface. NMR dispersion relaxation experiments of Halle and coworkers showed that three internal buried water molecules in the bovine trypsin inhibitor exchange with bulk solvent on time scales of milliseconds [21]. Therefore, the natural thermal dynamics of GFP should also allow for the transfer of these protons to the protein surface even more since the duration of this one-time process is not crucial.

However, Agmon and coworkers [22, 23] identified an extensive proton wire that connects the terminal chromophore residue Tyr66 via Glu222 to the surface of the protein. The new data now allowed for a detailed analysis of this wire in GFP on the basis of the ultra-high-resolution crystal structure [20]. The authors considered all backbone, side chain, and water oxygen atoms as well as the protein nitrogen atoms for the analysis of putative PT pathways. All O and N atoms within 3.0 Å distance were considered connected. It turned out that the GFP barrel contains two large internal clusters on both sides of the chromophore throughout the entire protein. Figure 2 shows one of these hydrogen-bonded clusters of putative proton-donating and -accepting groups. This "active-site wire" that connects to the hydroxyl oxygen of Tyr66 of the chromophore contains 147 atoms in the 2WUR crystal structure. Of these, 75 are protein atoms and 72 are water oxygens. This wire contains two exit segments on partly hydrophobic parts of the protein surface. According to the proton-antenna model, such surface patches are likely not involved in capturing protons from solution. A similar reasoning was put forward for the short Thr203-His148 exit [22]. A third exit connects the active-site wire to a cluster of neighboring carboxylate residues (Glu5, Asp36, Glu34, Asp117, Glu6) on the GFP surface. As discussed before, such an arrangement leads to a negative electrostatic potential on the protein surface that should thus be able to capture protons from solution.

The second large internal cluster (not shown in Fig. 2) located on the other face of the chromophore is disconnected from the first cluster and from the protein exterior. This wire does not seem to be involved in ESPT. It contains some key residues near the chromophore such as His181, Thr62, Thr108, Tyr145, and Arg96. Some of these residues are thought to be involved in chromophore biosynthesis. Consequently, this cluster was termed the "biosynthesis cluster". If Arg96 gets temporarily deprotonated during the dehydration step, this wire may serve as a proton storage. This is reminiscent of the situation at the exit pathway of the proton pump bacteriorhodopsin for which Gerwert and coworkers have argued that several buried water molecules that are held in place by two close-by carboxylate residues function as a proton storage [24].

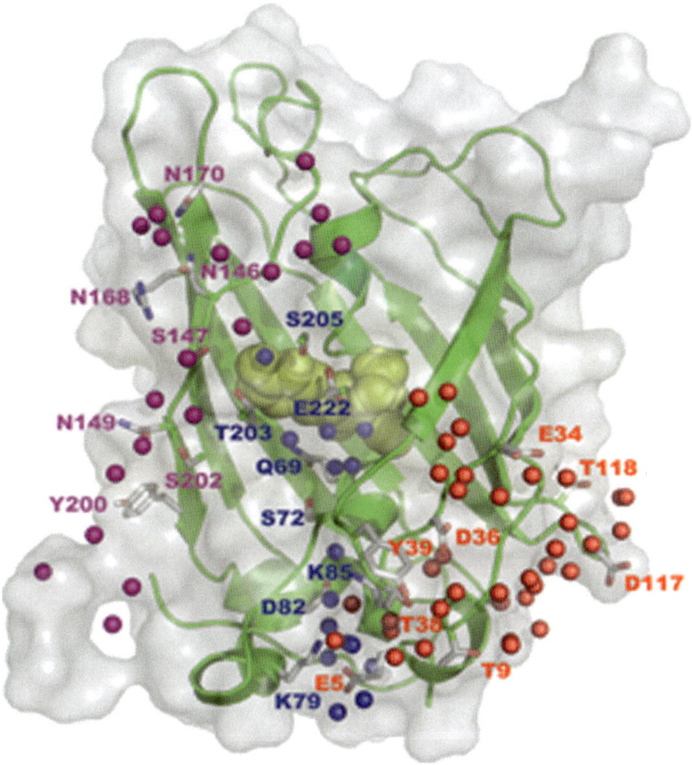

Fig. 2 Ribbon representation of the crystal structure 2WUR of a GFP triple mutant determined at 0.9 Å resolution. The chromophore atoms are shown as *yellow van der Waals spheres*. The protein ribbon is colored *green*, and the transparent *gray* surface illustrates the protein surface. *Colored spheres* indicate water oxygens. Different colors are used for different parts of the active-site hydrogen-bonding clusters identified in [20] according to the subcluster "entrance" on the right (*red*), "internal" in the middle (*blue*), and "exit" on the left (*magenta*). Amino acid residues involved in each cluster are also labeled with the corresponding color. Note that we did not draw residues for which only the backbone oxygen and nitrogen atoms are involved in a cluster

5 Computational Approaches to Study Biomolecular PT Dynamics

The simple algorithm applied in [20] that was discussed above did not consider the pK_a values of the residues involved nor the energy barriers for rotating their side chains. In contrast, Warshel and coworkers treated the dynamics of biomolecular PT by first computing the local pK_a values of titratable residues and then applying a Marcus-type approach to get transfer rates [25]. Taraphder and Hummer presented an algorithm that accounts for the conformational dynamics of proteins to a certain degree by considering all possible side-chain rotamer conformations and their torsion energies and identifying the pathway of lowest-cost for generating bridging conformations along which a proton may be translocated [26]. They applied this algorithm to the proton transfer from the surface of cytochrome P450cam to its buried active site and to carbonic anhydrase. We have implemented their algorithm in a software termed QVADIS, where we also considered the kinetic barriers for proton transfer in the Q-HOP framework [27]. Alternatively, Ullmann and coworkers have considered the degree of evolutionary conservation combined with analysis of hydrogen-bonding clusters in photosynthetic reaction centers [28].

Besides these approaches based on static conformations, also dynamic simulation techniques have been developed to model PT pathways based on the molecular dynamics technique [2]. Generally, molecular dynamics simulations have become a standard tool for characterizing the native protein dynamics on time scales from femtoseconds to microseconds [29]. For GFP, several authors have characterized the dynamics of the stiff protein matrix and the flexible internal water hydrogen bond network by molecular dynamics simulations [30–32]. On a nanosecond time scale, several water molecules near the chromophore were tightly locked into favorable coordinations, whereas others could almost freely rotate and alternate between alternative roles as hydrogen bond donors and acceptors [30]. These standard simulation techniques need to be modified to allow for bond breaking and formation that are required to describe PT reactions.

According to the "proton-antenna" model discussed by Gutman [8], patches of negatively charged residues on the surface of proteins may mark entry points for protons from bulk solution. We therefore compared the proton capturing tendencies of imidazole and acetic acid as side-chain analogs of histidine and of aspartic acid by our Q-HOP molecular dynamics method [33]. We found a qualitatively different protonation behavior of 4-methylimidazole compared to that of acetic acid. On one hand, deprotonated, neutral 4-methylimidazole cannot as easily attract a freely diffusing extra proton from solution. Once the proton is bound, however, it remains tightly bound on a time scale of tens of nanoseconds. In a linear chain composed of acetic acid, a separating water molecule, and 4-methylimidazole, an excess proton is equally shared between 4-methylimidazole and water. When a water molecule is linearly placed between two acetic acid molecules, the excess proton is always found on the central water. On the other hand, an excess proton in a 4-methylimidazole–water–4-methylimidazole chain is always localized on one of the two 4-methylimidazoles. This suggests that aspartic and glutamic acid function as attractors for protons from solution, whereas histidine residues may function as temporal reservoirs of proton storage.

We then mimicked the antenna effect of Gutman by a model system consisting of two transmembrane helices connected by a short loop fragment. 1–3 Asp residues were engineered into this loop and the helix-loop-helix fragment was embedded in a DOPC lipid bilayer. During molecular dynamics simulations with our hoppable proton model Q-HOP MD, we initially started all aspartate residues in the deprotonated form and placed an excess proton in bulk solution. Via normal Grotthuss-type diffusion, it is quite likely that the proton enters into the electrostatic capture radius of the negatively charged aspartates. Then, it may hop to one of the aspartates within a few picoseconds via a hydrogen-bonded water wire. We termed these wires "tightly connected water wires" (TCW) [34].

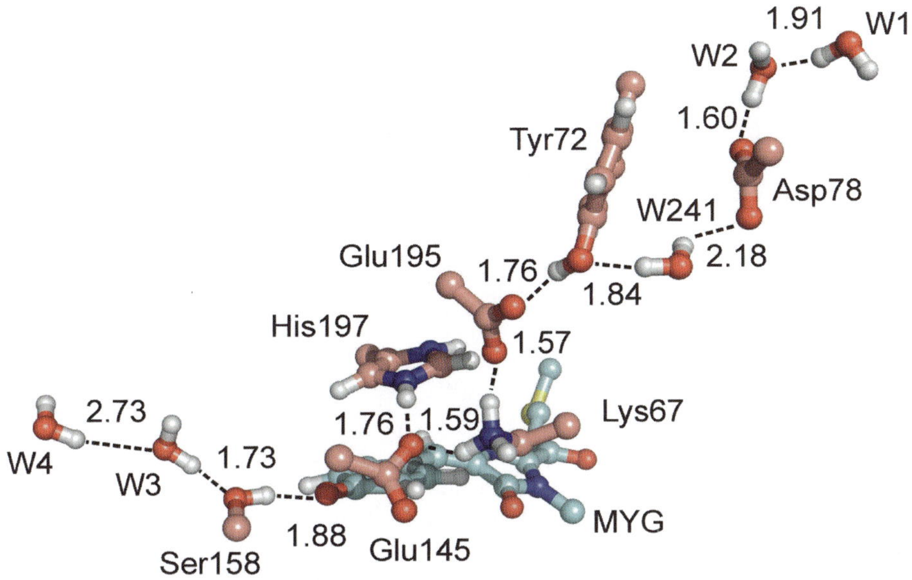

Fig. 3 Snapshot from a molecular dynamics simulation showing the proposed proton-entry and -exit wires in the asFP595 protein [35]. Figure altered following [35] with permission

Proton release to the bulk solvent has also been suggested to take place for the asFP595 protein [35]. Using molecular dynamics simulations, Schäfer and co-workers identified a pathway from the chromophore (termed MYG here) to the protein surface involving several titratable residues and buried water molecules (Fig. 3). This pathway leading to the terminal Asp78 residue was suggested to function as a proton-release pathway. During 20-ns long MD simulations, the hydrogen-bonding coordination along this pathway remained intact. The authors suggested that as the rate of proton migration along such wires is dominated by tunneling, the proton exchange may occur by rapid one-dimensional diffusion of protons along this wire.

6 Outlook

Unfortunately, rational mutation work in GFP has so far been focused on the vicinity of the chromophore with the aim of changing its fluorescence properties. None of the mutants generated by phage display genetic evolution has targeted the suggested proton-entry and -exit pathways. Likely, such mutations would provide no advantage to the protein stability or folding efficiency and therefore were not observed. However, GFP is clearly a very exciting model system for understanding the energetic and kinetic details of internal proton transfer. We hope that some future works will generate mutant fluorescent proteins that will affect the kinetics along proton-transfer pathways to or from the chromophore. Combined interpretation of structural, spectroscopic, and computational data may then converge to resolve the big picture of the "dynamic protome" in protein matrices.

Acknowledgments VH thanks Lars Schäfer (University of Groningen) for kindly making available Fig. 3 for this manuscript and Noam Agmon for a preprint of Shinobu et al. [20].

References

1. Marx D (2006) Proton transfer 200 years after van Grotthuss: insight from *ab initio* simulations. Chemphyschem 7:1848–1870. doi:10.1002/cphc.200600128
2. Swanson JMJ, Maupin CM, Chen H, Petersen MK, Xu J, Wu Y, Voth GA (2007) Proton salvation and transport in aqueous and biomolecular systems: insight from computer simulations. J Phys Chem B 111:4300–4314. doi:10.1021/jp070104x
3. Kemp DS (1995) How to promote proton transfer. Nature 373:196–197. doi:10.1038/373196a0
4. Tuckerman ME, Marx D, Parrinello M (2002) The nature and transport mechanism of hydrated hydroxide ions in aqueous solution. Nature 417:925–929. doi:10.1038/nature00797
5. Lu D, Voth GA (1998) Proton transfer in the enzyme carbonic anhydrase: an *ab initio* study. J Am Chem Soc 120:4006. doi:10.1021/ja973397o
6. Warshel A, Naray-Szabo G, Sussman F, Hwang JK (1989) How do serine proteases really work? Biochemistry 28:3629–3637. doi:10.1021/bi00435a002
7. Ädelroth P, Brzezinski P (2004) Surface-mediated proton-transfer reactions in membrane-bound proteins. Biochim Biophys Acta 1655:102–115. doi:10.1016/j.bbabio.2003.10.018
8. Gutman M, Nachliel E (1997) Time-resolved dynamics of proton transfer in proteinous systems. Annu Rev Phys Chem 48:329–356. doi:10.1146/annurev.physchem.48.1.329
9. Chattoraj M, King BA, Bublitz GU, Boxer SG (1996) Ultra-fast excited state dynamics in green fluorescent protein: multiple states and proton transfer. Proc Natl Acad Sci USA 93:8362–8367
10. Kötting C, Gerwert K (2005) Proteins in action monitored by time resolved FTIR spectroscopy. Chemphyschem 6:881–888. doi:10.1002/cphc.200400504
11. van Thor JJ, Pierik AJ, Nugteren-Roodzant I, Xie A, Hellingwerf KJ (1998) Characterization of the photoconversion of green fluorescent protein with FTIR spectroscopy. Biochemistry 37:16915–16921. doi:10.1021/bi981170f
12. Haupts U, Maiti S, Schwille P, Webb WW (1998) Dynamics of fluorescence fluctuations in green fluorescent protein observed by fluorescence correlation spectroscopy. Proc Natl Acad Sci USA 95:13573–13578
13. Widengren J, Terry B, Rigler R (1999) Protonation kinetics of GFP and FITC investigated by FCS – aspects of the use of fluorescent indicators for measuring pH. Chem Phys 249:259–271. doi:10.1016/S0301-0104(99)00256-6 DOI:dx.doi.org
14. Schwille P, Kummer S, Heikal AA, Moerner WE, Webb WW (2000) Fluorescence correlation spectroscopy reveals fast optical excitation-driven intramolecular dynamics of yellow fluorescent proteins. Proc Natl Acad Sci USA 97:151–156
15. Saxena AM, Udgaonkar JB, Krishnamoorthy G (2005) Protein dynamics control proton transfer from bulk solvent to protein interior: a case study with a green fluorescent protein. Prot Sci 14:1787–1799. doi:10.1110/ps.051391205
16. Gu W, Frigato T, Straatsma TP, Helms V (2007) Dynamic protonation equilibrium of solvated acetic acid. Angew Chem Int Ed 46:2939–2943. doi:10.1002/anie.200603583
17. Gutman M, Nachliel E, Kiryati S (1992) Dynamic studies of proton diffusion in mesoscopic heterogenous matrix II. The interbilayer space between phospholipid membranes. Biophys J 63:281–290. doi:10.1016/S0006-3495(92)81585-0

18. de Grotthuss CJT (1806) Sur la décomposition de l´eau et des corps qu´elle tient en dissolution à l´aide de l´électricité galvanique. Ann Chim Paris 58:54–73
19. Agmon N (1995) The Grotthuss mechanism. Chem Phys Lett 244:456–462. doi:10.1016/0009-2614(95)00905-J
20. Shinobu A, Palm GJ, Schierbeek AJ, Agmon N (2010) Visualizing proton antenna in a high-resolution green fluorescent protein structure. electronic version ahead of print. J Am Chem Soc. doi:10.1021/ja1010652
21. Denisov VP, Peters J, Hörlein HD, Halle B (1996) Using buried water molecules to explore the energy landscape of proteins. Nat Struct Biol 3:505–509. doi:10.1038/nsb0696-505
22. Agmon N (2005) Proton pathways in green fluorescence protein. Biophys J 88:2452–2461. doi:10.1529/biophysj.104.055541
23. Shinobu A, Agmon N (2009) Mapping proton-wires in proteins: carbonic anhydrase and GFP chromophore biosynthesis. J Phys Chem A 113:7253–7266. doi:10.1021/jp8102047
24. Garczarek F, Brown LS, Lanyi JK, Gerwert K (2005) Proton binding within a membrane protein by a protonated water cluster. Proc Natl Acad Sci USA 102:3633–3638
25. Sham YY, Muegge I, Warshel A (1999) Simulating proton translocations in proteins: probing proton transfer pathways in the rhodobacter sphaeroides reaction center. Protein Struct Funct Genet 36:484–500. doi:10.1002/(SICI)1097-0134(19990901)36:4<484::AID-PROT13>3.0.CO;2-R
26. Taraphder S, Hummer G (2003) Protein side-chain motion and hydration in proton-transfer pathways. Results for cytochrome P450cam. J Am Chem Soc 125:3931–3940. doi:10.1021/ja016860c
27. Kunz K, Helms V (2007) QVADIS: a package to compute proton transfer pathways in proteins. In: Falter C, Schliep A, Selbig J, Vingron M, Walther D (eds) GI-Edition – Lecture notes in informatics (LNI) – Proceedings 115
28. Krammer EM, Till MS, Sebban P, Ullmann GM (2009) Proton transfer pathways in photosynthetic reaction centers analyzed by profile hidden Markov models and network calculations. J Mol Biol 388:631–643. doi:10.1016/j.jmb.2009.03.020
29. Karplus M, McCammon JA (2002) Molecular dynamics simulations of macromolecules: a perspective. Nat Struct Biol 9:646–652. doi:10.1038/nsb0902-646
30. Helms V, Straatsma TP, McCammon JA (1999) Internal dynamics of green fluorescent proteins. J Phys Chem B 103:3263–3269. doi:10.1021/jp983120q
31. Nifosi R, Tozzini V (2003) Molecular dynamics simulations of enhanced green fluorescent proteins: effects of F64L, S65T and T203Y mutations on the ground-state proton equlibria. Proteins 51:378–389. doi:10.1002/prot.10335
32. Vallverdu G, Demachy I, Mérola F, Pasquier H, Ridard J, Lévy B (2010) Relation between pH, structure, and absorption spectrum of Cerulean: a study by molecular dynamics and TD DFT calculations. Proteins 78:1040–1054. doi:10.1002/prot.22628
33. Gu W, Helms V (2007) Different protonation equilibria of 4-methylimidazole and acetic acid. Chemphyschem 8:2445–2451. doi:10.1002/cphc.200700442
34. Gu W, Helms V (2009) Tightly connected water wires facilitate fast proton uptake at the proton entrance of proton pumping proteins. J Am Chem Soc 131:2080–2081. doi:10.1021/ja809301w
35. Schäfer LV, Groenhof G, Klingen AR, Ullmann GM, Boggia-Pasqua M, Robb MA, Grubmüller H (2007) Photoswitching of the fluorescent protein asFP595: mechanism, proton pathways, and absorption spectra. Angew Chem Int Ed 46:530–536. doi:10.1002/anie.200602315

Photoconversion of the Green Fluorescent Protein and Related Proteins

Jasper J. van Thor

Abstract This review focuses on the mechanistic details of photochromic reactions of the green fluorescent protein (GFP) and also of its mutant derivatives and related fluorescent proteins. A number of distinct photochromic processes have so far been identified that have entirely different photochemical and chemical basis, which will be reviewed. In addition to bright fluorescence, the GFP from the jellyfish *Aequorea victoria* undergoes photochromic transformation with blue or UV illumination. The associated change in electronic absorption provides a spectroscopic contrast that can be used in fluorescence microscopy application to tag and track the movement of populations that are photoconverted. Key to the successful use of photoconversion for such microscopy experiments is in fact the relatively low quantum yield of the irreversible process. In the wild-type GFP, photoconversion is triggered by light-induced electron transfer from the buried anionic carboxylate of Glu222 to the optically excited protonated chromophore. An unstable carboxylate radical subsequently cleaves off a CO_2 molecule in a "Kolbe" type reaction that has been trapped in a partially oriented site near the chromophore-binding site at 100K, as observed by low-temperature X-ray crystallography and cryo-infrared crystallography. Structural intermediates in the subsequent relaxation pathway involve motion of CO_2, amino acids and H-bonded waters both in the chromophore vicinity and at longer range. This review provides an overview of the molecular characterisation using structural and spectroscopy methods of this photoconversion reaction of GFP. In addition, the mechanisms of photochromic reactions of mutants of GFP and related fluorescent proteins will be summarised and discussed. These include the *cis–trans* isomerisation and protonation changes in Dronpa, asFP595 and IrisFP and related proteins, light-induced maturation in aceGFPL, and photoinduced beta-elimination and backbone cleavage that leads to "green-to-red" photoconversion in EosFP, Kaede, IrisFP and KikGR.

J.J. van Thor
Division of Molecular Biosciences, Imperial College London, South Kensington Campus, London SW7 2AZ, UK
e-mail: j.vanthor@imperial.ac.uk

G. Jung (ed.), *Fluorescent Proteins I*, Springer Ser Fluoresc (2012) 11: 183–216,
DOI 10.1007/4243_2011_20,
Published online: 7 April 2011

Keywords asFP595 · Dronpa · Fluorescent proteins · GFP · Green fluorescent protein · 'Green-to-red' photoconversion · Kaede · Oxidative decarboxylation · Photoconversion · Photoisomerisation · Photoswitching

Contents

1 Introduction

The green fluorescent protein (GFP) from the jellyfish *Aequorea victoria* was the first fluorescent protein to be fully characterised spectroscopically and structurally. Key to these successes was the molecular cloning of the *gfp* gene and its recombinant expression in bacterial *Escherichia coli* cells [1, 2]. Importantly, the light absorbing *p*-hydroxybenzylidene-imidazolidinone chromophore forms autocatalytically. It is this automatic maturation that allows the recombinant expression of the *gfp* gene *in vivo* to fluorescently tag protein populations in living cells [3]. The crystal structure of the S65T variant [4] and the wild type [5, 6] were subsequently solved. The examination of the molecular structures enabled the creation of mutants with altered spectroscopic properties, using both structure-based rational design and random mutagenesis approaches, and it also allowed sample preparation for detailed biophysical studies [7–9]. The important utility of GFP and other fluorescent proteins in fluorescent microscopy applications has revolutionised biological and medical studies. They have also impacted many other fields, particularly the biophysics and structural biology of light-induced reactions. With the number of available unique fluorescent (and non-fluorescent) proteins that are reported in the literature at the time of writing, there is a continuing research effort to chart and understand the diverse light and dark reactions that occur in the native and mutant derivative proteins. While the autocatalytic formation of a bright fluorescent gene product has already revolutionised biological research that uses fluorescence microscopy techniques, the additional photochemical characteristics of GFP and related fluorescent proteins have provided even more advanced tools for molecular biologists. Biophysical research of light-sensitive proteins has also made significant advances as a result. The discovery and development of the GFP was awarded with the 2008 Nobel Prize in Chemistry to Osamu Shimomura, Martin Chalfie and Roger Y. Tsien [10].

1.1 Molecular Structure Determination of GFP

An early breakthrough in GFP research was the successful crystallisation and structure solution. Very close together in time the structures of the S65T mutant [4], in which the anionic chromophore is stabilised, and the wild-type GFP [6], with a predominantly neutral chromophore, were reported. Interestingly, the first crystals of the native GFP isolated from *A. victoria* jellyfish were already reported in 1974 [11], and X-ray diffraction was recorded in 1988 [12]. However, it took many years before its structure was finally solved, using multiple isomorphous replacement and anomalous scattering [4, 6] and molecular replacement [5]. The results showed that GFP has a beta-barrel structure consisting of 11 beta strands with the light absorbing *p*-hydroxybenzylidene-imidazolidinone chromophore fully embedded within its core (Fig. 1). The availability of the structural coordinates started the rational design of directed mutations that alter the spectroscopic properties.

GFP exists predominantly in the neutral phenol state, with the phenolic *p*-hydroxybenzylidene-imidazolidinone chromophore protonated. The dominant but partial stabilisation of this species GFP_A ($\lambda_{max} = 398$ nm) at pH 8.0 results from electrostatic repulsion by an acidic group, identified from X-ray crystallography to be the carboxylate of Glu222 [5, 6]. The thermal equilibrium properties of proton transfer reactions include a plateau in the pH titration in the pH 7–10 region, which is predicted from a four-state system that describes the interaction of the two ionisable groups [13]. A comparison of the molecular interactions in the chromophore region of the wild-type (Fig. 2a) and the S65T mutant (Fig. 2b) highlights the stabilised neutral GFP_A and anionic GFP_B states, respectively. The stabilisation of the neutral and anionic forms in these proteins is central to understanding the spectroscopic details as well as the details of the photoconversion and the fluorescence reactions, which will be reviewed.

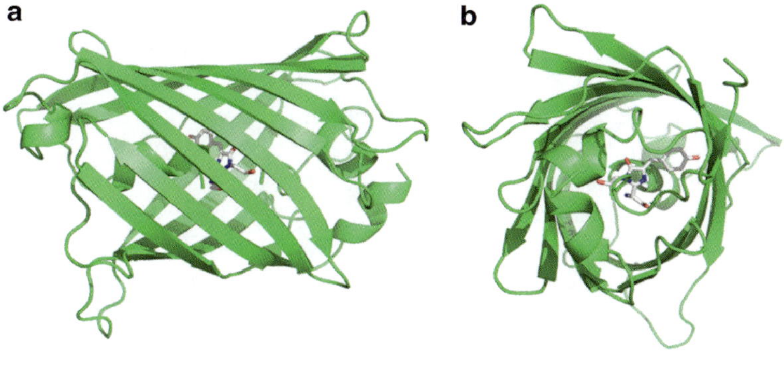

Fig. 1 (**a**) Crystal structure of the green fluorescent protein, drawn from 1GFL.pdb [6]. (**a**) A "side view" of the secondary structure cartoon representation and (**b**) "end-on view", showing the *p*-hydroxybenzylidene-imidazolidinone chromophore buried inside the beta-barrel

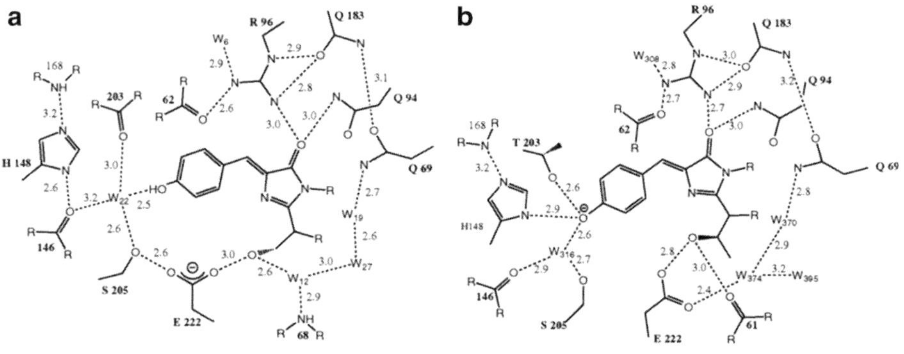

Fig. 2 Comparison of the chromophore environment of the wild-type (**a**) and the S65T mutant (**b**) of GFP. Reproduced with permission from [5]. In the wild type (**a**), the phenol group is predominantly protonated, identified as the GFP_A ($\lambda_{max} = 398$ nm) state. In the S65T mutant (**b**), the H-bonding network is different such that in contrast to the wild type the acidic Glu222 side chain is not in indirect H-bonding contact with the phenolic oxygen, which leads to stabilisation of the anionic GFP_B ($\lambda_{max} = 476$ nm) state

1.2 The Fluorescence Photocycle of GFP

It has been known since the early purification and characterisation of native GFP from jellyfish that fluorescence emission with blue excitation is green, and hence that the Stokes shift for the GFP_A state is very large, in contrast to that of the GFP_B state [11, 14, 15]. The absorption spectrum of the wild-type GFP shows peaks at 280, 398 and 476 nm, which are also observed in the fluorescence excitation spectrum (Fig. 3). The maxima at 398 and 476 nm belong to the neutral and the anionic chromophore, respectively, indicating that at pH 8.0 two species exist, the neutral form being the dominant population. The amplitude at 398 nm has about three times the amplitude of the 476 nm maximum, but the latter is associated with an extinction coefficient that is twice that of the neutral form. Therefore, the molar ratio of neutral to anionic chromophores is approximately 6:1 [16]. A structural basis for this heterogeneity was proposed by Sixma and co-workers [5]. It was noted that the neutral form is stabilised by indirect electrostatic repulsion from the acidic carboxylate group of Glu222, via an H-bonding network including Ser205 and a crystallographic water molecule (Fig. 2b). The green emission with blue excitation of the GFP_A neutral state is illustrated in Fig. 3, which indicates with double arrows both the Stokes shifts for the GFP_A and GFP_B states at room temperature (Fig. 3). While the fluorescence emits from the radiative deprotonated species (GFP_I*) in both cases, the species formed with GFP_A excitation is a more relaxed form (Fig. 3).

Pivotal work by Steven Boxer in 1996 studied the ultrafast fluorescence spectroscopy of the GFP_A state, which led to the discovery of excited state proton transfer (ESPT) in GFP [16]. Chattoraj et al. performed fluorescence upconversion measurements to show that after blue (400 nm) excitation of GFP_A, emission from the excited neutral species GFP_A* detected at 460 nm interconverts on picosecond

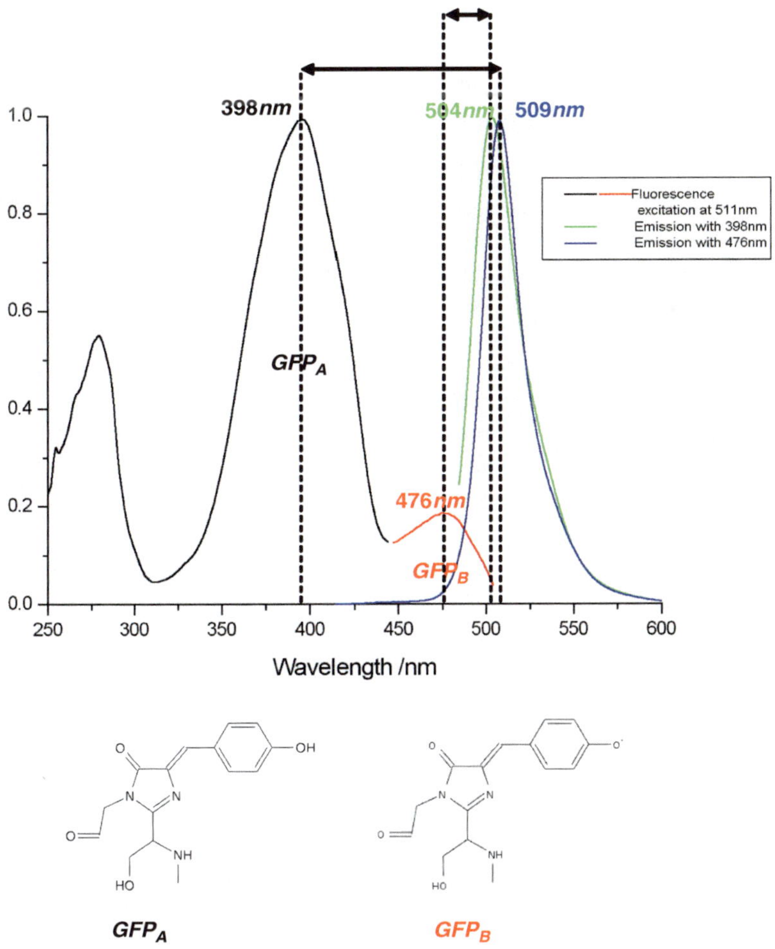

Fig. 3 Room temperature fluorescence properties of wild-type GFP at pH 8.0. The fluorescence excitation spectrum measured at 511 nm shows maxima at 280, 398 and 476 nm, the latter two being assigned to GFP_A and GFP_B species that have a neutral and anionic *p*-hydroxybenzylidene-imidazolidinone chromophore, respectively. The fluorescence emission of the GFP_A and GFP_B states is maximum at 509 and 504 nm, respectively. The *arrows* indicate the Stokes shifts for the GFP_A and GFP_B states. The spectra are normalised. The ratio of the excitation maxima at 476 and 398 nm with 511 nm detection is less than the ratio of the corresponding absorption maxima, reflecting primarily the difference of the fluorescent quantum yields of the GFP_A and GFP_B states, which are 0.8 and 0.65, respectively [3]

timescale to a redshifted species emitting in the green at 508 nm [16]. This ESPT reaction was shown to be biphasic in H_2O, and in D_2O, with both phases displaying a kinetic isotope effect (KIE) of about 5 [16]. The large Stokes shift as well as the KIE showed that GFP acts as a photoacid, similar to many photoacid compounds normally studied in solution. GFP is unique to show an ESPT reaction in a protein environment, and provides an ordered and directional environment for the proton transfer coordinate. The size of the very large Stokes shift, as an energy term, is taken as an approximation for the loss of proton affinity in the excited state, minus relaxation in the Frank Condon regions [17, 18]. The resulting drop of the pKa of the chromophore with optical excitation drives the ultrafast, picosecond timescale, ESPT reaction.

The ESPT reaction and the details of the fluorescence photocycle are not the focus of this review and will be discussed elsewhere in this volume by Steve Meech. However, the photoconversion reaction competes with the dominant fluorescence photocycle, and its properties must therefore be mentioned in the light of the essential characteristics of the fluorescence photocycle. Specifically, the charge transfer and structural response of both reactions will be discussed in this chapter.

Furthermore, mechanisms for photoconversion that were proposed in the literature are intricately linked with the fluorescence photocycle and the ESPT that occurs during it. Specifically, Sixma and co-workers proposed structural mechanisms for both the ESPT reaction in the fluorescence photocycle and one for the photoconversion reaction [5] (Fig. 4). A similar scheme was also proposed by Palm

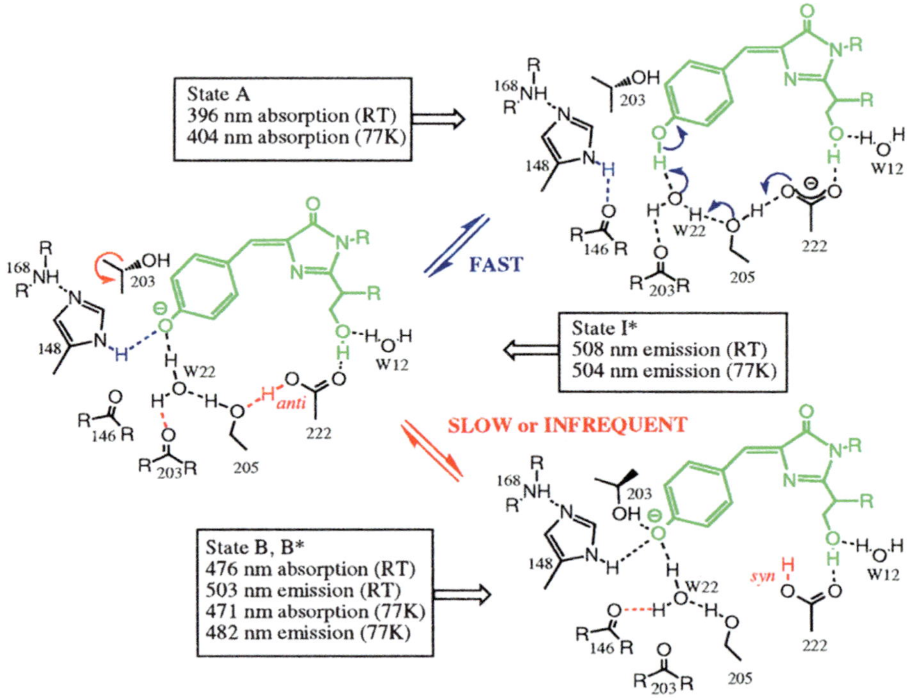

Fig. 4 Proposed mechanism for excited state proton transfer (*blue*) and photoconversion (*red*) in GFP made on the basis of X-ray structural analysis by Sixma and co-workers. Reproduced with permission from [5]. All available evidence agrees with the molecular mechanism proposed for ESPT (*blue*), as summarised here. However, the proposed syn-anti isomerisation of protonated Glutamic acid (*red*) was shown not to be the basis of photoconversion reaction [19, 20], but has since been invoked in the fluorescence photocycle of GFP instead [21–23]

et al. [24]. These proposals were based on the analysis of the X-ray structure and the details of the H-bonding interactions in the chromophore region. In 1997, Brejc et al. proposed that the fluorescence photocycle of the protonated chromophore involves the ultrafast proton transfer from the phenolic oxygen to the carboxylate of Glutamic acid 222 [5].

Evidence from ultrafast infrared spectroscopy later supported this mechanism in detail. Femtosecond optical excitation of the GFP_A state leads to absorption changes of an isolated band, in H_2O at 1,725 cm^{-1} [25–28] which is a specific frequency for protonated carboxylic acid. Based on the single wavelength measurements at a lower frequency of 1,706 cm^{-1}, the assignment of this transient to protonated Glutamate 222 was proposed [29] on the basis of the earlier proposal from Brejc et al. [5]. The assignment was made on the basis of the following four observations. (1) The isotope substitution effect of the mode observed at 1,712 cm^{-1} in D_2O and at 1,725 cm^{-1} in H_2O supports assignment to protonated glutamic acid [28]. (2) The corresponding H/D KIE of the observed ESPT equals 5 [28], which fits the KIE determined by fluorescence upconversion experiments for the ESPT reaction [16]. (3) Spectral and kinetic changes occur in a mutant Glu222Asp, in which the proton acceptor is replaced with a functional Aspartic acid 222 group [27, 28]. In this mutant, ESPT was delayed relative to the wild type with 13 and 125 ps time constants in H_2O and D_2O, respectively, and the COOH/D stretching mode of the carboxylate acceptor showed a reduced amplitude and downshifted frequency compared to the wild type [27, 28]. (4) The infrared anisotropy for the COOD band [27, 29], which was corrected for the finite bleach [27], agreed with the coordinates from the X-ray frame.

The results mentioned above are taken as strong evidence for the model put forward by Brejc et al. already in 1997. This is testimony for the important insight that was obtained from the functional analysis of X-ray structural information of GFP. It is interesting to note that an alternative proposal that is not supported by recent evidence invoking His 148 as the transient ESPT acceptor [6, 30] was also based on X-ray analysis, but made on the basis of different data.

In addition to predicting the identity of the excited state transient proton acceptor, Brejc et al. additionally proposed that subsequent syn-anti isomerisation of transiently protonated Glu222 would stabilise the anionic species and explain the photochromic behaviour [5] (Fig. 4; red). However, this was not supported by FTIR measurements which established the absence of protonation changes of COOH/D groups in GFP upon phototransformation [20]. It was later shown that electron transfer is at the molecular basis of photoconversion rather than structural rearrangements in the chromophore-binding site [19] (see Sect. 2). Instead, syn-anti isomerisation of Glu222 has been invoked in the fluorescence photocycle of GFP [21–23].

More recently, ESPT has also been reported in the red fluorescent protein mutant mKeima from the stony coral *Montipora* sp. [31] and in the LSSmKate1 and LSSmKate2 red fluorescent protein mutants from the sea anemone *Entacmaea quadricolor* [32].

2 Photoconversion of *A. victoria* GFP

2.1 Discovery and UV/VIS Spectroscopic Investigations of Photoconversion of GFP

Photoconversion of GFP was discovered shortly after the cloning of the *gfp* gene [1, 7]. Photoconversion is found to occur in a range of different optical regimes, using UV and visible light illumination, and with femtosecond and nanosecond pulses as well as continuous illumination. The report by Chalfie et al. [1] on recombinant expression of the gfp gene included a note: "Indeed, the fluorescence produced by 450- to 490-nm light appeared to be more intense after brief photobleaching by 340- to 390-nm light" [1]. A first spectroscopic characterisation of this photochromic reaction appeared in 1995 from the laboratory of Roger Tsien [7] (Fig. 5) This paper showed the fluorescence changes with UV illumination at 280 nm. Cubitt et al. [7] proposed photoisomerisation as a mechanism and found the spectroscopic changes to be irreversible. The fluorescence intensity changes in Fig. 5 indicate the occurrence of both photobleaching and photoconversion under the conditions used, considering that at pH 8.0

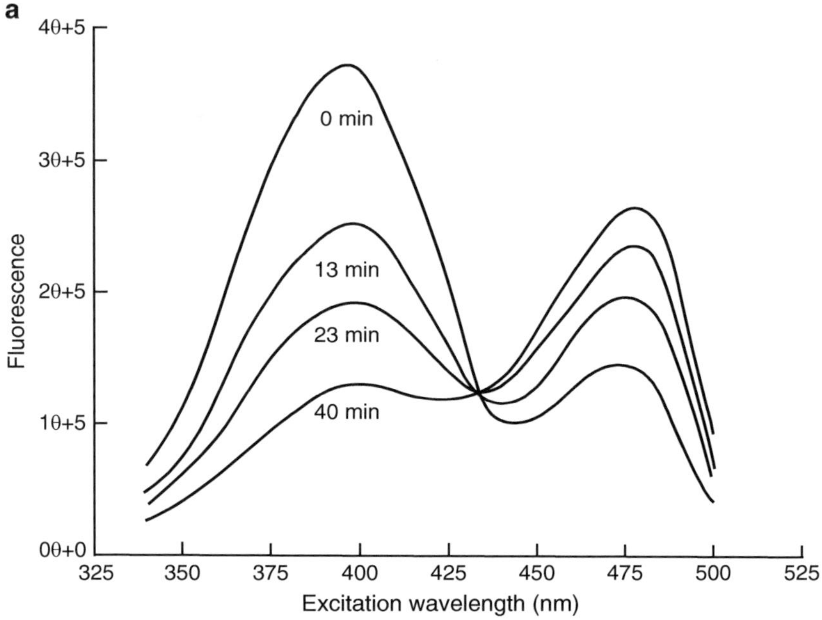

Fig. 5 First spectroscopic demonstration of photoconversion of GFP. Reproduced with permission from Cubitt et al. [7]. Selected figure caption: "Behavior of wild-type green fluorescent protein (GFP) upon progressive irradiation. GFP samples were illuminated at 280 nm from a xenon lamp and monochromator. Wild-type GFP suffered photoisomerization, decreasing the excitation amplitude at 395 nm while increasing the amplitude at 475 nm. This effect was not reversible upon placing the GFP in the dark" [7]

the cross-section of the phenolate species is twice that of the neutral species and taking into account the lower fluorescence quantum yield of the anionic species relative to the neutral state. The ratio between the photobleaching and photoconversion rate constants is pH dependent and showed a minimum at pH 8.0, from comparing the photogenerated absorption at 483 nm relative to ground state absorption decrease at 398 nm [19]. Under these conditions, a close to stoichiometric photoconversion of the neutral species to the anionic GFP_R species could be achieved with UV illumination at 254 nm [19].

Cubitt et al. already noted that the photoconversion can be used to monitor diffusion or migration of GFP fusion proteins after spatially localised photoisomerisation [7]. Indeed, such approaches have been taken very successfully [33–35].

The kinetics of photoconversion were investigated between 102K and 293K, but very little temperature dependence could be observed. Assuming Arrhenius behaviour, an activation energy of about 1 KJ/mol could be estimated [19]. However, performing photoconversion at cryogenic temperatures revealed thermal relaxation processes that are characterised by blue-shifted electronic absorption of the anionic photoproduct state. These relaxation processes were studied by time-resolved cryo X-ray crystallography and cryo-FTIR spectroscopy [26].

Both UV and visible (blue and green) light lead to photoconversion of GFP. However, photobleaching is significant under conditions of pulsed blue excitation. Figure 6 compares the absorption changes under visible (390 nm) and UV (254 nm) excitation. Kinetic analysis of the macroscopic photoconversion rate as a function of the power density of the nanosecond pulsed laser at 390 nm indicated a linear relationship with regard to the photoconversion rate [19, 36]. This shows that under the optical regime that was used, the reaction proceeds via a one photon process. Photoconversion also proceeds with intense continuous illumination at 476 nm [37], or femtosecond pulsed excitation at 478 nm [16], directly exciting the minor anionic GFP_B species. The quantum yield appears to be approximately 100 times below that with 254 nm excitation [37]. Two- or multiphoton photoconversion using 800-nm pulses was also reported [38].

An interesting and currently unresolved issue regarding blue light-induced photoconversion is that its rate constant may not depend linearly on the power density when continuous illumination is used, in contrast to nanosecond pulsed excitation. For this particular regime, the evidence available in the literature is less complete, and only single observations are reported. For example, Patterson and Lippincott-Schwarz observed second timescale photoconversion rates with 413 nm continuous excitation from an ion laser [35]. The power used was reported (1 mW), but not the spot size. When alternatively a 100 W Hg^+ lamp together with a 405-nm bandpass filter was used, photoconversion proceeded in ~60 s in their microscope setup [35]. Again, no power density was reported, but the power density effect of continuous illumination is clear. However, weaker blue illumination at 404 nm and 6 mW/cm^2 has been reported to be ineffective [37]. In the absence of a systematic investigation on the power dependence over these different regimes, no conclusion is presently possible. Clearly, two photon photoconversion with CW illumination at

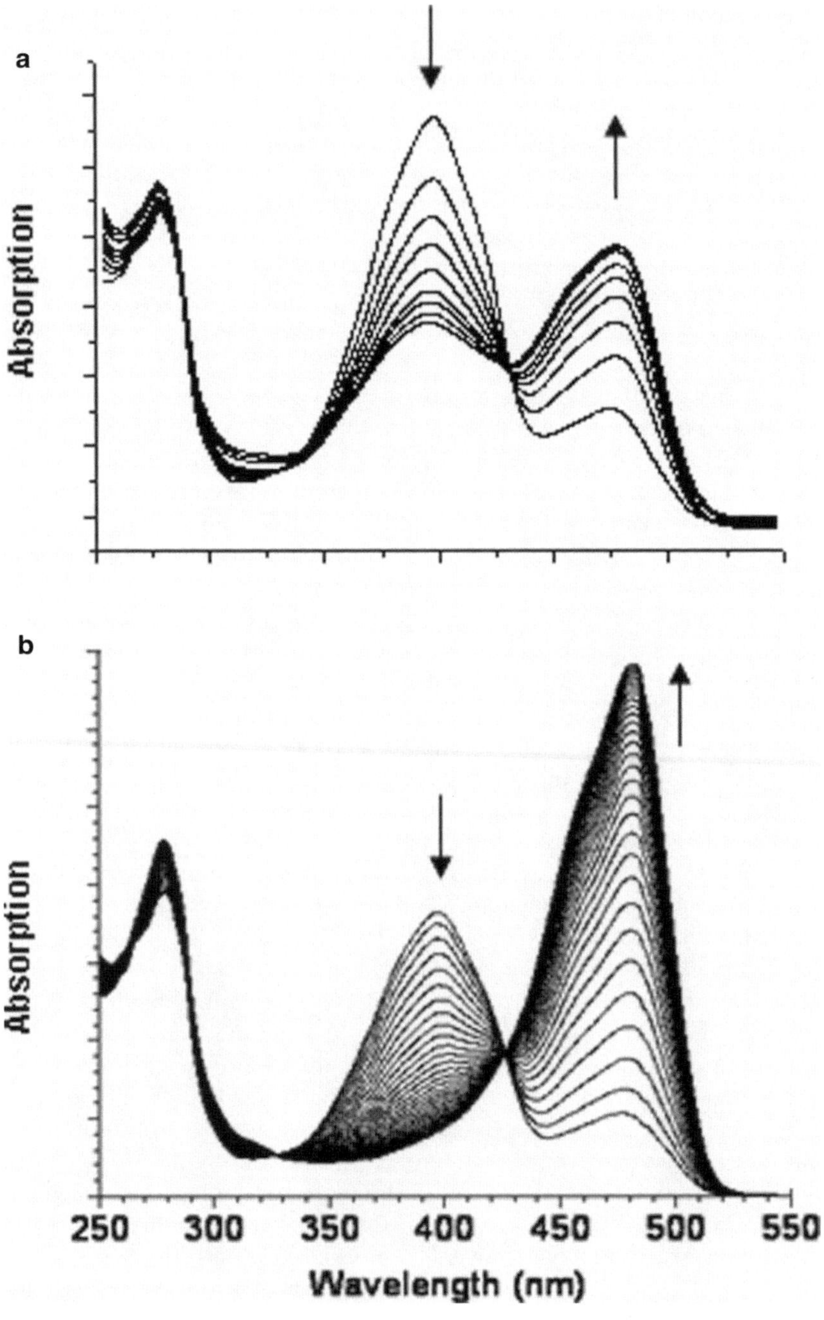

Fig. 6 Light-induced absorption changes of wild-type GFP under conditions of (**a**) nanosecond pulsed 390 nm excitation and (**b**) continuous 254 nm illumination, at pH 8.0 and 293 K. (**a**) Using nanosecond pulses at 5 Hz repetition rate and pulse energy of 1–9 mJ in an approximately 1 cm

400 nm would indicate the presence of a long-lived intermediate, possibly a triplet state, which has so far not been reported.

One study carefully characterised the photoconversion of a T203V mutant of GFP that lacks the GFP_B anionic species, with femtosecond excitation at 400 and 800 nm (via two photon absorption) [39]. In the range of 0–5 $\times$ 10^9 W cm^{-2} and 0–6 $\times$ 10^{11} W cm^{-2} for 400 and 800 nm 80 fs pulses, the power dependency for photoconversion was found to have reaction orders with exponents of 2.06 and 2.76, respectively. This quantitative information obtained in the high-power regime indicated that also with blue excitation a two photon process is more efficient than a one photon process. Given the large difference in quantum yield for continuous UV and blue illumination, this result at 400 nm appears to be in agreement with these observations. An alternate suggestion that for continuous illumination, in the low power regime, the wavelength dependence of photoconversion reflects two photon excitation of the chromophore [37] appears to be inconsistent with the linear power dependence at nanosecond pulsed excitation at 390 nm [19] and the observation of photoconversion with CW illumination at 413 nm [35]. A higher lying excited state, probably the S2 state, is considerably more efficient in triggering photoconversion than the S1 state [19]. The steady-state fluorescence anisotropy measurements identified the higher lying optical transition of the chromophore within the excitation envelope in the UV region and explain the high efficiency of photoconversion with 254 nm excitation via direct excitation of the chromophore [19]. The reaction order of 2.7 observed with 800 nm femtosecond pulses [39] could represent the effects of three photon excitation, at 267 nm, being more efficient than two photon excitation, at 400 nm [39]. In agreement with this proposal, it was observed that one 400 nm pulse in addition to one 800 nm pulse triggered photoconversion efficiently. Moreover, the photoconversion efficiency of the second pulse tracks the lifetime of the GFP_A* (S1) state, with scanning of the delay between the 400 and 800 nm pulses between 0 and 50 ps [39]. These results are fully consistent with the reported wavelength dependency with CW illumination [19], but do not explain the apparent lack of photoconversion reported with low power at 400 nm [37].

However, a similar behaviour was reported for the photoconversion of DsRed [40]. Although thought to involve both chromophore photoisomerisation and oxidative decarboxylation of Glu215 (see Sect. 2.3), evidence for a two photon process was provided from the power density analysis for pulsed as well as for CW irradiation. With regard to the latter observation, the involvement of an intermediate with microsecond life time would be needed to explain such behaviour, suggested to be either a triplet state, an isomerised form or a different ionic state [40].

Fig. 6 (continued) spot size, the observed rate constants were between 2 $\times$ 10^{-3} and 1.6 $\times$ 10^{-2} s^{-1}, showing a linear dependence of the rate constant with power density. The calculated photoconversion quantum yield at 390 nm was 1.6 $\times$ 10^{-3} [36]. Note that no isobestic point is present, indicating the contribution of photobleached product state in the mixture using pulsed blue excitation. (**b**) Using a 254 nm UV lamp, photoconversion proceeds with significantly higher quantum yield, estimated at 0.03 [19, 36]. Isobestic points at 325 and 426 nm are observed with UV illumination

Although photoconversion of DsRed may not be directly compared with that of GFP, the reported non-linear power density dependence of photoconversion with CW illumination in both cases may point at a similar mechanism. However, it should be emphasised that in the nanosecond pulsed regime, power density dependence is linear in the case of GFP [19] but not for DsRed [40].

Photoconversion is irreversible, as was already observed by Cubitt et al. [7]. However, after photoconversion very small (1–2%) absorption changes occur on hours–days time scales that are only a fraction of the phototransformation changes [19]. These occur at 293 K but not at cryogenic temperature, and are possibly caused by slow structural relaxations of the protein environment that have small effects on either the chromophore pKa or the cross-section [19]. In contrast, one report claims the occurrence of significant thermal reversion with 60% recovery of the lost absorption at 398 nm after 24 h dark incubation at 293 K [16]. This observation of reversibility has, however, not been confirmed since. It did, however, lead to a proposal for a specific structural mechanism for photoconversion that took the reversibility into account, which included the proposed syn-anti isomerisation of protonated Glu222 (Fig. 4) [5]. It is unlikely that non-photochemical reversal occurs after photoconversion [16]. It may be speculated that a contribution of slow mixing of unexposed material into the probed volume may have contributed to the measurements that were conducted in viscous samples containing glycerol [16]. Presently, there is agreement in the literature that photoconversion of GFP is fully irreversible.

With UV illumination, the photogenerated absorption at 483 nm is approximately twice the absorption decrease at 398 nm, which corresponds to a stoichoimetric conversion (Fig. 6b). With pulsed excitation at 390 nm several fold less product absorption at 483 nm is observed, resulting from a significant photobleaching process. Mass spectrometry analysis of fully photobleached samples after prolonged exposure to 390 nm nanosecond pulses indicated the presence of multiple fragments with reduced mass, showing several different cleavage reactions to be responsible for the photobleaching reaction. These observations could indicate the complete loss of the chromophore from the protein.

2.2 *Mechanism of Photoconversion of GFP*

Both UV and visible light transform the species with a neutral, phenolic chromophore that absorbs at 398 nm (GFP_A) into an ionic species absorbing at 483 nm (GFP_R), with a phenolate-containing chromophore [20]. Vibrational spectroscopy provided essential information about the chromophore changes during photoconversion. Difference FTIR measurements of the GFP_R and the GFP_A state confirmed the deprotonation of the chromophore as the molecular basis of the photochromic reaction [20] (Fig. 7). Prominent product bands in the mid-infrared region at 1,497 and 1,147 cm^{-1} are characteristic of the anionic phenolate group, which confirmed and established the deprotonation of the chromophore as the basis of

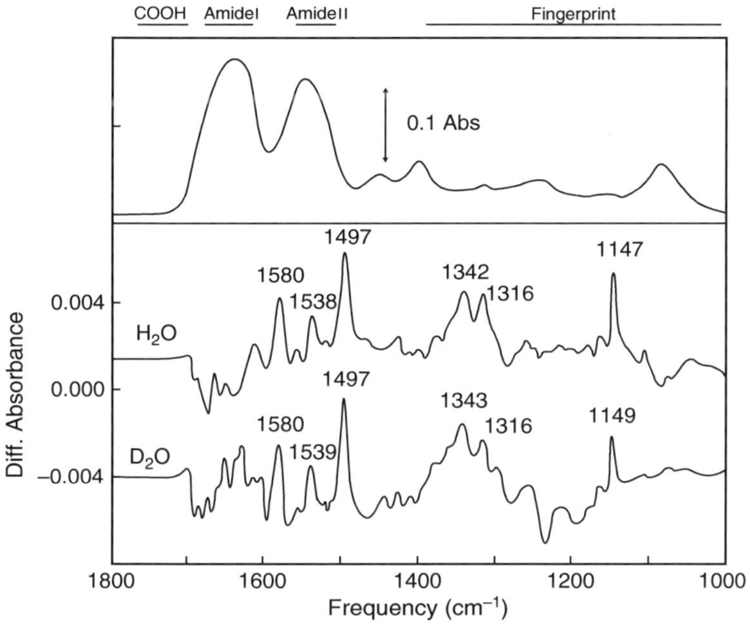

Fig. 7 Reaction-induced difference FTIR spectroscopy characterises the photoconversion of GFP. Reproduced with permission [20]. Figure legend: "Absolute FTIR spectrum of GFP in water (*upper panel*) and FTIR difference spectra (*lower panel*) between neutral and anionic species of GFP (i.e. GFP_A-GFP_R) in H_2O (*middle*) and D_2O (*bottom*), showing the carboxyl, amide I, amide II and fingerprint regions. An absolute spectrum (from 1,800 to 1,000 cm^{-1}) was obtained of the neutral form of GFP"

photoconversion and the associated strong red-shifted absorption [20]. Isotope substitutions and density functional theory calculations on a model compound for the GFP chromophore [41] suggest assignment of positive bands associated with vibrations of the phenol group (1,497 and 1,582 cm^{-1}), the imidazolidinone ring (1,537 cm^{-1}), and the bridging carbon (1,615 cm^{-1}) of the anionic chromophore in the room temperature IR difference spectrum [28, 41].

A direct test of stable proton rearrangements such as proposed by Brejc et al. (Fig. 4) by difference FTIR spectroscopy turned out negative, from direct measurement of absorption changes belonging to COOH stretching vibrations in the ~1,720 cm^{-1} region [20]. A quantification of the GFP_R–GFP_A difference spectrum showed that less than 0.1 carboxylate groups change protonation upon stable photoconversion [20].

Insight into the mechanism of photoconversion of GFP came from the discovery that covalent modification occurs upon UV or visible illumination. Specifically, illuminated samples were shown to be reduced in mass by 44 Da, suggesting the loss of a $-CO_2$ group [19] (Fig. 8). Structure solution of the purified photoproduct by X-ray crystallography revealed the loss of the carboxylate group of the Glutamate 222 side chain [19] (Fig. 9). In the GFP_R photoproduct, coordinate analysis suggests

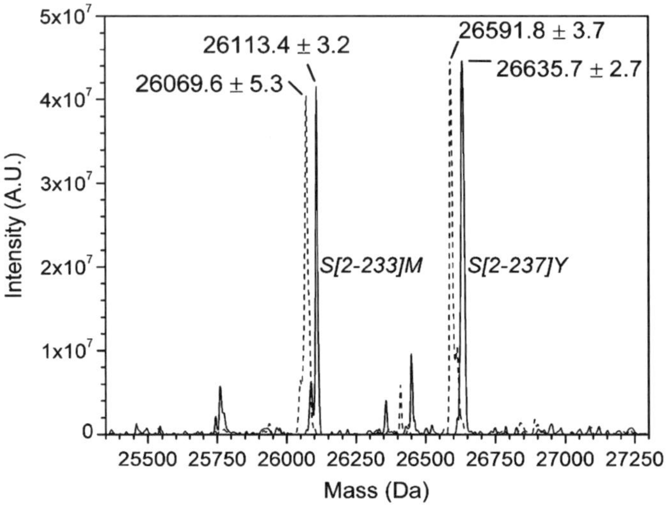

Fig. 8 Photoconversion results in a mass reduction of 44 Da. "Maximum entropy treated ESI mass spectrometry results are for GFP_A (*solid trace*) and the photoproduct GFP_R (*dotted trace*). The calculated values for the fragments Ser 2–Met 233 and Ser 2–Tyr 237 are 26,116.4 and 26,637.0 Da, respectively, taking into account a dehydration resulting from chromophore formation. A mass reduction of 43.8 and 43.9 Da, respectively, was observed for the fragments upon photoconversion" (reproduced with permission [19])

that a new H-bonding network replaces the H-bonding network in the ground state (Fig. 9). The removal of the acidic side chain in the photoproduct destabilises the protonated chromophore. This agrees with the model for the interactions in the GFP_A ground state proposed by Brejc et al. [5]. In order to test for the possible accumulation of a radical state, electron spin resonance (ESR) measurements were conducted. Under conditions of continuous illumination with 254 nm light at 200 and 293 K, no photochemically produced radical products or free spin intermediates could be detected by ESR spectroscopy, neither directly nor by spin-trapping using the spin traps 5,5-dimethyl-1-pyrroline-*N*-oxide (DPMO) and 2,2,6,6,-tetramethyl-4-piperidone (TEMP). This places a limit on the lifetime of any intermediate radical species from the rate of optical excitation, estimated at approximately 5 s^{-1}, multiplied by the photochemical quantum yield, 0.03, which equal ~0.15 s^{-1}.

The experimental observations show that light-induced electron transfer leads to oxidation of the anionic carboxylate of Glu222, which is proposed to proceed via direct transfer to the chromophore. Both the first (S1) and second (S2) electronically excited state are electron deficient singlet states, and the S2 state is expected to have the largest oxidising potential (see also below). Stark spectroscopy has shown that the imidazolidinone ring is electron deficient in the excited state of the neutral, phenolic chromophore [16, 42]. Assuming that the imidazolidinone ring, in particular the carbonyl (O2) oxygen atom of the phenolic chromophore, forms a strong oxidising moiety in the excited state, a mechanism can be proposed for the light-activated decarboxylation reaction: Upon excitation of the chromophore of GFP,

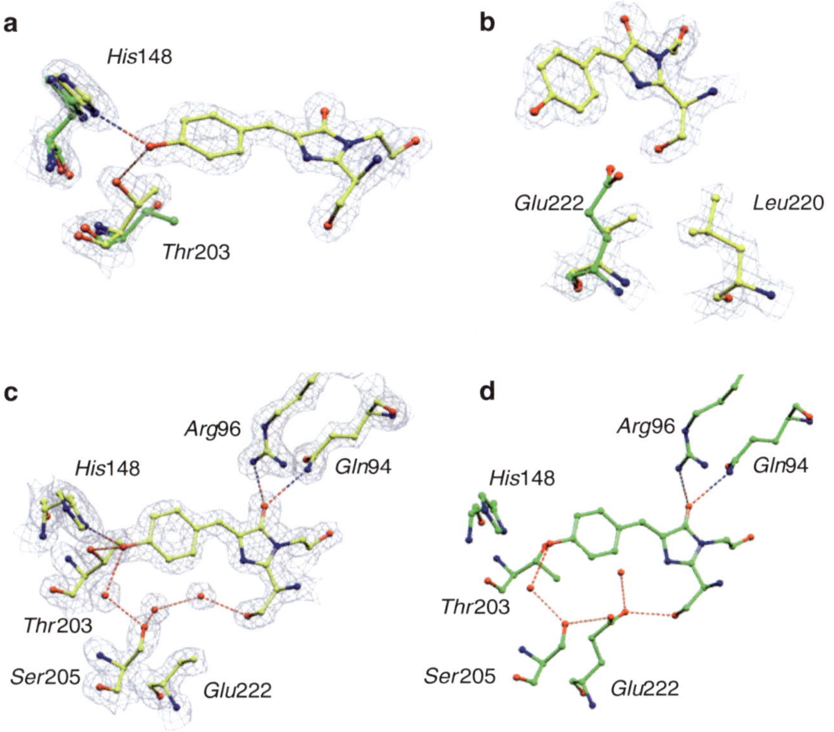

Fig. 9 "Photoinduced decarboxylation and conformational changes in the chromophore vicinity. Coordinates of the wild-type structure (PDB 1gfl) are shown with carbon atoms in *green*, whereas the structure of the photoproduct GFP_R is shown with carbon atoms in *yellow*. Electron density shown is the refined 2Fo–Fc map contoured at 1.6 σ for the 1.8 Å resolution structure. (**a**) Conformational changes of Thr 203 and His 148. (**b**) Absence of electron density for the $O\varepsilon1$, $O\varepsilon2$ and $C\delta$ atoms of Glu 222 in GFP_R. (**c**) Proposed hydrogen bonds in the chromophore vicinity of chain A in the GFP_{483} structure include solvent molecules 173[Z], 175[Z] and 203[Z]. (**d**) Chromophore structure in GFP_A. The hydrogen-bonding network includes solvent atoms 25 and 63" (reproduced with permission [19])

transfer of charge between the phenolic- and imidazolidinone rings of the chromophore will compete (efficiently) with electron transfer from Glu 222 to the chromophore. If an electron is transferred, the γ-carboxyl radical of Glu 222 will decarboxylate via the "Kolbe" mechanism [19]. This then leads to retro-transfer of an electron and a proton, or alternatively a hydrogen radical (Fig. 10).An experimental test for the accumulation of a radical intermediate was negative, which shows that at least at 200K, the lifetime of the carboxylate radical intermediate as well as the predicted chromophore radical intermediate is less than ~7 s. Considering the distances involved, these processes can proceed in the ns time domain, which is also in agreement with the low quantum efficiency of the reaction. Together with the low- and room-temperature photoconversion experiments, a two-step mechanism of the photoconversion of GFP_A to GFP_R was proposed.

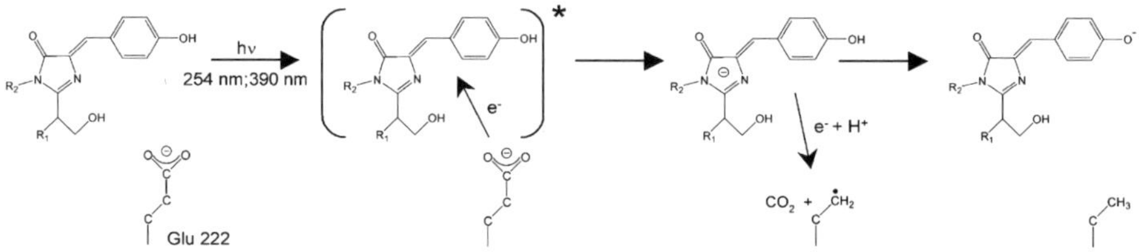

Fig. 10 Proposed mechanism for light-induced decarboxylation of Glu 222. R1 and R2 correspond to residues Phe 64 and Val 68. Asterisk denotes the short-lived ($\tau = 3.6$ and 12.0 ps[9]) excited state of the neutral, phenolic chromophore. The actual distances between the γ-carboxylic group of Glu 222 and the phenolic- and imidazolidinone rings are 3.7 and 3.8 Å, respectively. A CH_2·radical intermediate state is expected from the "Kolbe" mechanism [43, 44]. Recombination of this electron deficient carbon radical after decarboxylation may involve direct transfer of an electron in addition to transfer of a proton from the chromophore, or alternatively transfer of a hydrogen radical. Reproduced with permission [19]

First, decarboxylation of Glu 222 takes place (Fig. 10). This is followed by structural rearrangements of nearby amino acids Thr 203 and His 148 (Fig. 9). These thermal relaxation reactions blue-shift the absorption maximum [19]. Heating the stable photoproduct obtained at 92 to 293 K produced a photoproduct indistinguishable from that produced at 293 K. A slow transformation was already observed in real time (minutes) at 182 K. Re-cooling the product to 92 K, or cooling the photoproduct obtained at 293 to 92 K, produced a blue-shifted species absorbing at 452 and 477 nm. This indicates that at higher temperatures, a thermal relaxation process, such as a structural reorganisation, follows the primary photoconversion process, producing GFP_R with a more strongly hydrogen bonded phenolate chromophore as shown by the X-ray structure [19]. Structural changes associated with thermal transitions were revealed by X-ray crystallographic structure determination of intermediates after photoconversion of crystals of wild-type GFP [26]. In addition, spectroscopic changes in the visible and infrared region are observed, which has led to the identification of discrete intermediates in this pathway. The naming for the intermediates follows GFP_A (+hν) → GFP_L (100K) → GFP_M (200K) → GFP_R (293K), for "Lumi", "Meta" and "Relaxed". These intermediates are cryo-trapped, or accumulated, at the temperatures indicated. Detailed discussion of the spectral and structural characteristics of the intermediates [26] is beyond the scope of this review. In summary, structural changes of individual amino acids and solvent molecules occur in the discrete intermediates, and the available vibrational spectroscopy has led to assignment of modes to individual amino acids such as Gln69. Its C=O mode is observed at 1,696 cm^{-1} in the low-temperature product GFP_L, signalling structural perturbation of the amino acid side chain [26]. A key observation from ultrafast infrared spectroscopy was that the 1,696 cm^{-1} mode belonging to Gln69 perturbation is also observed, but reversibly, in the fluorescence photocycle. Interestingly, in H_2O the transient is developed within 1 ps, which demonstrates one of the most rapid amino acid structural reactions documented.

Small molecule activation and migration in proteins is an important topic, which has, for example, been addressed in detail in myoglobin [45–47] The photogenerated

CO_2 molecule in GFP provides a new opportunity to characterise the interactions of a single molecule of CO_2 in a specific protein environment [26]. After photoconversion, bound and unbound populations of photogenerated CO_2 have been observed and characterised by X-ray crystallography and cryo-infrared crystallography in the GFP_L, GFP_M and GFP_R products (at 100, 200 and 293 K). Electron density assigned to the CO_2 molecule in the GFP_L state at 100 K disappears at higher temperatures, indicating the loss of occupancy of the binding site identified, H-bonded to Ser205. The CO_2 vibrational signals evolve over the 100–200 K temperature range, coalescing into a single band at 2,338 cm^{-1} at 200 K. The appearance of multiple overlapping CO_2 bands suggests significant disorder for the photogenerated CO_2 already at 100 K. The observed disorder of the CO_2 molecule underlines the unexpected mobility of the chromophore environment at 100 K. Polarised IR data on oriented single crystals indicated a minimum of three CO_2 environments with distinct frequencies and orientations and additionally characterised the direction and changes of the Cys 70-Val 68 hydrogen bond during photoconversion [26].

Some conclusions can be drawn with regard to the energetics of the electron transfer reaction from Glu 222 to the chromophore. An in-depth characterisation of the thermodynamics of electron transfer is still not available, however [36]. Principally, the energy levels of the S1 and S2 states of the neutral and anionic chromophore in intact GFP relative to the anionic carboxylate of Glu222 are not known, from either experimental parameters or simulation and calculation [36]. This identifies an important area for future computational chemistry and theory. Considering the S1 state of the neutral chromophore, by making the assumption that electron transfer competes with the ESPT reaction in a parallel reaction scheme, the forward rate constant may be estimated. The singlet excited state of the protonated species decays with time-constants of 3 and 15 ps to form a longer lived (3 ns) deprotonated radiative state [16, 48]. From the apparent photoconversion quantum efficiency with excitation at 390 nm of 1.6×10^{-3}, a forward electron transfer rate of $2.4 \times 10^8\ s^{-1}$ is estimated [36]. For the anionic singlet state, a 3 ns lifetime would be considered, and the quantum yield is about ten times below that for the neutral state, leading to an estimated forward rate constant of ~$5 \times 10^4\ s^{-1}$. This suggests that the electron transfer rate is enhanced with increasing the free energy difference and/or reducing the reorganisation energy. However, the free energy difference cannot be determined as the back-transfer rate is not known.

Electron transfer in redox proteins is described by Marcus theory [49–51]. The non-adiabatic description of electron transfer uses "Fermi's golden rule" which provides the transfer rate. Both classical [50] and quantum mechanical [49] treatment of Marcus electron transfer theory predict that the dependence of the electron transfer rate, k_{et}, on the free energy ΔG leads to a maximum when the free energy equals the reorganisation energy ($\Delta G = -\lambda$). Empirical expressions have been derived for electron transfer in proteins that make predictions with regard to the, first order, forward electron transfer rate and the free energy and reorganisation energy of electron transfer [52]. For groups that are in van der Waals contact, the optimised k_{et} (when $\Delta G = -\lambda$) is expected to approach $10^{13}\ s^{-1}$ [52]. In GFP, the edge-to-edge distance between the Glu222 carboxylate and the phenolic ring of the chromophore is 3.5 Å, which is close to van der Waals distance. The electron transfer rate constant is estimated to be $2.4 \times 10^8\ s^{-1}$ which is five orders of magnitude below the optimal rate. This suggests that the free energy difference is very different in magnitude compared to the reorganisation energy. It is likely that the reaction has a very small ΔG or is endoergonic, in which case it would be expected that either a small decrease in ΔG or a small decrease in λ will result in an increase in the rate and in the photoconversion quantum yield [36]. This is consistent with the increased photoconversion quantum yield, and the electron transfer rate, with a shorter wavelength (254 nm UV illumination) that is resonant with a higher order optical transition of the chromophore [19].

2.3 Light-Induced Oxidative Decarboxylation in Other Fluorescent Proteins

GFP was the first known example showing oxidative decarboxylation of a carboxylic acid in a protein [19]. It has now been documented to occur also in mutants of GFP and in related fluorescent protein. Currently, besides the fluorescent proteins no other examples of oxidative decarboxylation are known in proteins. In GFP, oxidative decarboxylation of Glu222 results from the light-induced formation of an oxidative state. Similar reactions are well known in chemistry and are referred to as the "Kolbe reaction" [44]. Thermal oxidative decarboxylation of carboxylic acids can be catalysed by oxidants such as Mn(III) [43]. In redox proteins with strong oxidising centres, oxidative decarboxylation of nearby ionised carboxylic groups may also be expected to occur. Many proteins contain redox centres such as Mn(III) or Fe(III) that are coordinated by ionised γ-carboxylates. Depending on the specific redox potentials and lifetimes of the intermediate oxidative states and the γ-carboxylates, and resulting quantum efficiency, oxidative decarboxylation may limit the number of turnovers that such enzymes can make.

Since the discovery of light-induced decarboxylation in GFP, it has also been documented in mutants of *A. victoria* GFP and has been found in at least two other fluorescent proteins, DsRed and an aceGFP mutant. An *A. victoria* GFP mutant called yellow fluorescent protein (YFP) (10C variant, including mutations S65G, V68L, S72A and T203Y) undergoes photobleaching with strong CW illumination at 514 nm into the low energy absorption band ($\lambda_{max} = 527$ nm) [53]. Concomitant with bleaching, an absorption peak at 390 nm is produced which is a photoactive species. The report [53] mentions an additional product band at 470 nm, which could indicate that the 390-nm band is a higher optical transition. Subsequent excitation of this band at 390 nm causes photoactivation, recovering some of the fluorescence lost in the bleaching process. It is reported that the photobleached product recovers spontaneously in the dark on an hour timescale when CW green illumination was used, but when pulsed excitation was used the photoproduct was irreversibly bleached. Moreover, in this case a loss of 44 Da, indicating the likely decarboxylation of a carboxylate side chain, could be demonstrated [53]. The site of decarboxylation could also be at Glu222, as in the wild type. The fact that only pulsed excitation causes decarboxylation would suggest that light-induced electron transfer in YFP is triggered by a multi-photon process.

Another *A. victoria* GFP mutant has been reported to undergo classical $GFP_A \rightarrow GFP_R$ photoconversion involving decarboxylation. A T203H mutation shows no ground state population of the minor anionic species that is observed in the wild type and only shows the GFP_A neutral species absorbing at 398 nm [35], which is rationalised on the basis of an altered orientation of the Glu222 carboxylate [54]. Therefore, the spectroscopic contrast with photoconversion is enhanced when the anionic form is monitored. This improved contrast has obvious utility for fluorescence microscopy applications [35]. The mutant is referred to as "photoactivatable GFP" (PA-GFP) and includes secondary mutations F99S/M153T/V163A that do not appear to affect the spectroscopic properties [34, 35, 55, 56]. With regard to the photoconversion mechanism, it was shown to be the same as that of the wild type [54].

The *Aequorea coerulescens* aceGFP mutant "photoswitchable cyan fluorescent protein" (PS-CFP) was developed as an alternative to "PA-GFP" for protein tracking applications [57–59]. Photoconversion switches the fluorescence extinction and emission maxima from 402 and 468 nm to 490 and 511 nm, similar to PA-GFP. However, the fluorescence quantum yield of the neutral and anionic species is 0.16 and 0.19, respectively, and similar extinction coefficients are reported for both. While PA-GFP is significantly brighter, the difference of the fluorescence emission of neutral and anionic species can lead to improved contrast in emission [57, 59]. Although not fully characterised, light-induced electron transfer and oxidative decarboxylation is proposed to occur in PS-CFP as in PA-GFP and wild-type GFP. [57–59]. Although an estimate of the quantum yield is not available, conversion was reported to occur efficiently with 10 ns pulses of 15 μJ energy at 404 nm with no spot size reported [59]. Therefore, the photoconversion quantum yield of PS-CFP could possibly be higher than that for GFP at 400 nm.

The fluorescent protein DsRed from the coral *Discosoma* sp. undergoes photoconversion of the red ground state (R) absorbing at 559 nm and emitting at 583 nm with a pulsed laser at 532 nm into a red-shifted super red (SR) species, absorbing at 574 nm and emitting at 595 nm. The fluorescence quantum yield of the photoproduct is 0.012, in contrast to the bright R state which has a fluorescence quantum yield of 0.7. Photoconversion of DsRed is complex and thought to involve *cis–trans* isomerisation as well as oxidative decarboxylation, proposed to be of Glu215, the equivalent of avGFP Glu222. In addition, a minor photoproduct that has a protonated chromophore is observed [40]. Mass spectrometry and FTIR spectroscopy suggest the oxidative decarboxylation of a carboxylic acid in the anionic form from the loss of bands of 1,568 and 1,389 cm^{-1}, proposed to belong to the COO^- stretching modes [40], similar to those observed in GFP [20]. The proposed oxidative decarboxylation likely follows the same mechanism as in GFP, via the Kolbe mechanism [19]. A complex and interesting mechanism is proposed for the photoconversion of DsRed. The cavity created after decarboxylation of the nearby Glu215 is proposed to allow sufficient motion for *cis–trans* isomerisation of the DsRed chromophore to occur [40]. Nevertheless, in photoconverted samples, the decarboxylated species appears to be a minor species, and the majority of the sample has the native mass, suggesting it could occur as a side-reaction or that photoconversion is heterogeneous in nature [40].

Irreversible photoconversion was also reported for the A148G mutant of asFP595 (KFP) [60–62]. It was subsequently speculated that light-induced decarboxylation of Glu215 in this protein destabilises the *trans* zwitter-ionic chromophore, producing the fluorescent (0.45 quantum yield) neutral *trans* state [63, 64].

3 Irreversible Green-to-Red Photoconversion in Related Fluorescent Proteins

A distinct class of photoconvertable fluorescent proteins are the "green-to-red" convertible proteins, including *Anthozoa*-derived fluorescent proteins and variants having chromophores derived from the His-Tyr-Gly tripeptide. The mechanism of photoconversion is fundamentally different from that found in avGFP and involves covalent photochemical modification. In the dark, these proteins mature to fully fold and form a chromophore up to the green fluorescent state. Subsequent illumination with UV or blue violet light results in their irreversible conversion to a red fluorescent state. Members of this group are called Kaede [65], EosFP and mEosFP [66, 67], KikGR [68, 69, 71] and Dendra [70] (Fig. 11).

Kaede is from the stony coral *Trachyphyllia geoffroyi*, which changes the fluorescence from green at 518 nm to red at 582 nm upon irradiation with blue light at 400 nm [65]. Its chromophore is derived from the His-Tyr-Gly tripeptide. Another example in the same class of a photoconvertible "green-to-red" fluorescent proteins is EosFP from the coral *Lobophyllia hemprichii* [67]. EosFP has very similar spectroscopic and photoconversion characteristics as "Kaede".

Photoconversion of EosFP was shown to be initiated from the protonated green form of the chromophore [66]. The X-ray structures of the green and the red forms showed that cleavage occurred between His-62 N and C atoms[66], in agreement with the proposal based on mass spectrometry and NMR spectroscopy of peptides derived from photoconverted Kaede [72]. A model for the light-driven β-elimination reactions that causes reddening in this class of fluorescent proteins includes excited state chemistry that is proposed to be initiated by a ESPT reaction [66, 71, 72] (Fig. 12). Time-resolved spectroscopic evidence for the existence of an ultrafast ESPT reaction in Kaede or EosFP is not yet available, but the Stokes shift of fluorescence is similar to that of the neutral GFP_A state of avGFP. It is proposed that the resonance structures in the deprotonated excited state of the chromophore then lead to an intermediate with a cleaved N–C bond.

A number of different reaction models, involving different intermediates, have been proposed so far for the green-to-red photoconversion. Mizuno et al. [72] proposed that His62 in Kaede could become doubly protonated on the imidazole ring, leading to backbone cleavage via β-elimination [72] Nienhaus et al. [66] proposed an ESPT from the phenolic oxygen of the chromophore to the Nε of

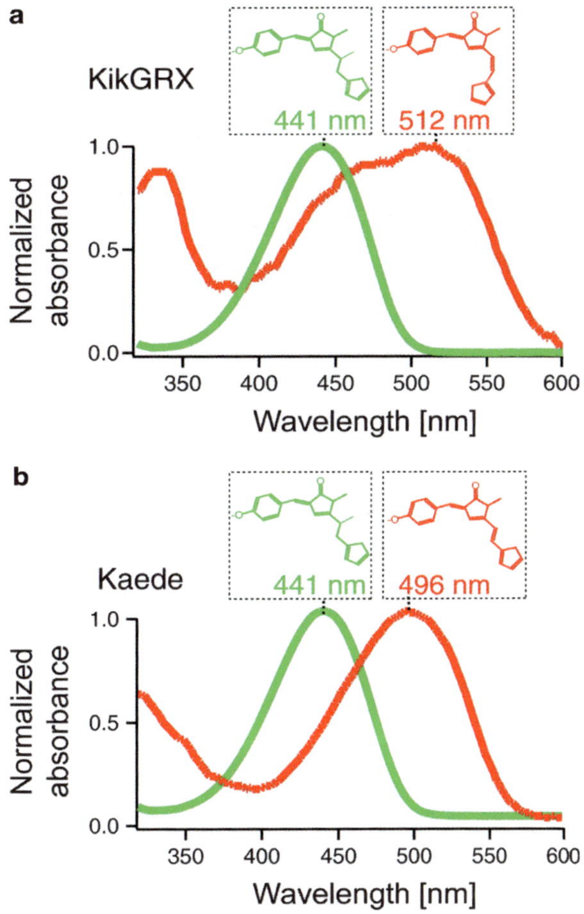

Fig. 11 Photoinduced backbond cleavage and β-elimination in "green-to-red" photoconvertable fluorescent proteins. Absorbance spectra of green and red forms of KikGRX (**a**) and Kaede (**b**) after denaturation (reproduced with permission [71])

His62 in EosFP, followed by a β-elimination step in which Glu212 acts as a proton acceptor [66]. Glu212 is known to be essential for photoconversion of EosFP, since the E212Q mutant does not photoconvert [66]. Hayashi et al. [73] suggested a "water-assisted" mechanism to explain the loss of a water molecule in the red photoproduct of Kaede [73]. Tsutsui et al. [71] discuss one specific detail of the β-elimination, which can follow either a E2-type elimination which is bimolecular, or an E1-type elimination which is unimolecular. Interestingly, the red photoproduct in KikGR(X) was found to be structurally different compared to Kaede and EosFP, having the C=C double bond between His62-Cα and His62-Cβ in a *cis* rather than a *trans* configuration [71]. This shows that free bond rotation occurs in a reaction intermediate that favours the "E1" mechanism of β-elimination.

Lelimousin et al. [74] proposed that intersystem crossing occurs in the reaction pathways, arguing that this explains the low quantum yield of photoconversion [74]. Hybrid QM/MM calculations were used to calculate the relative energies of

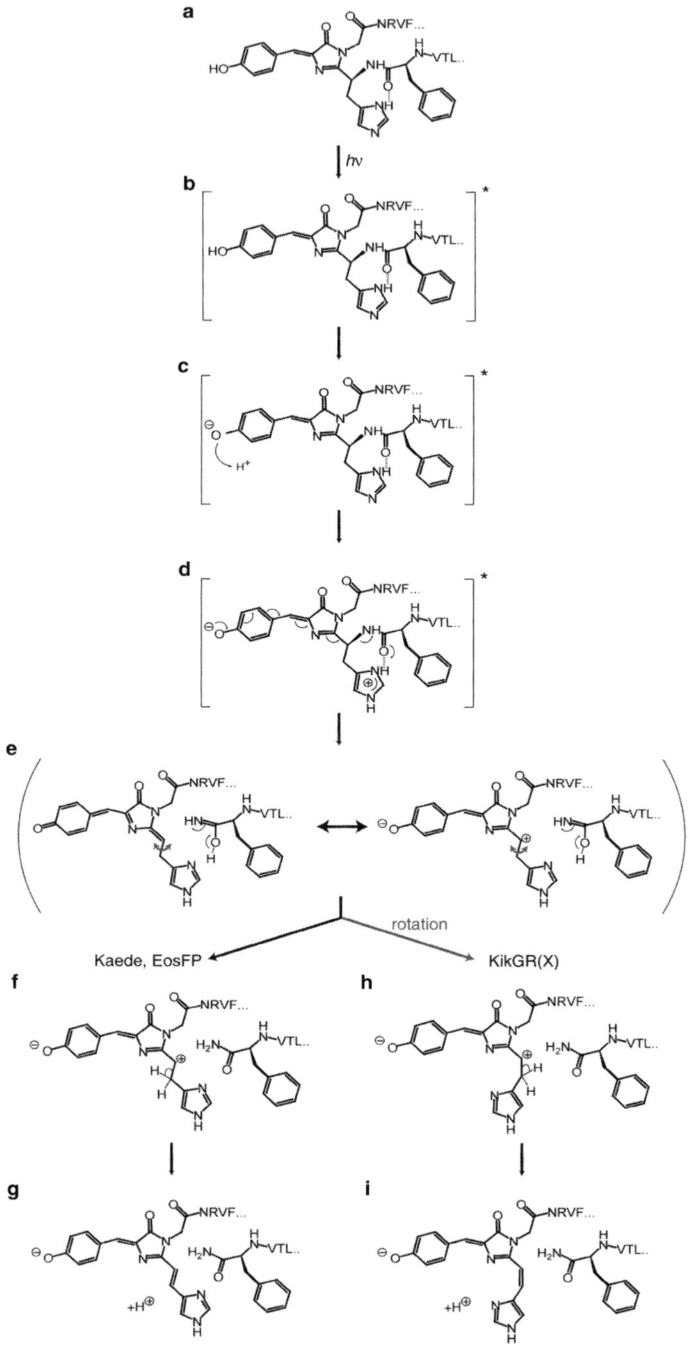

Fig. 12 Proposed photoconversion mechanism for Kaede, EosFP and KikGR(X). Reproduced with permission [71]. Figure caption from Tsutsui et al. [71]: "E1 Reaction Model for the Green-to-Red Photoconversion of Fluorescence Proteins. Structures derived from Phe61, His62, Tyr63, and

the predicted intermediates, involving ICS from S1 to T1 and finally to S0 [74]. These calculations have attempted to distinguish between the reaction models proposed by Nienhaus et al. [66] and Mizuno et al. [72] by comparing the energy minimised reaction trajectories. Interestingly, it was argued that the involvement of a triplet state may show that ESPT would not participate in the photoconversion reaction [74]. Awaiting ultrafast spectroscopy and cryo-trapping studies with this class of photoconvertable proteins there is currently no consensus in the literature with regard to the electronic configuration of reaction intermediates.

4 Light-Induced Maturation in Related Fluorescent Proteins

Oxidative decarboxylation of Glu222 was initially proposed to occur in a new type of UV-induced photoconversion that was reported for a mutant of a GFP homologue, a non-fluorescent jellyfish *A. coerulescens* chromoprotein which caries the same SYG motif for the chromophore, called "aceGFPL" [75]. Random mutagenesis identified an E222G mutation which leads to a fluorescent derivative, called "aceGFP". The GFP-like wild-type protein, or rather the "back-mutated" aceGFP-G222E variant, when expressed in *E. coli* is colourless and has only very minor absorption at 390 nm indicating a minor neutral chromophore species. Illumination with UV light in the 250–300 nm region resulted in photoconversion, generating an anionic species, with 480 nm excitation and 505 nm emission at room temperature [75]. Although further characterisation was not performed, it was suggested that light-induced electron transfer leading to decarboxylation of Glu222 could be part of the photochromic reaction. It appears, however, that UV light-induced photoconversion triggers maturation of the anionic chromophore specifically, whereas the minor neutral species does not convert. Recently the crystal structure of aceGFP-G222E was presented, which showed that the chromophore was not matured with Cα and Cβ atoms in the sp3 hybridisation [76]. The crystal structure of the photoconverted aceGFP-G222E, which has the anionic chromophore, did not indicate oxidative decarboxylation of Glu222. The observation is discussed in the light of the different proposals for chromophore maturation mechanisms [77, 78], particularly supporting a dehydrated, cyclised chromophore structure in the immature state in the aceGFP- G222E protein, as proposed by [77]. The photochemistry of UV-induced maturation starting with

Fig. 12 (continued) Gly64 are drawn; the neighbouring amino acids (single-letter code) are also shown. (**a**) The green chromophore in the initial protonated state. (**b**) Excitation of the chromophore by ultraviolet light. (**c**) Excited-state proton transfer. (**d**) The excited ionized phenolic hydroxyl group and protonated imidazole ring of His62. (**e**) Resonance stabilization of the intermediate structure. Allowed rotation of the imidazole ring along the bond between His62-Cα and His62-Cβ is indicated by *arrows*. The leaving group is carboxamidic acid, which becomes carboxamide through tautomerization. (F and H) Tautomerization with (H) and without (F) rotation of the imidazole ring (G and I). Loss of a proton from Cβ leads to completion of the red chromophore with either a *trans* (G) or *cis* (I) double bond" this state is, however, not yet fully understood in detail. Similar behaviour was reported for the photoactivatable red fluorescent protein PA-mRFP1s [62].

5 Reversible Photoconversion Reactions Resulting from Chromophore Photoisomerisation in Fluorescent Proteins

Yet another distinct class of photoconvertable fluorescent proteins concerns mutants of avGFP and natural and modified GFP homologues which undergo photoisomerisation reactions. These reactions are generally both thermally and photochemically reversible, and significant changes in the fluorescence quantum yield allows their use in protein-tracking applications in fluorescence microscopy. Additionally, the quantum yield of photoisomerisation can exceed that of any of the other photoconversion mechanisms that are known to date.

This class of proteins is called "reversibly switchable fluorescent proteins" (RSFPs). Examples displaying reversible photoconversion between dark and bright state are Dronpa [79–81] and its fast-switchable variant rsFastLime [81], asFP595 (asulCP, asCP) [82–85] and the kindling fluorescent protein (KFP; the A148G mutant of asFP595) [60], Padron [86] and IrisFP [87].It is established that *cis–trans* isomerisation of the chromophore is involved in switching, but protonation changes of the chromophore have also been implicated [63, 64, 88–90]. The role of photoisomerisation in RSFPs is currently well established, but recent studies, most of them computational, disagree on the protonation state and possible proton transfer reactions in the photoconversion reactions.

The "Dronpa" protein from the coral *Pectiniidae* undergoes photoswitching in the crystalline state. Andresen et al. [80], Stiel et al. [81] and Mizuno et al. [91] performed X-ray diffraction experiments with photoswitchable Dronpa crystals. Both in solution and in crystals, the dark resting state has a high fluorescence quantum yield of 0.85, with absorption and emission maxima at 503 and 522 nm [81]. Illumination with 500 nm light photoconverts the ground state to a non-fluorescent (fluorescence quantum yield of 0.02) protonated species absorbing at 380 nm, which can be re-transformed with illumination at ~400 nm [79]. The crystal structure of the fluorescent state showed the Cys-Tyr-Gly-derived chromophore to be in a *cis* conformation[81], as the avGFP chromophore [4, 6]. Subsequent illumination with 488 nm light switches the protein to a non-fluorescent state with an absorption maximum at 380 nm with a quantum yield of 3.2×10^{-4} [92] which decays back to ground state with an 18-h time constant [81]. Several studies found that the crystal structure of the photoproduct had the chromophore in a *trans* configuration, in addition to structural changes in the chromophore-binding site, possibly leading to stable protonation of the chromophore phenolic oxygen [80, 89, 91]. The protonation in the *trans* off-state is inferred from electrostatics calculations of the chromophore environment and the wavelength of electronic absorbance [80], but this currently needs to be confirmed from FTIR spectroscopy and/or pH titrations. Subsequent illumination of the *trans* off-state with ~400 nm light causes very efficient reformation of the *cis* "on" state (ground state), with a quantum yield of 0.37 [80]. This photochromic reaction is presently the most efficient photoactivation recorded, and as such is directly accessible for time-resolved pump-probe studies (Figs. 13 and 14).

It has been proposed that photoswitching of the *trans* "off" state to the bright *cis* "on" state of Dronpa involves ESPT as well as photoisomerisation [79, 92]. Photoswitching of Dronpa has been observed at the single molecule level [92] as well as in ensembles with picosecond time resolution [93]. Transient absorption measurements revealed a 4 ps component that was assigned to interconversion from the neutral "A2" excited state to the anionic I excited state. This phase was assigned to include ESPT from the twofold H/D KIE [93]. Recent QM/MM salutations addressed the possible reaction mechanisms in Dronpa, primarily the order of the ESPT and photoisomerisation processes [90]. Li et al. [90] invoke a twisted intramolecular charge transfer state in the photoisomerisation reaction, linking ESPT with *trans–cis* photoisomerisation in a concerted manner [90].

The GFP-like protein asFP595 from the sea anemone *Anemonia sulcata* is another example of an "RSFP", having a Met-Tyr-Gly-derived chromophore. Photoswitching from its non-fluorescent off-state to the fluorescent on-state is done with green light of 568 nm wavelength [82, 84, 85]. Spectroscopic measurements suggest that as for

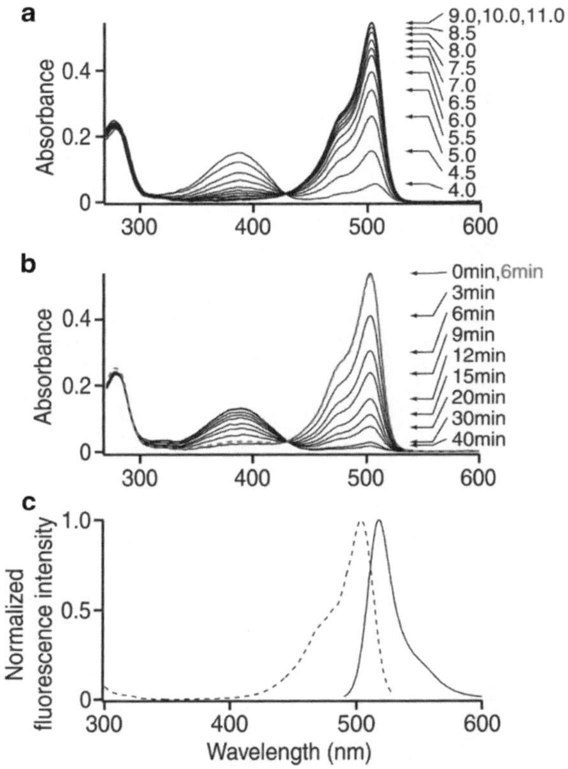

Fig. 13 Photochromic behaviour of Dronpa (22Gm3). Reproduced with permission from Ando et al. [79] "(**a**) The pH dependence of 22Gm3 absorbance. (**b**) Irradiation-dependent changes in 22Gm3 absorbance. Absorbance spectra obtained during irradiation at 490 nm (*black*) and after irradiation at 400 nm (*red*) are displayed. (**c**) Normalized excitation (*dotted line*) and emission (*solid line*) spectra of 22Gm3"

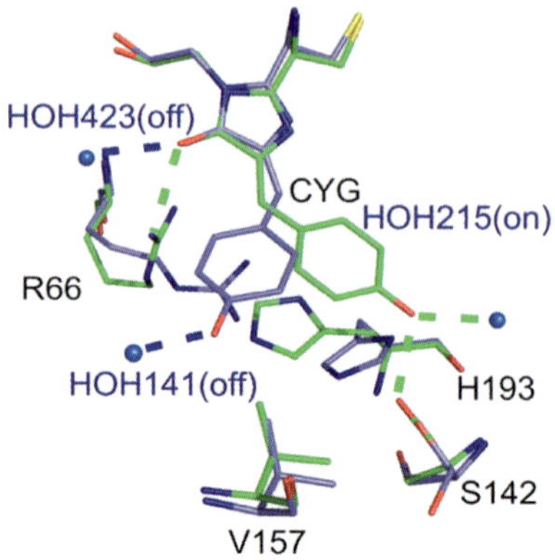

Fig. 14 Comparison of the on-state (*blue*) and off-state (*green*) structures of Dronpa in the chromophore vicinity (reproduced with permission from Andresen et al. [80])

Dronpa, photoswitching of asFP595 also involves protonation change of the chromophore [94]. However, in asFP595, the imidazolidinone nitrogen is proposed to form a zwitter-ionic species, on the basis of theoretical work [63, 64] (Fig. 15).

Several computational and simulation studies have focused on the photoswitching dynamics of "RSFPs" [63, 64, 90, 95, 96]. Schäfer et al. [63, 64] proposed that the chromophore is present as a zwitter-ion in the *trans* form, and that ESPT proceeds from the imidazolidinone nitrogen (rather than the phenolic oxygen). In GFP, the zwitter-ion was shown not to be present. For the free chromophore model compound ethyl 4-(4-hydroxyphenyl)methylidene-2-methyl-5-oxoimidazolacetate, Bell et al. [97] determined a pKa value of 8.0 for protonation of the imidazolidinone nitrogen [97] and showed from resonance Raman spectroscopy that in GFP at pH 8.0, where a mixture of anionic and neutral species is present, the position of the Raman bands exclude the possibility of cationic and zwitter-ionic forms [97]. In the case of asFP595, an acidic carboxylate from Glu215 is in hydrogen-bonding contact with the imidazolidinone nitrogen [82, 84, 85]. QM/MM simulations hitherto suggest that the zwitter-ion is the dominant species [63, 64], but this is yet to be confirmed by vibrational spectroscopy. Schäfer et al. [63] propose that ESPT deactivates the otherwise fluorescent zwitter-ionic *trans* chromophore, explaining the low fluorescence quantum yield [63, 64]. Indeed, ultrafast spectroscopy was used to determine the lifetime of the excited state, and found a dominant decay component of 320 fs, corresponding to the low fluorescence quantum yield $<10^{-4}$ [98]. The existence of a dark zwitter-ionic state was proposed early on from semi-empirical calculations of the GFP chromophore [99], providing a direct explanation for "blinking" behaviour observed at the single molecule level [100, 101]. Semi-empirical calculations predicted that in GFP in the ground state the zwitter-ion is not populated [102].Furthermore, the calculated excited-state free energy levels show the expected increased acidity of the phenol oxygen and increased basicity of the imino nitrogen [102]. This agrees with a picture of charge migration in the excited state typical of photoacid behaviour, and counters the results from Schäfer et al. [63, 64] who propose deactivation of the excited zwitter-ion state via proton transfer.

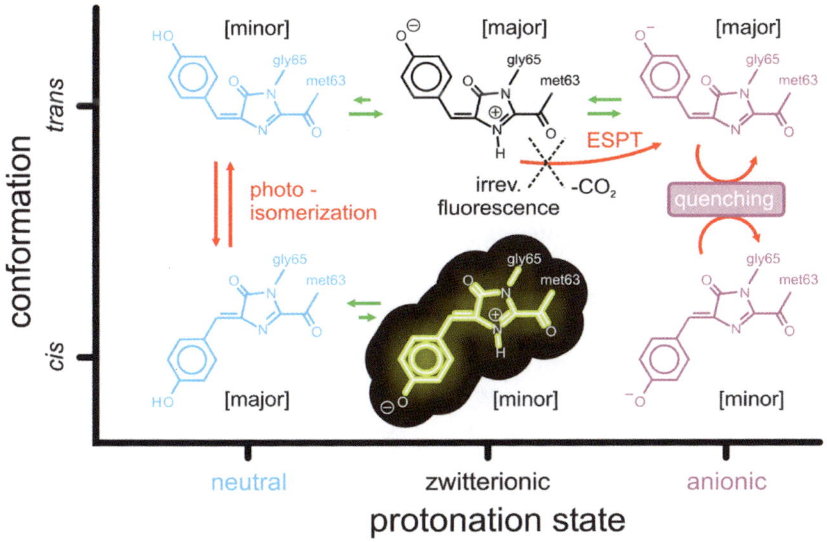

Fig. 15 Proposed mechanism of the reversible photoswitching mechanism of asFP595, reproduced with permission from Schäfer et al. [63]. QM/MM simulations have led to the proposal that protonation controls the photoswitching in asFP595. Legend from original figure: "The fluorescent state Zcis is *highlighted*. The *green arrows* indicate ground state equilibria, whereas the *red arrows* indicate excited state processes. The major protonation states are the zwitterionic and the anionic chromophores in the *trans* conformation, and the neutral chromophore in the *cis* conformation, as indicated in the *square bracket*"

In addition, the timescale of the deactivation 320 fs [98] would be very fast for the proposed ESPT reaction [63]. A number of theoretical calculations evaluated the potential energy of excited states of GFP chromophore models in the different protonation states [96, 99]. These suggest that the zwitterions have a conical intersection in a twisted geometry involving the rotation of only a single bond [96].

Li et al. [90] performed QM/MM calculations to study the photoswitching in the related Dronpa protein [90]. In contrast to the results for asFP595 [63], these calculations supported the anionic *cis* form for the "on" state and the neutral *trans* conformation for the "off" state of Dronpa [90]. The protonation is mentioned to be of critical importance for the non-fluorescent (dark) state of the *trans*-oriented chromophores in both asFP595 and Dronpa. It would appear that the general trend is that *cis* chromophores are bright states and *trans* chromophores are dark states. However, an exception is the eqFP611 fluorescent protein from *A. sulcata* which has a *trans*-oriented Met-Tyr-Gly-derived chromophore with a relatively high fluorescence quantum yield of 0.45 [103, 104]. It is possible that in the *trans* orientation of the eqFP611, chromophore is not in the zwitter-ionic state, which would support the conclusions reached for asFP595 [63, 64]. Vibrational spectroscopy information is not yet available to address this issue.

The possibility of the ESPT in asFP595 initiated from the imidazolidinone nitrogen [63] is unusual and suggests that the protein environment would have pronounced effects on the photophysical properties of the chromophore, since ESPT in photoacids is typically found to be initiated from a phenolate group. In the latter picture, the imidazolidinone nitrogen could act as a photobase rather than a photoacid [102]. Critical in the interpretation is the assignment of the transitions to the electronic species. From the QM/MM calculations, Schäfer et al. [64] assigned the low energy transition to the zwitter-ion in both *cis* and *trans* forms of asFP595 [64]. In contrast, Bravaya et al. [105] assign the low energy transition to the anionic state and predict the zwitter-ion at higher energy [105]. The calculations indicated that the S0–S1 transition energy for the zwitter-ion in the gas phase is particularly sensitive to the level of theory used and showed that multireference perturbation theory calculations [105] do not agree with TD-DFT [105] or ZINDO [64] results.

Taking a different approach, Olsen et al. [95] performed quantum mechanical calculations, which indicated that photoisomerisation occurs at different locations depending on the protonation state of the chromophore [95]. Specifically the calculations indicated excited state torsion of the phenoxy bridge in the neutral and the imidazolidinone bridge dihedral angles for the anionic chromophore [95].

The quantum yield of the *cis* on to *trans* off photoswitching is much lower, although mutations have led to increased rates of photoconversion [81, 106]. The possibility of intersystem crossing in the *cys* to *trans* photoisomerisation has been discussed [91, 92]. The involvement of a triplet state is a distinct possibility also in this photoconversion reaction, and has also previously been entertained in other types of (irreversible) photoconversion reactions (see Sects. 3 and 4). Because of the low quantum yield, spectroscopic studies of the *cis* on to *trans* off photoswitching are more challenging. In addition, the rate constant for thermal reversion differs substantially in different fluorescent proteins, from 7s in asFP595 and 50s in the A148G mutant of asFP595 (KFP) [60].

From the multitude of photoconversion reactions, there also exist fluorescent proteins that display more than one type. One example is IrisFP that undergoes both *cis–trans* photoisomerisation reactions as well as irreversible "green-to-red" photoconversion as found in "EosFP" (which IrisFP is derived of by mutagenesis) and related proteins (Sect. 4) [87].

6 General Conclusions

GFP and related fluorescent proteins not only provide a diverse palette of colours, but also display a very broad range of different types of photoreactions that can be exploited in fluorescence microscopy applications. These include irreversible and reversible photoconversion reactions that are associated with spectroscopic changes resulting in significant contrast. The photoconversion of GFP, which is the primary focus of this chapter, has been reviewed. While the principal molecular and spectroscopic details are known, there remain important questions to be answered, particularly with respect to the energetics. The low quantum yield of photoconversion is an obstacle for time-resolved spectroscopy. An important aim of this review has been not only to relate the historical development of significant research, but also to highlight and discuss the open and unanswered questions at the time of writing. In this fast developing field, an additional number of photoconversion processes have been found and have also briefly been discussed.

Acknowledgements Jasper van Thor is a Royal Society University Research Fellow. JvT acknowledges support from the European Research Council (Grant Agreement N° 208650) and EPSRC (Grant Ref EP/I003304/1).

References

1. Chalfie M, Tu Y, Euskirchen G, Ward WW, Prasher DC (1994) Green fluorescent protein as a marker for gene expression. Science 263:802–805
2. Prasher DC, Eckenrode VK, Ward WW, Prendergast FG, Cormier MJ (1992) Primary structure of the Aequorea victoria green-fluorescent protein. Gene 111:229–233
3. Tsien RY (1998) The green fluorescent protein. Annu Rev Biochem 67:509–544
4. Ormo M, Cubitt AB, Kallio K, Gross LA, Tsien RY, Remington SJ (1996) Crystal structure of the Aequorea victoria green fluorescent protein. Science 273:1392–1395
5. Brejc K, Sixma TK, Kitts PA, Kain SR, Tsien RY, Ormo M, Remington SJ (1997) Structural basis for dual excitation and photoisomerization of the Aequorea victoria green fluorescent protein. Proc Natl Acad Sci USA 94:2306–2311
6. Yang F, Moss LG, Phillips GN Jr (1996) The molecular structure of green fluorescent protein. Nat Biotechnol 14:1246–1251
7. Cubitt AB, Heim R, Adams SR, Boyd AE, Gross LA, Tsien RY (1995) Understanding, improving and using green fluorescent proteins. Trends Biochem Sci 20:448–455
8. Heim R, Cubitt AB, Tsien RY (1995) Improved green fluorescence. Nature 373:663–664
9. Heim R, Prasher DC, Tsien RY (1994) Wavelength mutations and posttranslational autoxidation of green fluorescent protein. Proc Natl Acad Sci USA 91:12501–12504
10. Nobelprize.org. (2008) The Nobel Prize in Chemistry 2008 – Press Release. "for the discovery and development of the green fluorescent protein, GFP". *Nobelprize.org*, http://nobelprize.org/nobel_prizes/chemistry/laureates/2008/press.html.
11. Morise H, Shimomura O, Johnson FH, Winant J (1974) Intermolecular energy transfer in the bioluminescent system of Aequorea. Biochemistry 13:2656–2662
12. Perozzo MA, Ward KB, Thompson RB, Ward WW (1988) X-ray diffraction and time-resolved fluorescence analyses of Aequorea green fluorescent protein crystals. J Biol Chem 263:7713–7716
13. Scharnagl C, Raupp-Kossmann R, Fischer SF (1999) Molecular basis for pH sensitivity and proton transfer in green fluorescent protein: protonation and conformational substates from electrostatic calculations. Biophys J 77:1839–1857
14. Bokman SH, Ward WW (1981) Renaturation of Aequorea gree-fluorescent protein. Biochem Biophys Res Commun 101:1372–1380
15. Ward WW, Cody CW, Hart RC, Cormier MJ (1980) Spectrophotometric identity of the energy transfer chromophores in Renilla and Aequorea green-fluorescent proteins. Photochem Photobiol 31:611–615

16. Chattoraj M, King BA, Bublitz GU, Boxer SG (1996) Ultra-fast excited state dynamics in green fluorescent protein: multiple states and proton transfer. Proc Natl Acad Sci USA 93: 8362–8367
17. Förster T (1949) Fluorezenzspektrum und Wasserstoffionenkonzentration. Naturwiss 36:186
18. Förster T (1950) Die pH-abhangigkeit der fluoreszenz von naphthalinderivaten. Z Electrochem 54:531
19. van Thor JJ, Gensch T, Hellingwerf KJ, Johnson LN (2002) Phototransformation of green fluorescent protein with UV and visible light leads to decarboxylation of glutamate 222. Nat Struct Biol 9:37–41
20. van Thor JJ, Pierik AJ, Nugteren-Roodzant I, Xie A, Hellingwerf KJ (1998) Characterization of the photoconversion of green fluorescent protein with FTIR spectroscopy. Biochemistry 37:16915–16921
21. Lill MA, Helms V (2002) Proton shuttle in green fluorescent protein studied by dynamic simulations. Proc Natl Acad Sci USA 99:2778–2781
22. van Thor JJ (2009) Photoreactions and dynamics of the green fluorescent protein. Chem Soc Rev 38:2935–2950
23. Vendrell O, Gelabert R, Moreno M, Lluch JM (2008) Operation of the proton wire in green fluorescent protein. A quantum dynamics simulation. J Phys Chem B 112:5500–5511
24. Palm GJ, Zdanov A, Gaitanaris GA, Stauber R, Pavlakis GN, Wlodawer A (1997) The structural basis for spectral variations in green fluorescent protein. Nat Struct Biol 4:361–365
25. Stoner-Ma D, Jaye AA, Matousek P, Towrie M, Meech SR, Tonge PJ (2005) Observation of excited-state proton transfer in green fluorescent protein using ultrafast vibrational spectroscopy. J Am Chem Soc 127:2864–2865
26. van Thor JJ, Georgiev GY, Towrie M, Sage JT (2005) Ultrafast and low barrier motions in the photoreactions of the green fluorescent protein. J Biol Chem 280:33652–33659
27. van Thor JJ, Ronayne KL, Towrie M, Sage JT (2008) Balance between ultrafast parallel reactions in the green fluorescent protein has a structural origin. Biophys J 95:1902–1912
28. van Thor JJ, Zanetti G, Ronayne KL, Towrie M (2005) Structural events in the photocycle of green fluorescent protein. J Phys Chem B 109:16099–16108
29. Stoner-Ma D, Melief EH, Nappa J, Ronayne KL, Tonge PJ, Meech SR (2006) Proton relay reaction in green fluorescent protein (GFP): polarization-resolved ultrafast vibrational spectroscopy of isotopically edited GFP. J Phys Chem B 110:22009–22018
30. Youvan DC, Michel-Beyerle ME (1996) Structure and fluorescence mechanism of GFP. Nat Biotechnol 14:1219–1220
31. Henderson JN, Osborn MF, Koon N, Gepshtein R, Huppert D, Remington SJ (2009) Excited state proton transfer in the red fluorescent protein mKeima. J Am Chem Soc 131(37): 13212–13213
32. Piatkevich KD, Malashkevich VN, Almo SC, Verkhusha VV (2010) Engineering ESPT pathways based on structural analysis of LSSmKate red fluorescent proteins with large stokes shift. J Am Chem Soc 132:10762–10770
33. Lippincott-Schwartz J, Manley S (2009) Putting super-resolution fluorescence microscopy to work. Nat Methods 6:21–23
34. Lippincott-Schwartz J, Patterson GH (2008) Fluorescent proteins for photoactivation experiments. Methods Cell Biol 85:45–61
35. Patterson GH, Lippincott-Schwartz J (2002) A photoactivatable GFP for selective photolabeling of proteins and cells. Science 297:1873–1877
36. van Thor JJ, Sage JT (2006) Charge transfer in green fluorescent protein. Photochem Photobiol Sci 5:597–602
37. Bell AF, Stoner-Ma D, Wachter RM, Tonge PJ (2003) Light-driven decarboxylation of wildtype green fluorescent protein. J Am Chem Soc 125:6919–6926
38. Schneider M, Barozzi S, Testa I, Faretta M, Diaspro A (2005) Two-photon activation and excitation properties of PA-GFP in the 720-920-nm region. Biophys J 89:1346–1352

39. Langhojer F, Dimler F, Jung G, Brixner T (2009) Ultrafast photoconversion of the green fluorescent protein studied by accumulative femtosecond spectroscopy. Biophys J 96:2763–2770
40. Habuchi S, Cotlet M, Gensch T, Bednarz T, Haber-Pohlmeier S, Rozenski J, Dirix G, Michiels J, Vanderleyden J, Heberle J, De Schryver FC, Hofkens J (2005) Evidence for the isomerization and decarboxylation in the photoconversion of the red fluorescent protein DsRed. J Am Chem Soc 127:8977–8984
41. He X, Bell AF, Tonge P (2002) Isotopic labeling and normal-mode analysis of a model green fluorescent protein chromophore. J Phys Chem B 106:6056–6066
42. Bublitz GU, Boxer SG (1997) Stark spectroscopy: applications in chemistry, biology, and materials science. Annu Rev Phys Chem 48:213–242
43. Anderson JM, Kocji JK (1970) Manganese(III) complexes in oxidative decarboxylation of acids. J Am Chem Soc 92:2450–2460
44. Kolbe H (1849) Untersuchungen über die Elektrolyse organischer Verbindungen. Ann Chem Pharm 69:257–294
45. Cao W, Ye X, Sjodin T, Christian JF, Demidov AA, Berezhna S, Wang W, Barrick D, Sage JT, Champion PM (2004) Investigations of photolysis and rebinding kinetics in myoglobin using proximal ligand replacements. Biochemistry 43:11109–11117
46. Ye X, Yu A, Georgiev GY, Gruia F, Ionascu D, Cao W, Sage JT, Champion PM (2005) CO rebinding to protoheme: investigations of the proximal and distal contributions to the geminate rebinding barrier. J Am Chem Soc 127:5854–5861
47. Zeng W, Silvernail NJ, Wharton DC, Georgiev GY, Leu BM, Scheidt WR, Zhao J, Sturhahn W, Alp EE, Sage JT (2005) Direct probe of iron vibrations elucidates NO activation of heme proteins. J Am Chem Soc 127:11200–11201
48. Lossau H, Kummer A, Heinecke R, Pollinger-Dammer F, Kompa C, Bieser G, Jonsson T, Silva CM, Yang MM, Youvan DC, Michel-Beyerle ME (1996) Time-resolved spectroscopy of wild-type and mutant green fluorescent proteins reveals excited state deprotonation consistent with fluorophore-protein interactions. Chem Phys 213:1–16
49. Hopfield JJ (1974) Electron transfer between biological molecules by thermally activated tunneling. Proc Natl Acad Sci USA 71:3640–3644
50. Marcus RA, Sutin N (1985) Electron transfers in chemistry and biology. Biochim Biophys Acta 811:265–322
51. Moser CC, Keske JM, Warncke K, Farid RS, Dutton PL (1992) Nature of biological electron transfer. Nature 355:796–802
52. Page CC, Moser CC, Chen X, Dutton PL (1999) Natural engineering principles of electron tunnelling in biological oxidation-reduction. Nature 402:47–52
53. McAnaney TB, Zeng W, Doe CF, Bhanji N, Wakelin S, Pearson DS, Abbyad P, Shi X, Boxer SG, Bagshaw CR (2005) Protonation, photobleaching, and photoactivation of yellow fluorescent protein (YFP 10C): a unifying mechanism. Biochemistry 44:5510–5524
54. Henderson JN, Gepshtein R, Heenan JR, Kallio K, Huppert D, Remington SJ (2009) Structure and mechanism of the photoactivatable green fluorescent protein. J Am Chem Soc 131:4176–4177
55. Lippincott-Schwartz J, Altan-Bonnet N, Patterson GH (2003) Photobleaching and photoactivation: following protein dynamics in living cells. Nat Cell Biol Suppl:S7–S14
56. Lippincott-Schwartz J, Patterson GH (2003) Development and use of fluorescent protein markers in living cells. Science 300:87–91
57. Chudakov DM, Lukyanov S, Lukyanov KA (2007) Tracking intracellular protein movements using photoswitchable fluorescent proteins PS-CFP2 and Dendra2. Nat Protoc 2:2024–2032
58. Chudakov DM, Lukyanov S, Lukyanov KA (2007b) Using photoactivatable fluorescent protein Dendra2 to track protein movement. Biotechniques 42:553, 555, 557 passim
59. Chudakov DM, Verkhusha VV, Staroverov DB, Souslova EA, Lukyanov S, Lukyanov KA (2004) Photoswitchable cyan fluorescent protein for protein tracking. Nat Biotechnol 22:1435–1439

60. Chudakov DM, Belousov VV, Zaraisky AG, Novoselov VV, Staroverov DB, Zorov DB, Lukyanov S, Lukyanov KA (2003) Kindling fluorescent proteins for precise in vivo photolabeling. Nat Biotechnol 21:191–194
61. Lukyanov KA, Chudakov DM, Lukyanov S, Verkhusha VV (2005) Innovation: photoactivatable fluorescent proteins. Nat Rev Mol Cell Biol 6:885–891
62. Verkhusha VV, Sorkin A (2005) Conversion of the monomeric red fluorescent protein into a photoactivatable probe. Chem Biol 12:279–285
63. Schäfer LV, Groenhof G, Boggio-Pasqua M, Robb MA, Grubmuller H (2008) Chromophore protonation state controls photoswitching of the fluoroprotein asFP595. PLoS Comput Biol 4(3):e1000034
64. Schäfer LV, Groenhof G, Klingen AR, Ullmann GM, Boggio-Pasqua M, Robb MA, Grubmuller H (2007) Photoswitching of the fluorescent protein asFP595: mechanism, proton pathways, and absorption spectra. Angew Chem Int Ed Engl 46:530–536
65. Ando R, Hama H, Yamamoto-Hino M, Mizuno H, Miyawaki A (2002) An optical marker based on the UV-induced green-to-red photoconversion of a fluorescent protein. Proc Natl Acad Sci USA 99:12651–12656
66. Nienhaus K, Nienhaus GU, Wiedenmann J, Nar H (2005) Structural basis for photo-induced protein cleavage and green-to-red conversion of fluorescent protein EosFP. Proc Natl Acad Sci USA 102:9156–9159
67. Wiedenmann J, Ivanchenko S, Oswald F, Schmitt F, Rocker C, Salih A, Spindler KD, Nienhaus GU (2004) EosFP, a fluorescent marker protein with UV-inducible green-to-red fluorescence conversion. Proc Natl Acad Sci USA 101:15905–15910
68. Habuchi S, Tsutsui H, Kochaniak AB, Miyawaki A, van Oijen AM (2008) mKikGR, a monomeric photoswitchable fluorescent protein. PLoS ONE 3(12):e3944
69. Tsutsui H, Karasawa S, Shimizu H, Nukina N, Miyawaki A (2005) Semi-rational engineering of a coral fluorescent protein into an efficient highlighter. EMBO Rep 6:233–238
70. Gurskaya NG, Verkhusha VV, Shcheglov AS, Staroverov DB, Chepurnykh TV, Fradkov AF, Lukyanov S, Lukyanov KA (2006) Engineering of a monomeric green-to-red photoactivatable fluorescent protein induced by blue light. Nat Biotechnol 24:461–465
71. Tsutsui H, Shimizu H, Mizuno H, Nukina N, Furuta T, Miyawaki A (2009) The E1 mechanism in photo-induced beta-elimination reactions for green-to-red conversion of fluorescent proteins. Chem Biol 16:1140–1147
72. Mizuno H, Mal TK, Tong KI, Ando R, Furuta T, Ikura M, Miyawaki A (2003) Photo-induced peptide cleavage in the green-to-red conversion of a fluorescent protein. Mol Cell 12:1051–1058
73. Hayashi I, Mizuno H, Tong KI, Furuta T, Tanaka F, Yoshimura M, Miyawaki A, Ikura M (2007) Crystallographic evidence for water-assisted photo-induced peptide cleavage in the stony coral fluorescent protein Kaede. J Mol Biol 372:918–926
74. Lelimousin M, Adam V, Nienhaus GU, Bourgeois D, Field MJ (2009) Photoconversion of the fluorescent protein EosFP: a hybrid potential simulation study reveals intersystem crossings. J Am Chem Soc 131:16814–16823
75. Gurskaya NG, Fradkov AF, Pounkova NI, Staroverov DB, Bulina ME, Yanushevich YG, Labas YA, Lukyanov S, Lukyanov KA (2003) A colourless green fluorescent protein homologue from the non-fluorescent hydromedusa Aequorea coerulescens and its fluorescent mutants. Biochem J 373:403–408
76. Pletneva NV, Pletnev VZ, Lukyanov KA, Gurskaya NG, Goryacheva EA, Martynov VI, Wlodawer A, Dauter Z, Pletnev S (2010) Structural evidence for a dehydrated intermediate in green fluorescent protein chromophore biosynthesis. J Biol Chem 285:15978–15984
77. Barondeau DP, Tainer JA, Getzoff ED (2006) Structural evidence for an enolate intermediate in GFP fluorophore biosynthesis. J Am Chem Soc 128:3166–3168
78. Pouwels LJ, Zhang L, Chan NH, Dorrestein PC, Wachter RM (2008) Kinetic isotope effect studies on the de novo rate of chromophore formation in fast- and slow-maturing GFP variants. Biochemistry 47:10111–10122

79. Ando R, Mizuno H, Miyawaki A (2004) Regulated fast nucleocytoplasmic shuttling observed by reversible protein highlighting. Science 306:1370–1373
80. Andresen M, Stiel AC, Trowitzsch S, Weber G, Eggeling C, Wahl MC, Hell SW, Jakobs S (2007) Structural basis for reversible photoswitching in Dronpa. Proc Natl Acad Sci USA 104:13005–13009
81. Stiel AC, Trowitzsch S, Weber G, Andresen M, Eggeling C, Hell SW, Jakobs S, Wahl MC (2007) 1.8 A bright-state structure of the reversibly switchable fluorescent protein Dronpa guides the generation of fast switching variants. Biochem J 402:35–42
82. Andresen M, Wahl MC, Stiel AC, Grater F, Schafer LV, Trowitzsch S, Weber G, Eggeling C, Grubmuller H, Hell SW, Jakobs S (2005) Structure and mechanism of the reversible photoswitch of a fluorescent protein. Proc Natl Acad Sci USA 102:13070–13074
83. Lukyanov KA, Fradkov AF, Gurskaya NG, Matz MV, Labas YA, Savitsky AP, Markelov ML, Zaraisky AG, Zhao X, Fang Y, Tan W, Lukyanov SA (2000) Natural animal coloration can be determined by a nonfluorescent green fluorescent protein homolog. J Biol Chem 275:25879–25882
84. Quillin ML, Anstrom DM, Shu X, O'Leary S, Kallio K, Chudakov DM, Remington SJ (2005) Kindling fluorescent protein from Anemonia sulcata: dark-state structure at 1.38 A resolution. Biochemistry 44:5774–5787
85. Wilmann PG, Petersen J, Pettikiriarachchi A, Buckle AM, Smith SC, Olsen S, Perugini MA, Devenish RJ, Prescott M, Rossjohn J (2005) The 2.1A crystal structure of the far-red fluorescent protein HcRed: inherent conformational flexibility of the chromophore. J Mol Biol 349:223–237
86. Brakemann T, Weber G, Andresen M, Groenhof G, Stiel AC, Trowitzsch S, Eggeling C, Grubmuller H, Hell SW, Wahl MC, Jakobs S (2010) Molecular basis of the light-driven switching of the photochromic fluorescent protein Padron. J Biol Chem 285:14603–14609
87. Adam V, Lelimousin M, Boehme S, Desfonds G, Nienhaus K, Field MJ, Wiedenmann J, McSweeney S, Nienhaus GU, Bourgeois D (2008) Structural characterization of IrisFP, an optical highlighter undergoing multiple photo-induced transformations. Proc Natl Acad Sci USA 105:18343–18348
88. Faro AR, Adam V, Carpentier P, Darnault C, Bourgeois D, de Rosny E (2010) Low-temperature switching by photoinduced protonation in photochromic fluorescent proteins. Photochem Photobiol Sci 9:254–262
89. Henderson JN, Ai HW, Campbell RE, Remington SJ (2007) Structural basis for reversible photobleaching of a green fluorescent protein homologue. Proc Natl Acad Sci USA 104:6672–6677
90. Li X, Chung LW, Mizuno H, Miyawaki A, Morokuma K (2010) A theoretical study on the nature of on- and off-states of reversibly photoswitching fluorescent protein Dronpa: absorption, emission, protonation, and Raman. J Phys Chem B 114:1114–1126
91. Mizuno H, Mal TK, Walchli M, Kikuchi A, Fukano T, Ando R, Jeyakanthan J, Taka J, Shiro Y, Ikura M, Miyawaki A (2008) Light-dependent regulation of structural flexibility in a photochromic fluorescent protein. Proc Natl Acad Sci USA 105:9227–9232
92. Habuchi S, Ando R, Dedecker P, Verheijen W, Mizuno H, Miyawaki A, Hofkens J (2005) Reversible single-molecule photoswitching in the GFP-like fluorescent protein Dronpa. Proc Natl Acad Sci USA 102:9511–9516
93. Fron E, Flors C, Schweitzer G, Habuchi S, Mizuno H, Ando R, De Schryver FC, Miyawaki A, Hofkens J (2007) Ultrafast excited-state dynamics of the photoswitchable protein dronpa. J Am Chem Soc 129:4870–4871
94. Chudakov DM, Feofanov AV, Mudrik NN, Lukyanov S, Lukyanov KA (2003) Chromophore environment provides clue to "kindling fluorescent protein" riddle. J Biol Chem 278:7215–7219
95. Olsen S, Lamothe K, Martinez TJ (2010) Protonic gating of excited-state twisting and charge localization in GFP chromophores: a mechanistic hypothesis for reversible photoswitching. J Am Chem Soc 132:1192–1193

96. Polyakov IV, Grigorenko BL, Epifanovsky EM, Krylov AI, Nemukhin AV (2010) Potential energy landscape of the electronic states of the GFP chromophore in different protonation forms: electronic transition energies and conical intersections. J Chem Theory Comput 6:2377–2387
97. Bell AF, He X, Wachter RM, Tonge PJ (2000) Probing the ground state structure of the green fluorescent protein chromophore using Raman spectroscopy. Biochemistry 39:4423–4431
98. Schuttrigkeit TA, von Feilitzsch T, Kompa CK, Lukyanov KA, Savitsky AP, Voityuk AA, Michel-Beyerle ME (2006) Femtosecond study of light-induced fluorescence increase of the dark chromoprotein asFP595. Chem Phys 323:149–160
99. Weber W, Helms V, McCammon JA, Langhoff PW (1999) Shedding light on the dark and weakly fluorescent states of green fluorescent proteins. Proc Natl Acad Sci USA 96: 6177–6182
100. Blum C, Subramaniam V (2009) Single-molecule spectroscopy of fluorescent proteins. Anal Bioanal Chem 393:527–541
101. Dickson RM, Cubitt AB, Tsien RY, Moerner WE (1997) On/off blinking and switching behaviour of single molecules of green fluorescent protein. Nature 388:355–358
102. Scharnagl C, Raupp-Kossmann RA (2004) Solution pK(a) values of the green fluorescent protein chromophore from hybrid quantum-classical calculations. J Phys Chem B 108: 477–489
103. Petersen J, Wilmann PG, Beddoe T, Oakley AJ, Devenish RJ, Prescott M, Rossjohn J (2003) The 2.0-A crystal structure of eqFP611, a far red fluorescent protein from the sea anemone Entacmaea quadricolor. J Biol Chem 278:44626–44631
104. Wiedenmann J, Schenk A, Rocker C, Girod A, Spindler KD, Nienhaus GU (2002) A far-red fluorescent protein with fast maturation and reduced oligomerization tendency from Entacmaea quadricolor (Anthozoa, Actinaria). Proc Natl Acad Sci USA 99:11646–11651
105. Bravaya KB, Bochenkova AV, Granovsky AA, Savitsky AP, Nemukhin AV (2008) Modeling photoabsorption of the asFP595 chromophore. J Phys Chem A 112:8804–8810
106. Andresen M, Stiel AC, Folling J, Wenzel D, Schonle A, Egner A, Eggeling C, Hell SW, Jakobs S (2008) Photoswitchable fluorescent proteins enable monochromatic multilabel imaging and dual color fluorescence nanoscopy. Nat Biotechnol 26:1035–1040

Spectral Versatility of Fluorescent Proteins Observed on the Single Molecule Level

Christian Blum and Vinod Subramaniam

Abstract The photophysics of visible fluorescent proteins (VFPs) remains a topic of intense research, driven by the widespread use of these proteins as reporters and sensors in living cells. The photophysical complexity of these markers originates from the multistep chemical reaction that forms the chromophore, and from the embedding of the chromophore within the protein nanoenvironment. To accurately interpret the biological and biochemical processes illuminated by the VFPs, it is essential to understand the details of their photophysics. Certain aspects of VFP photophysics can only be observed and understood at the single molecule level, which removes the averaging effect inherent to ensemble studies. Here, we review how spectrally resolved single molecule emission detection at room temperature has helped to understand the complex photophysics of VFPs. We focus on the detection of spectrally distinct subensembles, the spontaneous or light-induced transition between these subensembles, and on subtle spectral variations induced by changes in the local nanoenvironment of VFP chromophores.

Keywords Fluorescence spectra · Photophysics · Photochromism · Single molecule spectroscopy · Spectral diffusion · Spectral forms

Contents

C. Blum (✉) and V. Subramaniam
Nanobiophysics, MESA+ Institute for Nanotechnology, Faculty of Science and Technology, University of Twente, P.O. Box 217, 7500 AE Enschede, The Netherlands
e-mail: c.blum@tnw.utwente.nl

G. Jung (ed.), *Fluorescent Proteins I*, Springer Ser Fluoresc (2012) 11: 217–240,
DOI 10.1007/4243_2011_22,
Published online: 31 March 2011

1 Introduction

The discovery, further development, and application of visible fluorescent proteins (VFPs) as tools for visualizing biological processes have had tremendous impact on cellular biology. In combination with optical microscopy techniques, VFPs allow for the visualization of the subcellular localization of proteins and the dynamics of their transport, trafficking, and interactions with other molecules and cellular components. Indeed, the breakthrough advances enabled by VFPs have been recognized by the award of the Nobel Prize in Chemistry in 2008. Furthermore, the exciting recent advances in optical super-resolution techniques that allow for near-molecular resolution optical imaging of biological systems [1–7] strongly rest on the unique properties of switchable fluorescent dyes and switchable fluorescent proteins.

To date, a wide palette of fluorescent proteins has been discovered or engineered for genetically encoded in vivo labeling. Besides proteins derived from the first discovered green fluorescent protein (GFP) in the *Aequoria* jellyfish, the palette of VFPs has been enormously extended by the discovery of new intrinsically fluorescent VFPs from other marine organisms [8–14] and by their optimization by protein engineering [15–20]. The combination of these genetically encodable markers with advanced microscopic and spectroscopic techniques has enabled the quantitative measurement of protein–protein interactions at high spatial and temporal resolutions. Variants of VFPs exhibiting different colors and photophysical properties such as photoactivation [21–28], photobleaching [29, 30], and phototoxicity [31, 32] have provided new windows into the cell.

However, many studies have established that VFPs exhibit intrinsically complex photophysical behavior. To effectively and correctly exploit the enormous power of VFPs to visualize biological processes, it is of paramount importance to comprehensively understand the intrinsic photophysical properties of these remarkable fluorophores in detail.

In addition to the ensemble characterization of emitters, rapid advances in ultrasensitive optical detection and spectroscopy have made it possible to visualize and characterize emitters at the single molecule level [33–36]. In contrast to ensemble measurements that yield information about the averaged properties of the sample, single molecule studies yield information about individual molecular entities and hidden subensembles that are not observable due to the ensemble averaging effect. These molecular properties vary from molecule to molecule and with time for individual molecules. As a result, single molecule studies on a statistically relevant number of molecules yield distributions of parameters that in the limit of very large numbers of molecules approach ensemble data, but which contain a great deal of more detailed information about subensembles and individual properties than ensemble measurements can hope to provide. A particular strength of single molecule studies is the ability to identify rare forms and subensembles that are hidden or easily overlooked in ensemble studies, and to visualize the evolution of these forms in time, which is impossible in ensemble studies. To access the details of the photophysical complexity of VFPs, single molecule studies are of great value. In particular, many aspects of the spectral versatility of fluorescent proteins can only be analyzed on the single molecule level using spectrally resolved single molecule spectroscopy.

2 Basis for Photophysical Complexity of VFPs

VFPs show an extraordinary spectral versatility and photophysical complexity compared to most commonly used chemical marker molecules. The reasons for this spectral complexity are manifold and have their origin, on the one hand, in the general structure of VFPs and, on the other hand, in the details of the formation of the light-emitting chromophore.

All VFPs now known have been isolated from different marine life forms or were created by using the toolbox of protein engineering to modify these proteins found in nature. Remarkably, all these proteins share the same basic structure elucidated for *Aequoria* GFP [37, 38]. GFP has a molecular weight of 26.9 kDa and forms a barrel-like structure with a diameter of about 2.4 nm and a height of 4.2 nm (Fig. 1). 11 β-sheets form the outer wall of the barrel structure, while an α-helix runs diagonally through this barrel. The fluorescent chromophore is enclosed in the center of the β-barrel.

The formation of this light-emitting chromophore is a complex, multistage chemical reaction within the protein (see Fig. 1), with significant potential for the formation of side products that complicate the emission behavior of the VFPs, e.g., by resulting in a mixture of different end product chromophores of different emission colors [19, 40, 43, 44]. A prime example of the parallel formation of different chromophores starting from identical precursors is the tetrameric reef coral fluorescent protein DsRed, in which a green- or a red-emitting chromophore can be formed [40]. Variants generated from this protein show a very different spectral appearance, depending on the exact ratio of occurrence of these two chromophores within the tetramer. From a chemistry point of view, the appearance of different end products in a complex chemical reaction over many steps is not unusual, and is in fact expected. When synthesizing chemical marker chromophores, this problem of unwanted side products is solved by a number of purification steps that result in essentially pure emitter molecules of choice. Clearly, this approach cannot be followed using genetically encoded fluorescent markers such as the VFPs that are created in vivo. As a result, when using VFPs as marker molecules, one has to accept the presence of different end product chromophores that are formed within the proteins.

In addition to the spontaneous formation of different chromophores, a number of photoinduced modifications of the chromophore and its nanoenvironment have

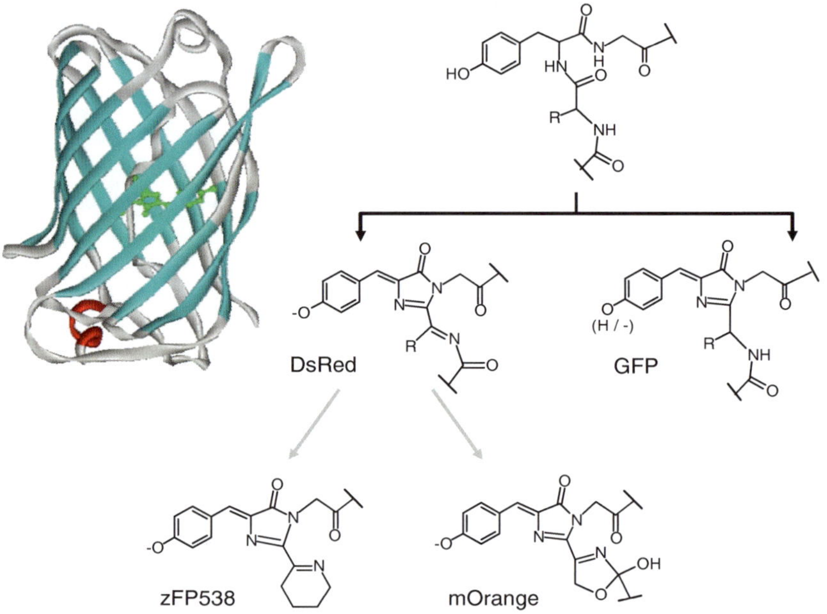

Fig. 1 *Left*: VFPs share a universal "beta-can" structure of ~2.4 nm in diameter and 4.2 nm in height. The light-emitting chromophore is formed in a multistep autocatalytic reaction within the protein cylinder. *Right*: Schematic outlining chromophore formation in VFPs. The red-emitting chromophore in DsRed and the green-emitting chromophore in GFP are different end products in a multistep chemical reaction [39, 40]. The red-emitting chromophore is an intermediate in the formation of the chromophores in zFP538 [41] and mOrange [42]

been identified [45–48]. While this possibility to modify the emission of a fluorescent protein by means of applied light adds to the spectral complexity and versatility of VFPs, the photoactivation and photoswitching properties have made these proteins an important tool in a number of applications, including the revolutionary super resolution microscopy techniques.

The detailed photophysics and spectral appearance of an emitter is further dependent on its embedding matrix. By virtue of its embedding within the protein scaffold, the chromophore interacts with its local nanoenvironment defined by the surrounding protein. The exact nature of the nanoenvironment of the chromophore can be modified by mutating surrounding amino acid residues, resulting in changes in the photophysical properties of the chromophore [42, 49–52]. Although VFPs do not show major structural rearrangements as seen in many enzymes, they are not static structures, so that thermally driven or photoinduced reorientations in the chromophore vicinity can lead to a modification of photophysical parameters [29, 30, 51].

Finally, the tendency of many VFPs to form dimers and tetramers [17, 53, 54] can lead to complex energy transfer interactions within the specific oligomeric species formed.

3 Ultrasensitive Spectroscopy at the Single Molecule Level

3.1 Technical Principles of Single Molecule Studies

To access and resolve the photophysical complexity of VFPs as described above, single molecule spectroscopy is a very powerful method of choice. Standard ensemble spectroscopy yields an average of a given observable for a large number of presumably identical molecules that are sampled in parallel. By applying single molecule spectroscopy – hence analyzing one molecule at a time – this averaging effect is removed and variations of parameters characteristic of the individual single molecule become visible. Sampling statistically relevant numbers of single molecules gives access not only to the observable averaged over a large number of molecules, but also to histograms of the observable that describe the distribution of the respective parameter. The width and shape of such distributions can be analyzed to gain even deeper insight into the analyzed systems. Especially for complex systems where a multitude of different interactions between the emitting chromophore and its direct environment are possible, the distributions of parameters contain a wealth of information not accessible by conventional ensemble spectroscopy. As outlined above, for VFPs, such heterogeneity arises due to differing chromophores or chromophore conformations, or to variations in the specific chromophore nanoenvironment.

Furthermore, single molecule detection allows the observation of a single molecule over time, thereby giving access to dynamical changes, either spontaneous or arising from photophysics or photochemistry, without any need for synchronization.

For the practical realization of single molecule fluorescence spectroscopy, a number of approaches are by now well established [34, 55–57]. In short, to realize single molecule detection the sampled volume is reduced until there is only one target molecule in the observation volume left at a time. The minimum sample volume in optical microscopy and spectroscopy is defined by the optical diffraction limit, although recently some techniques have been demonstrated to circumvent this limit [3]. However, the sampled volume is in any case much larger than the size of the target molecule. It is hence necessary to have the target molecule embedded in a surrounding that is nonfluorescent to realize the condition to have only one target molecule in the detection volume at a time. This is realized by working with exceedingly high dilutions of target molecules in a nonfluorescent matrix or solvent in the observation volume at any given time.

For applications discussed here that are targeted at the characterization of emitters, the most commonly used technique is confocal raster-scanning microscopy. To localize the single emitters, the sample is raster-scanned, and for further analysis the localized emitters are selectively positioned in the observation volume. To detect the photons emitted by the single molecule highly sensitive avalanche photo diodes (APDs) or highly sensitive intensified or back illuminated and cooled CCD cameras are used. APDs can be used either to study the evolution of the single molecule emission intensity over time, or, when pulsed excitation is used, to determine the emission lifetimes of single emitters. Sensitive CCD cameras in combination with a dispersing element such as a grating or a prism are used to detect full fluorescence spectra of a single molecule.

To characterize the photophysical properties of VFPs, the fluorescent proteins are usually immobilized by embedding them at very high dilutions (~10^{-11} M VFP in polymer solution) in a very thin film of a nonfluorescent polymer. In this way, the proteins are separated laterally, so that on average less than one protein can be found within the diffraction limited observation volume. The immobilization in the polymer ensures that the VFPs are not diffusing out of the detection volume during the characterization, which typically takes some seconds per single protein.

3.2 Single VFP Intensity Trajectories

The intensity trajectory of a single emitter describes the evolution of its emission intensity over time. The majority of single fluorescent protein studies analyze the intensity trajectory, gaining insights into processes affecting the brightness of the single molecule, such as changes in molecular orientation or quantum yield, transient transitions into dark states (often called "blinking"), or photobleaching. Many of these phenomena can be observed for any single emitter including chemical fluorophores, quantum dots, or the VFPs discussed here. Since the intensity trajectory is the most accessible parameter on the single molecule level, the analysis of single VFP intensity trajectories already started in 1997 [58]. Dickson et al. recorded intensity trajectories that showed evidence of repeated cycles of fluorescent emission and transitions into dark, nonemitting states on a timescale of several seconds, behavior clearly not observable in ensemble studies due to the averaging of the emission from different molecules. Comparable blinking of the emission was observed from different fluorescent proteins [59–62] (Fig. 2). The duration of the on-times was found to decrease with increasing excitation power [35], which points toward an excitation-driven, photoinduced transition to the dark state. The nature of these dark states can vary, including changes in the protonation state of the chromophore [58, 62, 63], efficient *cis/trans* photoisomerization that quenches the fluorescence [64], or rearrangements in the chromophore environment defined by the protein backbone leading to other effective deactivation pathways.

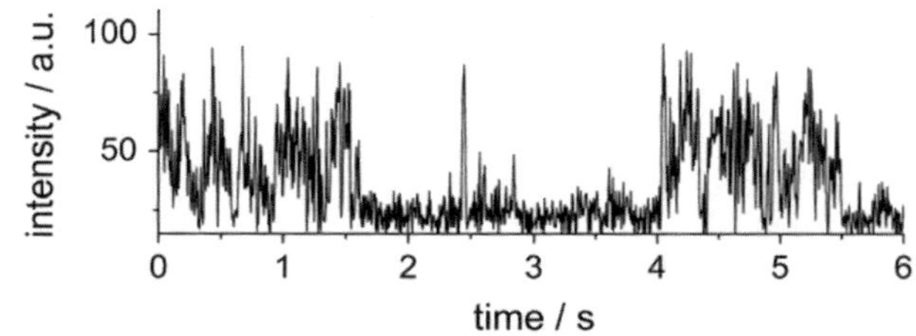

Fig. 2 Typical single molecule intensity trajectories of the enhanced Green Fluorescent Protein (EGFP). The emission is interrupted by numerous dark intervals

In addition, a long lived dark state found for many GFP group proteins can be depopulated by illumination with light around 400 nm [58, 62, 65]. This switching is induced by a change of the protonation state of the GFP chromophore. The protonated, neutral chromophore is essentially nonemitting due to its low quantum efficiency and blue-shifted absorbance, while illumination with light of around 400 nm results in deprotonation of the chromophore and thereby the reconstitution of the efficiently emitting chromophore [58, 62, 65, 66]. This switching between different states induced by light and the possibility to observe the emission from single fluorescent proteins has played an important role in the development of super-resolution microscopy techniques [1–7].

3.3 Spectrally Resolved Single Molecule Detection

Intensity trajectories give fascinating insights into the details of single emitters, but for photophysically complex systems such as the VFPs, it is often necessary to sample additional parameters. The emission spectrum of a single molecule is characteristic of the molecule, and direct observation of the spectra and changes in the spectral shape and position is a powerful means to gain further insights into the photophysics of complex emitters. Techniques to record full emission spectra from single molecules with high spectral resolution have been established [67–70].

The emission spectrum is a robust parameter since it does not suffer from changes by reorientation of the analyzed molecule with respect to the excitation light and detection, which change the detected emission intensity. Also spectra can be used to effectively discriminate between target molecules and unwanted but unavoidable contaminations. Further, emission spectra can be analyzed in great detail, enabling access to parameters such as the emission maximum position, the spectral shape, the spacing of the vibronic progression, or the intensity ratio between different transitions of the vibronic progression.

Spectrally resolved analysis of the emission from single fluorophores unfolds its full potential when series of emission spectra from a statistically relevant number of single emitters are measured (Fig. 3). The identification of rare spectral forms becomes directly accessible since the influence from other, dominating forms is simply removed by looking at one emitter at a time. Recording the evolution of the emission spectrum over time in spectral series makes possible the identification of connected spectral species by direct observation of transitions of the single emitter between these forms. These transitions between spectrally different forms originate from rather large-scale modifications of the chromophore, such as conformational reorganizations or chemical modifications of the chromophore itself. Besides these pronounced changes in spectral signature, subtle changes in the spectral position and shape can also be observed. These smaller spectral variations are termed "spectral diffusion" and are attributed to fluctuations in the nanoenvironment of the emitter that defines the exact photophysical properties of the emitter [71]. Precisely these capabilities of spectrally resolved single molecule spectroscopy

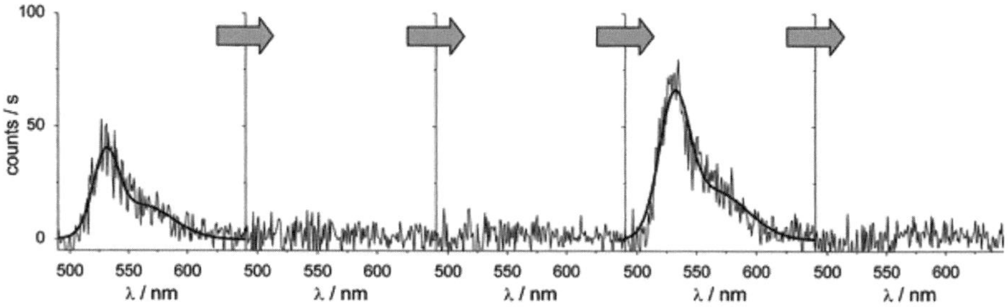

Fig. 3 Typical spectral series recorded from a single EYFP with integration time of 1 s per spectrum. A transition to a dark state and recurrence of emission is evident, the spectra before and after the dark period show only minor differences (Reprinted from [36], with permission from Elsevier)

to yield insights into the appearance and dynamics of spectral forms and the possibility to analyze the chromophore–nanoenvironment interaction make the method a powerful tool to analyze the complex emission behavior of VFPs, as will be discussed below.

However, the gain in information by recording emission spectra is accompanied by a loss of time resolution and a substantial increase in experimental complexity compared to spectrally integrated detection. To record single molecule emission spectra without wasting any photons by scanning the detected wavelength, the emitted fluorescence is dispersed by a spectrometer and imaged onto a highly sensitive, usually cooled, CCD camera. In this way, emission spectra can be recorded with integration times in the order of hundreds of milliseconds to seconds – depending on the experimental conditions used, on the excitation powers, on the overall sensitivity of the setup, and on the extinction coefficient and quantum efficiency characteristic of the molecule being studied.

3.4 Single VFP Emission Spectra for Target Molecule Identification

Single emitter fluorescence spectra are especially rich in information about complex emitting systems showing spectral diversity such as VFPs. However, to record emission spectra of high spectral resolution with acceptable signal to noise ratio rather large numbers of photons need to be collected. VFPs are known to be exquisitely sensitive emitters and their photostability is often far worse than that of many synthetic fluorescent dyes [72], which makes the recording of emission spectra or even series of emission spectra from single VFPs an especially challenging task.

These experimental challenges long hampered the systematic use of spectrally resolved single molecule spectroscopy to analyze VFP properties. The first single molecule emission spectrum recorded from VFPs was hence not used to analyze the emission of the protein, but to merely confirm that indeed the targeted proteins and not impurities were being studied. The single protein emission spectrum was embedded in a study analyzing intensity trajectories of single GFP variants [73]. In a similar way, the single protein emission spectra from the GFP variant EGFP and from DsRed were used to prove that the immobilization of the proteins in a film of polyvinylalcohol for single molecule analysis did not change their photophysical properties [74].

3.5 Spectral Diffusion: Sampling the Parameter Distribution

Classical bulk spectroscopy averages over all emitters in a sample and thereby results in one value, e.g., the emission maximum position, for the ensemble. Clearly, if there are deviations from the average value attributable to individual emitters, these are not discernible within the broad distribution of the ensemble. Single molecule spectroscopy, on the contrary, yields detailed insights into the spatial and temporal heterogeneity of single emitters. Instead of one value for a specific photophysical parameter, the full distribution of this parameter within the ensemble becomes accessible. Thus, in the case of the emission maximum position, single molecule spectroscopy gives access to the distribution of emission maximum positions of the individual emitters within the sampled ensemble rather than one value for the peak of the emission. In a sample composed of molecules of identical chemical forms, the different emitters can experience variations in their immediate nanoenvironments. The heterogeneity of spectral properties of single molecules originates exactly from this fact that any non-gas-phase chromophore interacts with its immediate environment; it is this interaction between the chromophore and its environment that determines the exact photophysical properties of the emitter. These local variations in the nanoenvironment of a single chromophore lead to spatial and temporal heterogeneity of the spectral properties of the chromophores. These small variations in spectral properties induced by varying chromophore–environment interactions are termed “spectral diffusion” since the variations are typically small and occur randomly [71]. Spectral diffusion generally shows a unimodal distribution of the respective parameter around one peak value. The unimodal nature of the distribution discriminates spectral diffusion from transitions between distinct spectral forms, since the transition between two forms will show a bimodal distribution around two peak values. Each of the two spectral forms will be characterized by a central value of the distribution (the most likely value of the respective observable), and a distribution width that is characteristic of the spectral diffusion associated with each form.

The heterogeneity of different single molecule emission parameters has been shown for single emitters at cryogenic temperatures and at room temperature. For polymer embedded dyes, variations in the shape as well as the exact spectral position of the emission spectrum have been observed [69], although it was impossible to conclusively describe the mechanisms underlying the observed variations since the polymer matrix embedding the chromophore is not exactly defined. In contrast to chromophores embedded in a polymer, the local environment of the chromophore of a VFP is precisely defined by the amino-acid composition of the protein scaffold encapsulating the chromophore. However, this strict definition of the chromophore environment is not sufficient to ensure invariant spectral characteristics of the VFPs. Indeed at the single molecule level, VFP photophysical parameters have been found to be distributed as in any other system of chromophores embedded in a matrix. Thus, even in these protein systems that possess a chemically well-defined nanoenvironment around the chromophore, the observed spectral heterogeneity reflects the subtle structural variations within the protein that influence the chromophore nanoenvironment and lead to distinct chromophore–matrix interactions.

By site-directed mutagenesis in and around the chromophore the chemical environment of the chromophore in a VFP can be precisely altered. Genetic engineering has yielded a wide range of mutant VFPs, such that a whole palette of proteins with different chromophore environments is available. Based on these variants it was possible to analyze the effect on spectral diffusion of interactions between the chromophore and its environment. Since some VFPs are capable of forming chemically different chromophores emitting at different wavelengths within one and the same scaffold, one can not only analyze changes in the environment of the chromophore defined by the protein, but also characterize spectral diffusion of different chromophores within identical nanoenvironments.

Spectral diffusion of VFPs was researched in detail in a comparative study on the tetrameric reef coral fluorescent protein DsRed and a number of its variants. As the first discovered red-emitting VFP, DsRed has been widely studied [17, 19, 20, 40, 75–79]. DsRed was extensively engineered to yield variants with changed and optimized properties. In DsRed, the autocatalytic reaction that forms the chromophore within the protein can branch and result in two different end products, yielding either a green-emitting or red-emitting chromophore. Different variants of DsRed exhibiting different green to red chromophore ratios have been generated, yielding proteins that are essentially all green emitting, all red emitting, or exhibiting mixed green and red emission. For this study, a large number of single molecule emission spectra were collected from DsRed, DsRed2, and Fluorescent timer (all predominantly red emitting), DsRed_N42H (mixed green and red emission), and AG4 (mainly green emitting) (see Fig. 4) [51].

Spectral diffusion was characterized by measuring the emission maximum position. The exact position of the emission spectrum is easy to determine and the emission maximum position has been found to be especially sensitive to changes in the local chromophore environment [42, 49–52]. To precisely determine the emission maximum position, a double Gaussian was fitted to each emission band in the single molecule spectra. From the individual emission maximum positions, histograms were constructed that showed the distribution of emission maximum positions for each protein variant (Fig. 5).

The width of the distribution was found to be characteristic for each chromophore-matrix combination and independent of experimental parameters such as the

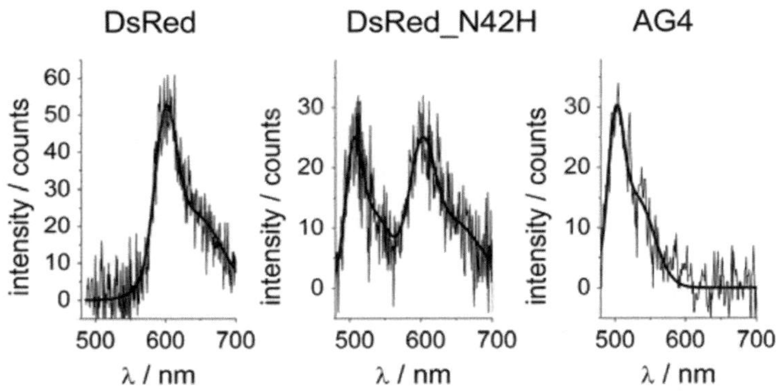

Fig. 4 Single oligomer spectra from DsRed group proteins. Different chromophores can form within the protein, a property, which make this group of proteins ideal for studying the influence of the chromophore embedding on the spectral diffusion (Reprinted with permission from [51]. Copyright 2006 American Chemical Society)

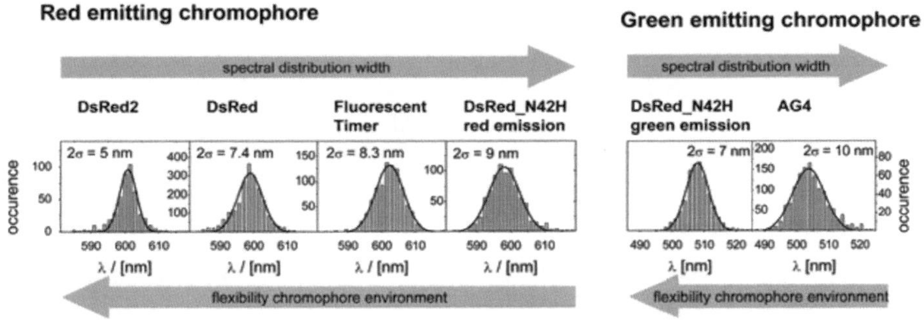

Fig. 5 Distribution of single molecule emission maximum positions from the different chromophores of the analyzed DsRed variants. The distribution width and the flexibility of the chromophore nanoenvironment are anti-correlated (Adapted with permission from [51]. Copyright 2006 American Chemical Society)

excitation power. For each variant, the red-emitting chromophore showed a different width of the determined maximum position distribution; the same trend was observed for the green-emitting chromophore. Clearly, this observation reflected the differences in the local environment within different variants. Further differences in the widths of the distribution were found for green- and for red-emitting chromophores residing within the same nanoenvironment of one specific variant (DsRed_N42H), reflecting the differences in sensitivity of chemically different chromophores to variations in the embedding.

Since the exact protein sequence for each variant was known and structural data for DsRed are available, it was possible to analyze the data further to discriminate between interactions due to changes in the chemical environment of the chromophores, e.g., different residues in the nanoenvironment of the chromophore, and changes due to differences in the flexibility of the chromophore nanoenvironment.

To estimate changes in the flexibility of the chromophore environment, the exact nature of the mutations in the chromophore environment and a model that linked the flexibility of the chromophore environment with the speed of chromophore formation was considered for each variant. The narrowest distribution of maximum positions was found for the protein variant DsRed2, which had the largest chromophore environment flexibility. For the other protein variants, the increase in distribution width of maximum positions corresponded to decreasing chromophore environment flexibility. Clearly, a more rigid chromophore environment results in a broader distribution of the emission maximum positions, that is, exhibits increased spectral diffusion. It was postulated that the correlation between the width of the distribution of single molecule emission maximum positions and the flexibility of the chromophore environment reflects the ability of the protein to return to the energetically most favorable conformation after, for example, a thermally induced reorientation. Flexible systems can rapidly reach the global energy minimum while in more rigid systems the return to the global energy minimum goes via local minima that act as traps and that would have different emission spectra.

Although the chemical environment of the chromophore showed clear differences from variant to variant, these differences in the chemical environment did not break the correlation between distribution width and rigidity of the chromophore environment. It was concluded that in this fluorescent protein system these chemical variations close to the chromophore only played a minor role for the observed spectral diffusion.

3.6 Identification and Characterization of Distinct Spectral Forms

Chemical dyes are purified before use to ensure an ensemble of chemically identical emitters. The purification of VFPs in a similar fashion is not possible. Due to the complex autocatalytic processes forming the fluorescent chromophore potentially resulting in different chromophores, the influences of chromophore maturation, and the varying effects of the protein that is encapsulating the chromophore, VFP ensembles are essentially complex mixtures of photophysically different subensembles. Additionally, these forms might be in equilibrium with each other and exhibit interchanges between the different forms.

To discriminate between spectral diffusion and different spectral forms, the most straightforward approach is to analyze a statistically relevant number of molecules and to subsequently create histograms of the relevant observable. In the histogram spectral diffusion results in a unimodal distribution. Different spectral forms are apparent as distinct peaks in a multimodal distribution. Each spectral form is defined by a characteristic central value, which is the most likely value of the relevant observable, and a distribution width, which is defined by spectral diffusion. Since each spectral form is subject to spectral diffusion, large spectral distributions generally point toward the presence of different spectral forms, while variations originating solely from spectral diffusion are usually small [69].

The properties of different subensembles and the transitions between these forms were first addressed for ensembles of GFP group proteins using low temperature spectroscopy. At cryogenic temperatures, three spectroscopically different, interconvertible forms were found for a number of GFP variants [48, 80, 81]. These forms were named A-form with neutral chromophore, B-form with deprotonated, anionic chromophore, and I-form which was found to be an intermediate involved in the transition between these forms. The B-form with anionic chromophore is strongly fluorescent and the predominant form in brightly emitting proteins such as the Enhanced Yellow Fluorescent Protein (EYFP). Experiments at cryogenic temperatures can establish limits on the heights of the energy barriers between the various forms, but it is not possible to conclude from these experiments if the different forms are of relevance at room temperature conditions typical for the use of these proteins.

To address the question of spectrally different forms, spectrally resolved single molecule detection was used to analyze the fluorescent protein EYFP in detail [36] by recording a large number of single protein spectral series. Once again the emission maximum peak position was the chosen analysis parameter, and was determined for each spectrum and assembled into the histogram presented in Fig. 6.

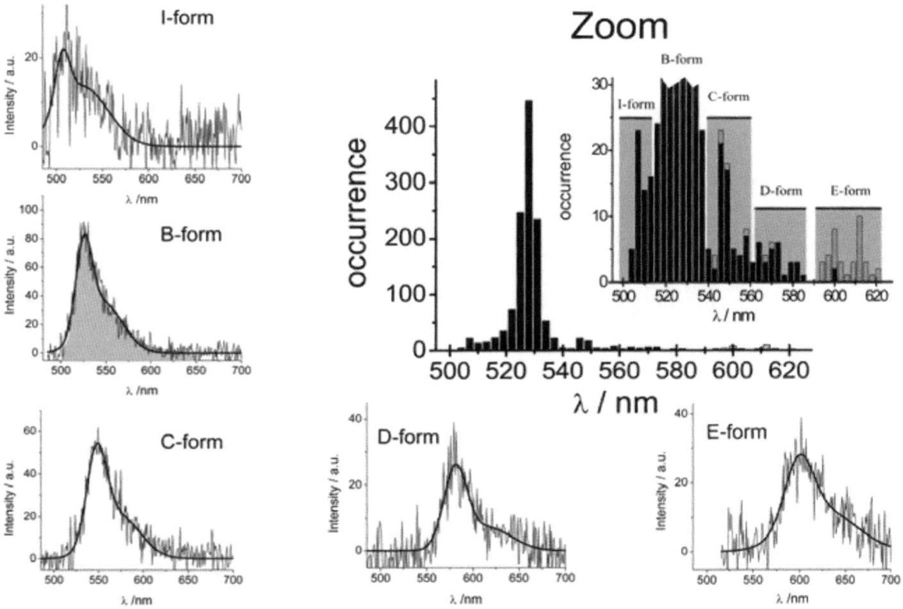

Fig. 6 Characteristic single protein emission spectra of all spectrally different forms observed for the enhanced yellow fluorescent protein (EYFP); ensemble spectrum in solid grey. The distribution of emission maximum positions from all analyzed single EYFP molecules is clearly not unimodal. Besides the main distribution around 527 nm, side maxima on the blue as well as on the red side of the main distribution are clearly discernible (Adapted from [36], with permission from Elsevier)

The histogram of the emission maximum positions from single EYFP molecules was clearly not a unimodal Gaussian distribution. The distribution is very widely stretched with a dominating peak around ~527 nm. This peak was expected, since the predominant form in EYFP was known from ensemble studies to be the B-form with emission maximum at 527 nm. More interesting were the secondary distributions on the blue as well as on the red side of the main distribution that suggested the existence of different, coexisting spectral forms. On the blue side, a secondary distribution centered at ~508 nm was found that was identified to originate from the blue-shifted intermediate I-form that had heretofore only been observed at cryogenic temperatures. Using single molecule spectroscopy does not only provide evidence of the presence of spectrally different forms but also allows the characterization of these forms without being disturbed by the overwhelming contribution of the dominating forms as in ensemble studies. In case of EYFP, the single molecule study resulted in the first undisturbed emission spectra of the I-form (Fig. 6), which yielded interesting hints about the underlying molecular basis of this form. Interestingly, the EYFP I-form spectrum was found to be congruent with the well-known emission spectrum from the anionic GFP chromophore. The EYFP emission is red shifted compared to the emission of the anionic GFP chromophore due to π-stacking interactions between the aromatic ring of the tyrosine residue introduced at position 203 and the chromophoric π-system [50]. Based on the observation that the EYFP I-form spectrum resembles the emission of the chromophore lacking π-stacking, it was concluded that a disturbed interaction of the π-stacking by a conformational change in the chromophore surrounding, e.g., by a turn or tilt of the phenolic group of the tyrosine residue most likely results in the spectral signature observed for the I-form.

Since series of emission spectra from each molecule were sampled, it was also possible to observe direct dynamics between the different forms – an observation impossible in ensemble spectroscopy. Interestingly, the transition between the B- and the I- form was always accompanied by a dark period between the disappearance of one and the appearance of the other form (Fig. 7). This spectral signature

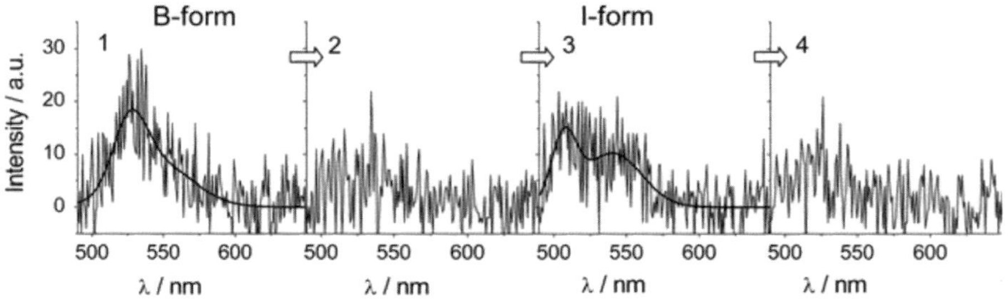

Fig. 7 Spectral sequence showing the transition from the predominant EYFP B-form to the blue-shifted I-form. The transitions were always accompanied by a dark period between the observations of emission from the two different forms (Reprinted from [36], with permission from Elsevier)

clearly points towards an at least two-step mechanism with a nonradiative – possibly very effectively quenched – intermediate.

Besides the I-form, which is blue shifted relative to the dominant B-form, a number of emission spectra (6% of all sampled single molecules) were found for EYFP that showed emission peaks red shifted well beyond 540 nm. The wide spread of the maximum positions and the comparatively low occurrence made the assignment of different forms difficult since no clear distribution of maximum positions could be identified. Nevertheless, it was postulated that the observed emission spectra belonged to three, so far unknown, spectral forms (Fig. 6). The recorded spectral series showed transitions from the dominating B-form to the red-shifted forms, evidencing the connection between these forms. Upon closer inspection, the recorded emission spectra from these red-shifted forms showed striking similarities to emission spectra recorded from DsRed group of proteins (for details see below). Further, comparable red-shifted forms occurring with comparable frequencies were found for EGFP, indicating a generic origin of all these forms.

Besides the VFPs that belong to the GFP group of proteins, the proteins of the DsRed group of proteins were intensely researched on the single molecule level using spectrally resolved spectroscopy to identify different forms. The photophysical complexity and multimeric nature is shared by many of the known red-emitting proteins to date [10, 16, 82–84], which makes DsRed and its variants a valuable model system for this group of proteins.

DsRed forms obligate tetramers [85] even at concentrations used for single molecule spectroscopy, and the constituent monomeric subunits contain either a green-emitting chromophore analogous to that of GFP or a red-emitting chromophore [86]. Within the tetramers, the different chromophores are likely to be coupled by fluorescence resonance energy transfer. This energy transfer coupling, together with the intrinsic complexity seen for all VFPs, results in markedly rich photophysics.

The first single molecule emission spectra from DsRed were presented as evidence that immobilization of the proteins in a film of PVA does not change the characteristic emission [74]. Already the first study of a statistically relevant number of DsRed single molecule emission spectra highlighted the photophysical complexity of DsRed [77]. The histogram of the single molecule emission maximum positions yielded the expected Gaussian distribution originating from spectral diffusion. However, the distribution was clearly red shifted with respect to the bulk emission maximum position, providing evidence for the fast formation of a red-shifted, so-called super-red form under typical single molecule detection conditions. The super-red form of DsRed was found to be photoinduced [47, 87] and single molecule experiments yielded emission spectra of this form for the first time (Fig. 8).

These first pioneering studies were followed by systematic studies of the emission spectra of different DsRed variants at the single tetramer level that further highlighted the spectral complexity of this group of proteins [88]. DsRed variants with different amino acid substitutions (DsRed2: Arg2Ala, Lys5Glu, Lys9Thr,

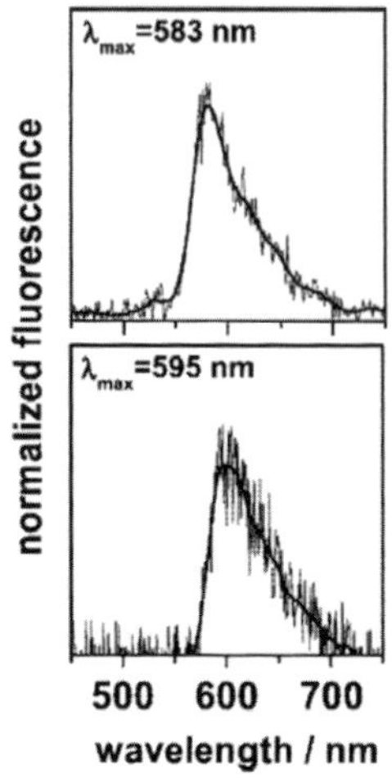

Fig. 8 Emission spectra from single DsRed tetramers. Emission with maximum at 583 nm as observed for ensemble samples was only rarely observed. Most of the emission spectra showed a distinct shift of the emission spectra to longer wavelengths. This red-shifted emission was attributed to the photoinduced formation of a "super-red form" (From [77], Copyright (2001) National Academy of Sciences, USA)

Val105Ala, Ile161Thr, Ser197Ala; Fluorescent Timer: Val105Ala, Ser197Thr; DsRed_N42H: Asn42His; AG4: Val71Met, Val105Ala, Ser197Thr) exhibiting altered spectral and maturation properties were thoroughly analyzed. In these studies, single molecule emission spectra from the green-emitting DsRed chromophore were measured for the first time. Full sets of emission spectra consisting of spectra from the green-emitting chromophore and from the red-emitting chromophore were presented for DsRed as well as for all sampled variants (Fig. 9). Also single protein tetramers were found that showed mixed emission from the green as well as from the red-emitting chromophore, providing direct evidence of the existence of mixed tetramers consisting of tetramers containing the different chromophores.

To identify different spectral forms and especially to establish whether the super-red form observed for DsRed can be found for all DsRed variants, the emission maximum positions were determined and assembled into histograms. The analysis of the emission maximum positions of all single tetramer spectra of all variants showed that the emission maximum positions of the green-emitting chromophores are distributed around ~503 nm, consistent with bulk spectroscopy (Fig. 9). The emission maximum positions of the red chromophores were found to be distributed ~600 nm (with slight differences for the different variants), a value that is ~16 nm red shifted relative to the bulk emission maxima. This observation verified the rapid formation of a high quantum efficiency super-red form for all DsRed variants.

Although the super-red form has been observed for DsRed and a number of variants, its underlying molecular origins are still unclear. It has been suggested that the photoconversion to the super-red form originates from structural changes in the vicinity of the chromophore due to a *cis–trans* isomerization of the chromophore and a decarboxylation of a glutamate (Glu-215) in the chromophore vicinity [46]. These changes result in a red-shifted, low quantum efficiency form.

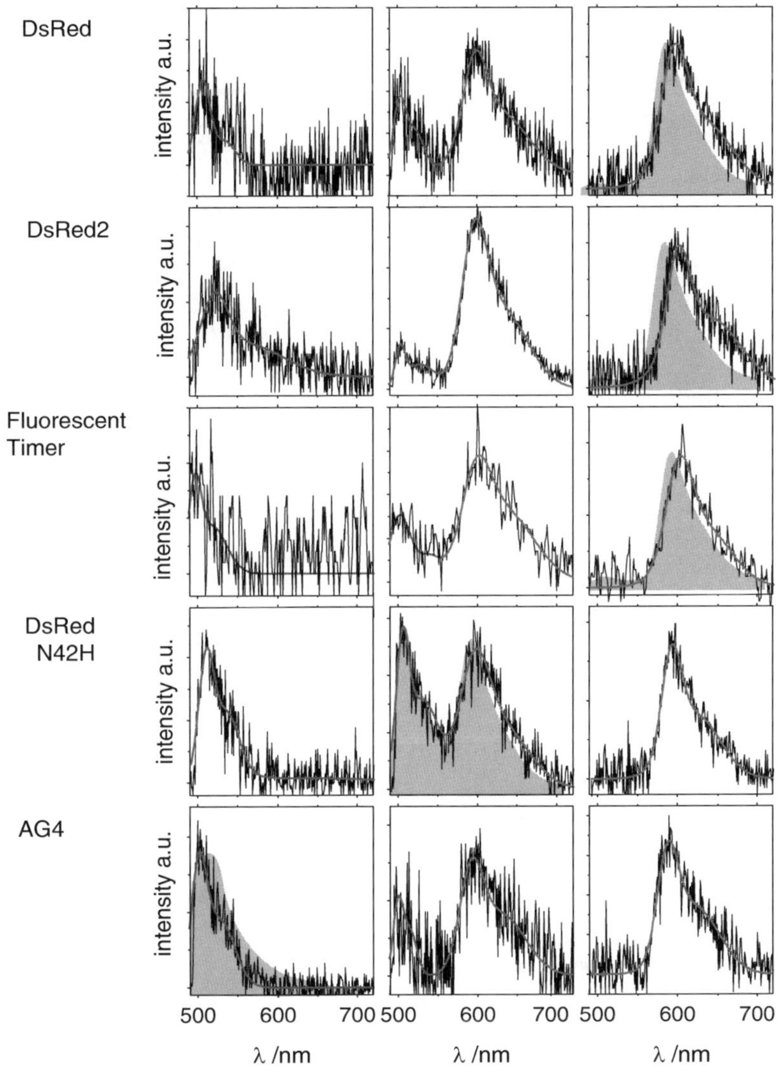

Fig. 9 Single tetramer spectra of DsRed and the analyzed variants. For all proteins emission spectra showing solely green emission (*first column*), mixed emission (*second column*), or solely red emission (*third column*) were found. The emission from the red-emitting chromophore was systematically shifted to the red, consistent with the fast formation of a super-red form. The frequencies of occurrence of the respective spectra were consistent with the bulk spectra, displayed in grey behind the predominantly observed spectrum (From [88], Copyright Wiley-VCH Verlag GmbH & Co. KGaA. Reproduced with permission)

Indeed, all variants for which the super-red form was found contain a glutamate residue at position 215; thus, the universal observation of the super-red form is consistent with this proposed mechanism. However, the super-red form was readily observed at the single molecule level, which suggests a high quantum efficiency. Additionally, experiments at cryogenic temperatures revealed that reversible photoconversion between a super-red form and other spectral forms is possible [47], which contradicts the proposed mechanism since photoinduced decarboxylation is usually irreversible. Considering all the information gathered from ensemble spectroscopy, cryogenic spectroscopy, and single molecule spectroscopy, it appears likely that two different super-red-emitting forms exist, one with high fluorescence quantum efficiency, which reversibly photoconverts to other forms, and which is observed at the single molecule level and in the experiments at cryogenic temperatures. The second super-red form would then be of low quantum efficiency and is potentially formed by irreversible decarboxylation as suggested [46].

The histograms of single molecule emission maximum positions revealed emission spectra with maximum positions between 530 and 570 nm that were found for DsRed and all the variants sampled (Fig. 10). It was impossible to assign these spectra (Fig. 11a) to any known form of the green- or the red-emitting chromophore. The percentage of tetramers showing this rare form emission is generally low but not negligible (DsRed ~3%, Fluorescent Timer ~1%, DsRed2 ~6%, AG4 ~2% and DsRed_N42H ~2%). Spectral series showed transitions from the known main forms to the rare forms and thus demonstrated that these rare forms are connected to the main forms. (Fig. 11b). It is also striking that these forms appeared at similar wavelengths as the rare forms found for proteins from the GFP group of proteins.

As demonstrated before, it is often helpful to combine results obtained from the single molecule experiments with results obtained from conventional ensemble spectroscopy to understand the details of complex emitting systems. Comparing the emission observed from the rare forms of DsRed and its variants (Fig. 11) to the dominant emission from the protein zFP538 from coral *Zoanthus sp.* and mOrange reveals clear similarities. The emission maximum of zFP538 is at 538 nm, the emission maximum of mOrange is at 562 nm, which is in the same wavelength region as the observed DsRed rare forms. zFP538 and mOrange embody a chromophore that resembles a truncated red-emitting DsRed chromophore [41, 42]. This similarity is suggestive of an analogous, possibly photoactivated, modification of the red-emitting chromophore from an acylimine to an imine resulting in the truncated chromophore. In spectral series only transitions from the main forms to the rare spectral forms were observed, but never transitions back from rare forms to one of the main forms. This observation supports the hypothesis of a modification of the chromophore since the creation of the imine is not likely to be reversible. As discussed above, not only the chemical structure of the chromophore determines the spectral properties, but also the chromophore nanoenvironment contributes to the exact emission position and distribution of the emission maximum positions observed at the single molecule level. It is likely that these differences in the chromophore embedding between the DsRed variants and zFP538 and mOrange accounts for the variances in the emission maximum position of the DsRed rare forms compared to zFP538 and mOrange.

Fig. 10 Histograms of emission peak positions for DsRed and variants. The emission from the red-emitting chromophore is systematically red-shifted for all variants compared to the bulk emission, and is attributed to the rapid formation of a super-red form of the chromophore. The emission maximum position from the green-emitting chromophore agrees with the ensemble peak of this chromophore for all variants. The histogram further shows the emission maximum positions of rare forms emitting in the wavelength area between the typical emission of the green- and the red-emitting chromophore (see inset in the *top panel*) (From [88], Copyright Wiley-VCH Verlag GmbH & Co. KGaG. Reproduced with permission)

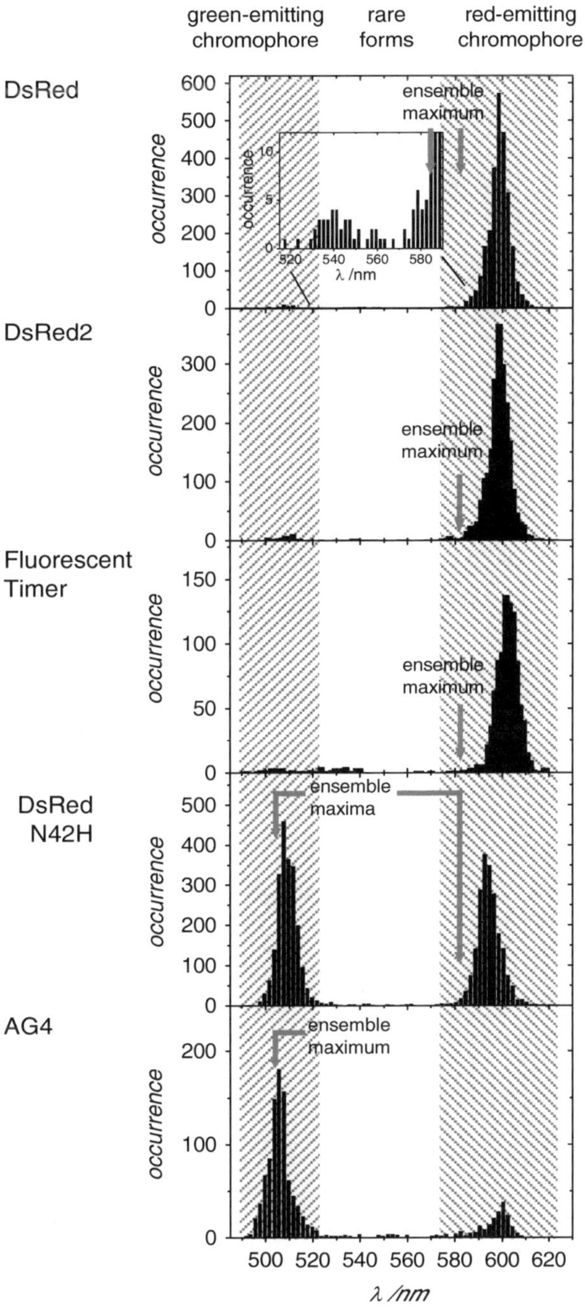

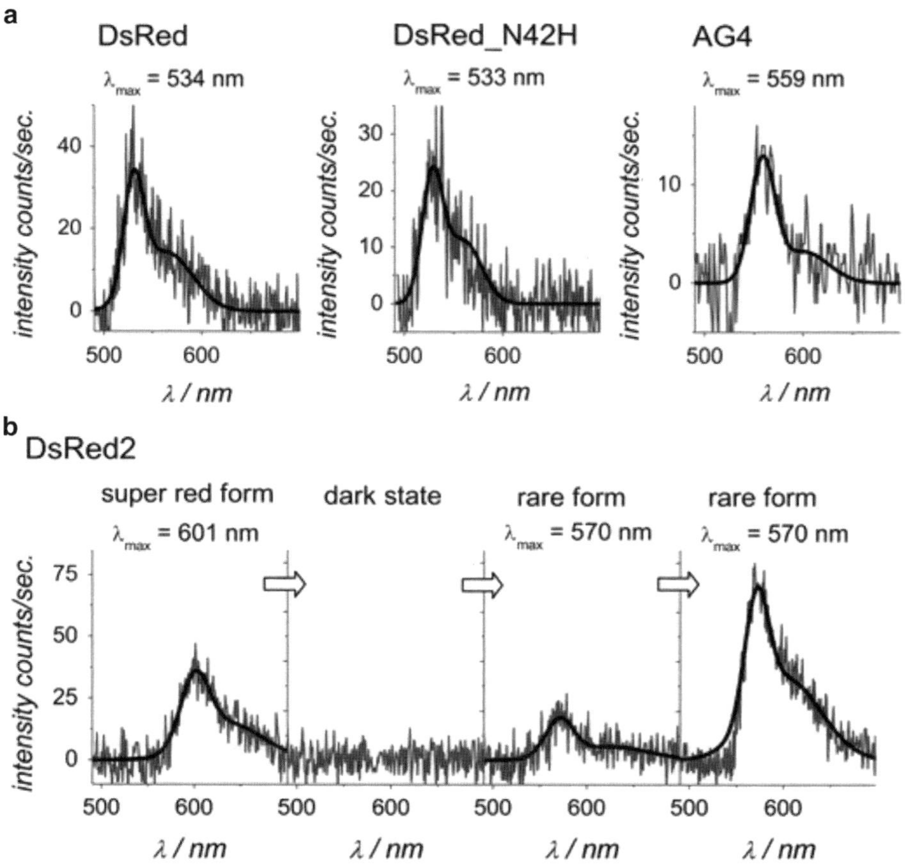

Fig. 11 (**a**) For all protein variants rare spectral forms with emission maximum positions between ~530 nm and 575 nm were observed. (**b**) Transitions between the predominant form and a rare spectral form with emission maximum at 570 nm via a dark state provide evidence that the rare forms are connected to the main forms [88] (Copyright Wiley-VCH Verlag GmbH & Co. KGaG. Reproduced with permission)

4 Summary

It is essential to understand VFP emission properties in depth, since the applications in which VFPs are used to quantitatively elucidate biological and biochemical processes are becoming more and more complex. Single molecule emission spectroscopy is a powerful tool to access aspects of VFP photophysics, which are hidden to the averaging of properties inherent to conventional ensemble spectroscopy.

In particular, techniques that record more parameters than simply the change of emission intensity over time give detailed insights into the photophysics of VFPs. These techniques even give access to the dynamics of single molecules when the evolution of the respective parameter is followed in time.

Single molecule spectroscopy of VFPs is experimentally still challenging because of the low photostability of VFPs, but the insights that can be gained have immense potential, and make the challenge well worth the effort. Spectrally resolved single molecule spectroscopy yielding full emission spectra from individual VFPs has proven to be an excellent method to access the amazing spectral heterogeneity of VFPs. Single molecule emission spectroscopy has added a lot to understanding the emission properties of VFPs by allowing the identification of hidden subensembles, giving valuable information about the possible molecular origins of these subensembles, by shedding light on the transitions and dynamics between these subensembles and by allowing quantitative insights into the spectral diffusion and its origins in VFPs. To date, single molecule analysis of VFPs has focused on the most commonly used proteins from the GFP and the DsRed group of proteins, but the rapidly growing number of VFPs, and in particular the photoactivatable or photoswitchable proteins that underpin optical super-resolution methods, holds further promise for exciting insights into the spectral heterogeneity of these complex emitters.

References

1. Betzig E, Patterson GH, Sougrat R, Lindwasser OW, Olenych S, Bonifacino JS, Davidson MW, Lippincott-Schwartz J, Hess HF (2006) Science 313(5793):1642–1645
2. Donnert G, Keller J, Wurm CA, Rizzoli SO, Westphal V, Schonle A, Jahn R, Jakobs S, Eggeling C, Hell SW (2007) Biophys J 92(8):L67–L69
3. Hell SW (2007) Science 316(5828):1153–1158
4. Hess ST, Girirajan TP, Mason MD (2006) Biophys J 91(11):4258–4272
5. Schermelleh L, Carlton PM, Haase S, Shao L, Winoto L, Kner P, Burke B, Cardoso MC, Agard DA, Gustafsson MG, Leonhardt H, Sedat JW (2008) Science 320(5881):1332–1336
6. Shroff H, Galbraith CG, Galbraith JA, Betzig E (2008) Nat Methods 5(5):417–423
7. Willig KI, Kellner RR, Medda R, Hein B, Jakobs S, Hell SW (2006) Nat Methods 3(9):721–723
8. Matz MV, Fradkov AF, Labas YA, Savitsky AP, Zaraisky AG, Markelov ML, Lukyanov SA (1999) Nat Biotechnol 17(10):969–973
9. Gurskaya NG, Fradkov AF, Terskikh A, Matz MV, Labas YA, Martynov VI, Yanushevich YG, Lukyanov KA, Lukyanov SA (2001) FEBS Lett 507(1):16–20
10. Wiedenmann J, Schenk A, Rocker C, Girod A, Spindler KD, Nienhaus GU (2002) Proc Natl Acad Sci USA 99(18):11646–11651
11. Karasawa S, Araki T, Yamamoto-Hino M, Miyawaki A (2003) J Biol Chem 278 (36):34167–34171
12. Kogure T, Karasawa S, Araki T, Saito K, Kinjo M, Miyawaki A (2006) Nat Biotechnol 24 (5):577–581
13. Chan MC, Karasawa S, Mizuno H, Bosanac I, Ho D, Prive GG, Miyawaki A, Ikura M (2006) J Biol Chem 281(49):37813–37819
14. Beddoe T, Ling M, Dove S, Hoegh-Guldberg O, Devenish RJ, Prescott M, Rossjohn J (2003) Acta Crystallogr D Biol Crystallogr 59(Pt 3):597–599
15. Merzlyak EM, Goedhart J, Shcherbo D, Bulina ME, Shcheglov AS, Fradkov AF, Gaintzeva A, Lukyanov KA, Lukyanov S, Gadella TW, Chudakov DM (2007) Nat Methods 4(7):555–557
16. Shcherbo D, Merzlyak EM, Chepurnykh TV, Fradkov AF, Ermakova GV, Solovieva EA, Lukyanov KA, Bogdanova EA, Zaraisky AG, Lukyanov S, Chudakov DM (2007) Nat Methods 4(9):741–746

17. Campbell RE, Tour O, Palmer AE, Steinbach PA, Baird GS, Zacharias DA, Tsien RY (2002) Proc Natl Acad Sci USA 99(12):7877–7882
18. Terskikh A, Fradkov A, Ermakova G, Zaraisky A, Tan P, Kajava AV, Zhao XN, Lukyanov S, Matz M, Kim S, Weissman I, Siebert P (2000) Science 290(5496):1585–1588
19. Bevis BJ, Glick BS (2002) Nat Biotechnol 20(1):83–87
20. Shaner NC, Campbell RE, Steinbach PA, Giepmans BN, Palmer AE, Tsien RY (2004) Nat Biotechnol 22(12):1567–1572
21. Lippincott-Schwartz J, Patterson GH (2008) Methods Cell Biol 85:45–61
22. Lippincott-Schwartz J, Altan-Bonnet N, Patterson GH (2003) Nat Cell Biol 5(Suppl):S7–S14
23. Patterson GH, Lippincott-Schwartz J (2002) Science 297(5588):1873–1877
24. Chudakov DM, Belousov VV, Zaraisky AG, Novoselov VV, Staroverov DB, Zorov DB, Lukyanov S, Lukyanov KA (2003) Nat Biotechnol 21(2):191–194
25. Chudakov DM, Verkhusha VV, Staroverov DB, Souslova EA, Lukyanov S, Lukyanov KA (2004) Nat Biotechnol 22(11):1435–1439
26. Lukyanov KA, Chudakov DM, Lukyanov S, Verkhusha VV (2005) Nat Rev Mol Cell Biol 6 (11):885–891
27. Chudakov DM, Chepurnykh TV, Belousov VV, Lukyanov S, Lukyanov KA (2006) Traffic 7 (10):1304–1310
28. Flors C, Hotta J, Uji-i H, Dedecker P, Ando R, Mizuno H, Miyawaki A, Hofkens J (2007) J Am Chem Soc 129(45):13970–13977
29. Sinnecker D, Voigt P, Hellwig N, Schaefer M (2005) Biochemistry 44(18):7085–7094
30. Henderson JN, Ai HW, Campbell RE, Remington SJ (2007) Proc Natl Acad Sci USA 104 (16):6672–6677
31. Bulina ME, Chudakov DM, Britanova OV, Yanushevich YG, Staroverov DB, Chepurnykh TV, Merzlyak EM, Shkrob MA, Lukyanov S, Lukyanov KA (2006) Nat Biotechnol 24 (1):95–99
32. Bulina ME, Lukyanov KA, Britanova OV, Onichtchouk D, Lukyanov S, Chudakov DM (2006) Nat Protoc 1(2):947–953
33. Weiss S (1999) Science 283(5408):1676–1683
34. Moerner WE, Fromm DP (2003) Rev Sci Instrum 74(8):3597–3619
35. Garcia-Parajo MF, Veerman JA, Bouwhuis R, Vallee R, van Hulst NF (2001) Chemphyschem 2(6):347–360
36. Blum C, Meixner AJ, Subramaniam V (2004) Biophys J 87(6):4172–4179
37. Yang F, Moss LG, Phillips GN (1996) Nat Biotechnol 14(10):1246–1251
38. Ormo M, Cubitt AB, Kallio K, Gross LA, Tsien RY, Remington SJ (1996) Science 273 (5280):1392–1395
39. Verkhusha VV, Chudakov DM, Gurskaya NG, Lukyanov S, Lukyanov KA (2004) Chem Biol 11(6):845–854
40. Strack RL, Strongin DE, Mets L, Glick BS, Keenan RJ (2010) J Am Chem Soc 132(24): 8496–8505
41. Remington SJ, Wachter RM, Yarbrough DK, Branchaud B, Anderson DC, Kallio K, Lukyanov KA (2005) Biochemistry 44(1):202–212
42. Shu X, Shaner NC, Yarbrough CA, Tsien RY, Remington SJ (2006) Biochemistry 45 (32):9639–9647
43. Tubbs JL, Tainer JA, Getzoff ED (2005) Biochemistry 44(29):9833–9840
44. Terskikh AV, Fradkov AF, Zaraisky AG, Kajava AV, Angres B (2002) J Biol Chem 277 (10):7633–7636
45. van Thor JJ, Gensch T, Hellingwerf KJ, Johnson LN (2002) Nat Struct Biol 9(1):37–41
46. Habuchi S, Cotlet M, Gensch T, Bednarz T, Haber-Pohlmeier S, Rozenski J, Dirix G, Michiels J, Vanderleyden J, Heberle J, De Schryver FC, Hofkens J (2005) J Am Chem Soc 127 (25):8977–8984
47. Bonsma S, Gallus J, Konz F, Purchase R, Volker S (2004) J Lumin 107(1–4):203–212

48. Creemers TMH, Lock AJ, Subramaniam V, Jovin TM, Volker S (2000) Proc Natl Acad Sci USA 97(7):2974–2978
49. Tsien RY (1998) Annu Rev Biochem 67:509–544
50. Wachter RM, Elsliger MA, Kallio K, Hanson GT, Remington SJ (1998) Structure Fold Des 6 (10):1267–1277
51. Blum C, Meixner AJ, Subramaniam V (2006) J Am Chem Soc 128(26):8664–8670
52. Jung G, Wiehler J, Zumbusch A (2005) Biophys J 88(3):1932–1947
53. Jung G, Ma YZ, Prall BS, Fleming GR (2005) Chemphyschem 6(8):1628–1632
54. Wiedenmann J, Ivanchenko S, Oswald F, Nienhaus GU (2004) Mar Biotechnol (NY) 6 (3):270–277
55. Roy R, Hohng S, Ha T (2008) Nat Meth 5(6):507–516
56. Schneckenburger H (2005) Curr Opin Biotechnol 16(1):13–18
57. Nie SM, Zare RN (1997) Annu Rev Biophys Biomol Struct 26:567–596
58. Dickson RM, Cubitt AB, Tsien RY, Moerner WE (1997) Nature 388(6640):355–358
59. Peterman EJG, Brasselet S, Moerner WE (1999) J Phys Chem A 103(49):10553–10560
60. Garcia-Parajo MF, Segers-Nolten GMJ, Veerman JA, Greve J, van Hulst NF (2000) Proc Natl Acad Sci USA 97(13):7237–7242
61. Steinmeyer R, Noskov A, Krasel C, Weber I, Dees C, Harms GS (2005) J Fluoresc 15 (5):707–721
62. McAnaney TB, Zeng W, Doe CFE, Bhanji N, Wakelin S, Pearson DS, Abbyad P, Shi XH, Boxer SG, Bagshaw CR (2005) Biochemistry 44(14):5510–5524
63. Habuchi S, Cotlet M, Gronheid R, Dirix G, Michiels J, Vanderleyden J, De Schryver FC, Hofkens J (2003) J Am Chem Soc 125(28):8446–8447
64. Niwa H, Inouye S, Hirano T, Matsuno T, Kojima S, Kubota M, Ohashi M, Tsuji FI (1996) Proc Natl Acad Sci USA 93(24):13617–13622
65. Cinelli RAG, Pellegrini V, Ferrari A, Faraci P, Nifosi R, Tyagi M, Giacca M, Beltram F (2001) Appl Phys Lett 79(20):3353–3355
66. Chirico G, Diaspro A, Cannone F, Collini M, Bologna S, Pellegrini V, Beltram F (2005) Chemphyschem 6(2):328–335
67. Stracke F, Blum C, Becker S, Mullen K, Meixner AJ (2004) Chem Phys 300(1–3):153–164
68. Jung C, Hellriegel C, Michaelis J, Brauchle C (2007) Adv Mater 19(7):956–960
69. Blum C, Stracke F, Becker S, Mullen K, Meixner AJ (2001) J Phys Chem A 105(29): 6983–6990
70. Xie XS (1996) Acc Chem Res 29(12):598–606
71. Moerner WE, Orrit M (1999) Science 283(5408):1670–1676
72. Moerner WE (2002) J Chem Phys 117(24):10925–10937
73. Jung G, Wiehler J, Gohde W, Tittel J, Basche T, Steipe B, Brauchle C (1998) Bioimaging 6 (1):54–61
74. Cotlet M, Hofkens J, Kohn F, Michiels J, Dirix G, Van Guyse M, Vanderleyden J, De Schryver FC (2001) Chem Phys Lett 336(5–6):415–423
75. Lounis B, Deich J, Rosell FI, Boxer SG, Moerner WE (2001) J Phys Chem B 105 (21):5048–5054
76. Garcia-Parajo MF, Koopman M, van Dijk E, Subramaniam V, van Hulst NF (2001) Proc Natl Acad Sci USA 98(25):14392–14397
77. Cotlet M, Hofkens J, Habuchi S, Dirix G, Van Guyse M, Michiels J, Vanderleyden J, De Schryver FC (2001) Proc Natl Acad Sci USA 98(25):14398–14403
78. Heikal AA, Hess ST, Baird GS, Tsien RY, Webb WW (2000) Proc Natl Acad Sci USA 97 (22):11996–12001
79. Baird GS, Zacharias DA, Tsien RY (2000) Proc Natl Acad Sci USA 97(22):11984–11989
80. Creemers TMH, Lock AJ, Subramaniam V, Jovin TM, Volker S (1999) Nat Struct Biol 6 (6):557–560
81. Creemers TMH, Lock AJ, Subramaniam V, Jovin TM, Volker S (2002) Chem Phys 275 (1–3):109–121

82. Tsutsui H, Karasawa S, Shimizu H, Nukina N, Miyawaki A (2005) EMBO Rep 6(3):233–238
83. Lessard GA, Habuchi S, Werner JH, Goodwin PM, De Schryver F, Hofkens J, Cotlet M (2008) J Biomed Opt 13(3):031212
84. Wiedenmann J, Ivanchenko S, Oswald F, Schmitt F, Rocker C, Salih A, Spindler KD, Nienhaus GU (2004) Proc Natl Acad Sci USA 101(45):15905–15910
85. Yarbrough D, Wachter RM, Kallio K, Matz MV, Remington SJ (2001) Proc Natl Acad Sci USA 98(2):462–467
86. Gross LA, Baird GS, Hoffman RC, Baldridge KK, Tsien RY (2000) Proc Natl Acad Sci USA 97(22):11990–11995
87. Malvezzi-Campeggi F, Jahnz M, Heinze KG, Dittrich P, Schwille P (2001) Biophys J 81 (3):1776–1785
88. Blum C, Meixner AJ, Subramaniam V (2008) Chemphyschem 9(2):310–315

Structure–Function Relationships in Fluorescent Marker Proteins of the Green Fluorescent Protein Family

G. Ulrich Nienhaus, Karin Nienhaus, and Jörg Wiedenmann

Abstract GFP-like proteins, originally cloned from marine animals, are genetically encoded fluorescence markers that have become indispensable tools for the life sciences. The search for GFP-like proteins with novel and improved properties is ongoing, driven by the persistent need for advanced and specialized fluorescence labels for cellular imaging. The 3D structures of these proteins are overall similar. However, considerable variations have been found in the covalent structures and the stereochemistry of the chromophore, which govern essential optical properties such as the absorption/emission wavelengths. A detailed understanding of the structure and dynamics of GFP-like proteins greatly aids in the rational engineering of advanced fluorescence marker proteins. In this chapter, we summarize the present knowledge of the structural diversity of GFP-like proteins and discuss how structure and dynamics govern their optical properties.

Keywords Chromophore · Fluorescent protein · GFP

Contents

G.U. Nienhaus (✉)
Institute of Applied Physics and Center for Functional Nanostructures, Karlsruhe Institute of Technology (KIT), Wolfgang-Gaede-Str. 1, 76131 Karlsruhe, Germany
and
Department of Physics, University of Illinois at Urbana-Champaign, 1110 West Green Street, Urbana, IL 61801, USA
e-mail: uli@illinois.edu

K. Nienhaus
Institute of Applied Physics and Center for Functional Nanostructures, Karlsruhe Institute of Technology (KIT), Wolfgang-Gaede-Str. 1, 76131 Karlsruhe, Germany

J. Wiedenmann
National Oceanography Centre, University of Southampton, Southampton SO14 3ZH, UK

G. Jung (ed.), *Fluorescent Proteins I*, Springer Ser Fluoresc (2012) 11: 241–264,
DOI 10.1007/4243_2011_15,
Published online: 31 March 2011

1 Introduction

The green fluorescent protein (GFP) and the related fluorescent proteins (FPs) of the GFP family have revolutionized life sciences research by enabling a vast array of novel approaches to study biomolecular processes, especially in living cells and organisms. In recognition of GFP's enormous impact on the life sciences, the Nobel Committee for Chemistry awarded the Prize in 2008 to Osamu Shimomura, Martin Chalfie, and Roger Y. Tsien "for the discovery and development of the green fluorescent protein, GFP" [1]. Shimomura isolated the protein already in 1962 from tissue extracts of the jellyfish *Aequorea victoria* [2]. But only in 1994, when the GFP gene was expressed recombinantly in *Escherichia coli* and *Caenorhabditis elegans*, researchers realized that GFP can be used as a genetically encoded fluorescence marker because the fluorophore forms spontaneously in a reaction that only requires O_2 [3, 4]. To extend its range of applications, color variants of *A. victoria* GFP were developed by mutagenesis, with emission peaks ranging from blue (BFP) to yellow (YFP) [5]. A major step forward was the discovery of GFP homologues in anthozoa animals [6–8], so that a large number of fluorescent proteins (FPs) became available as potential fluorescent marker tools including the long-sought orange and red FPs. They extended the color palette for multicolor labeling or FRET experiments, and were also highly welcomed for live-cell and tissue imaging applications because of the reduced cellular autofluorescence and scattering in the red spectral range. In recent years, so-called photoactivatable or optical highlighter FPs have emerged as powerful new tools for cellular imaging [9–12]. Upon irradiation with light of specific wavelengths, these FPs can either be switched reversibly between a fluorescent and a nonfluorescent state (photoswitching), or they change their fluorescence emission intensity or color irreversibly (photoconversion).

However, FPs obtained from natural sources frequently show poor performance as fluorescence markers, including a tendency to aggregate or oligomerize, incomplete (or even completely lacking) chromophore formation (maturation), especially at physiological temperatures (37°C), fluctuating emission (photodynamics, flickering), and fast photobleaching. Detailed structural and spectroscopic investigations of GFP-like proteins have furthered our understanding of structure–dynamics–function relationships in recent years up to a point that, in many instances, rational development of optimized variants using genetic engineering approaches has become feasible.

2 Molecular Structure of GFP-Like Proteins

2.1 The β-Can Fold

The X-ray structures of GFP, first solved by Remington's and Phillips's groups in 1996 [13, 14], revealed that its polypeptide chain of 238 amino acids folds into a rigid, 11-stranded β-can with a central helix running along its axis (Fig. 1a). It is capped at either end by short helical sections and loops. The fluorescent chromophore interrupts the central helix and resides close to the geometric center of the protein. Surrounding residues and structural water molecules anchor the chromophore and hold it tightly in a planar conformation, which is essential for a high fluorescence quantum yield. Indeed, the isolated chromophore is nonfluorescent in solution because it rotates around its exocyclic bonds and undergoes fast radiationless decay by internal conversion upon reaching an avoided crossing of the S_0 and S_1 surfaces [15, 16]. Over the years, X-ray structures of FPs from many different species have been solved. They all share the β-can structure, suggesting that this

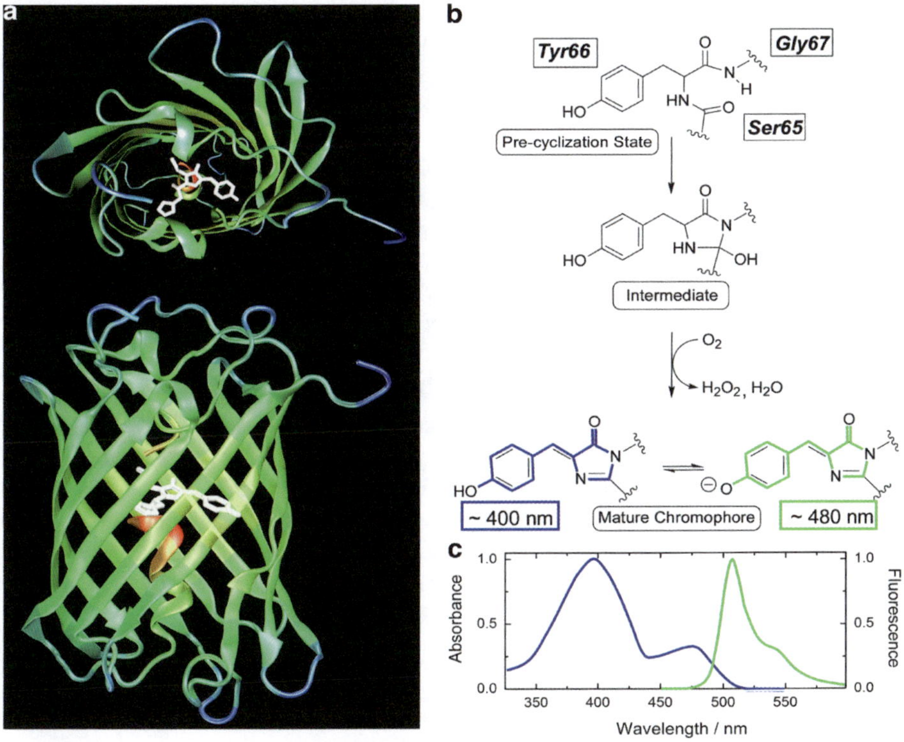

Fig. 1 Green fluorescent protein, GFP. (**a**) Cartoon models of the molecular structure. (**b**) Maturation scheme. (**c**) Absorption (*blue*) and emission (*green*) spectra of GFP

particular tertiary structure plays a crucial role both in formation/maturation of the chromophore and in regulating its photophysical properties.

2.2 *The Chromophore*

The chromophore forms autocatalytically from a tripeptide (in GFP: Ser65-Tyr66-Gly67), requiring nothing else but molecular oxygen. In general, the first amino acid may be any one, but the second and third amino acids, tyrosine and glycine, are strictly conserved in nature. However, while glycine is absolutely crucial for chromophore self-synthesis, tyrosine can be replaced, e.g., by histidine, phenylalanine, or tryptophan [5].

The green GFP chromophore, 4-(*p*-hydroxybenzylidene)-5-imidazolinone (*p*-HBI), results from a sequential cyclization–oxidation–dehydration reaction, with characteristic time scales ranging from minutes to hours (Fig. 1b). In a first step during protein folding, a peptide cyclization and possibly proton transfers occur, and a heterocyclic intermediate forms within a few minutes. In a second step, the protein reacts with O_2 to oxidize the tyrosine Cα–Cβ bond, and to generate a cyclic imine, hydrogen peroxide is released. The final step again involves a proton transfer reaction that induces bond rearrangements and leads to the fully conjugated, mature chromophore [17]. In the planar *p*-HBI chromophore, the conjugated π-electron system extends from the *p*-hydroxybenzyl ring of the tyrosine to the imidazolinone ring. With respect to the double bond of the methene bridge (Cα–Cβ of Tyr66), both rings are arranged in a *cis* conformation. Real-time visualization of de novo synthesis and folding of GFP molecules using single-molecule fluorescence microscopy has demonstrated that the fluorescence of the fastest GFP molecules appeared already within 1 min [18]. Because of the technique used, GFP molecules become detectable only after chromophore formation. Therefore, the characteristic time obtained from these experiments is related to the overall sequence of polypeptide synthesis, protein folding, and chromophore formation. In bulk measurements, transition midpoints of ~30 to ~90 min were observed for the maturation of GFP [5].

The absorption spectrum of wild-type GFP displays two bands associated with the chromophore (Fig. 1c). The dominant A band at 395 nm and the weaker B band at 475 nm represent the neutral and anionic tyrosine moieties, respectively. Upon excitation of the B band, the fluorescence emission peaks at 508 nm in the green (Fig. 1c). Interestingly, excitation of the neutral chromophore also results in green emission of an anionic species: in the excited state, the phenolic proton is released and transferred to Glu222 (excited state proton transfer, ESPT), so that the anionic species forms before photon emission [19–21].

2.3 *Variations of the Chromophore Motif*

In recent years, a number of modifications of the *p*-HBI chromophore have been found in GFP-like proteins from natural sources. In orange and red-fluorescent

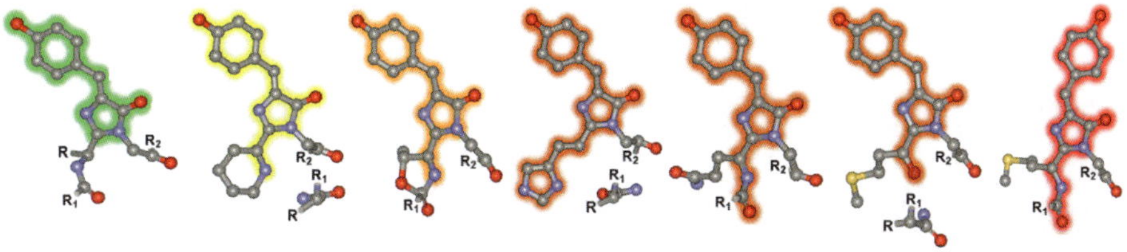

Fig. 2 The GFP chromophore and its autocatalytically produced variants (adopted from [24])

proteins such as DsRed [22] and eqFP611 [23], the single bond between the amide nitrogen and the Cα atom of the first of the three chromophore-forming amino acids is oxidized to an acylimine group in-plane with the p-HBI chromophore (Fig. 2). Consequently, the conjugated π-electron system of the resulting 2-imino-5-(4-hydroxybenzylidene)-imidazolinone chromophore is further extended, and the emission shifts bathochromically to ~600 nm. The delocalized π-electron system may extend even further to include the carbonyl group of the preceding amino acid in FPs emitting further to the red [25].

In mOrange, a monomeric DsRed variant emitting at 562 nm, an oxazole heterocycle (Fig. 2) is created from the side chain of Thr66, the first amino acid of the chromophore triad in this protein [26]. Its hydroxyl group apparently attacks the preceding carbonyl carbon. The yellow zFP538 from *Zoanthus* sp. likewise features a third heterocycle, formed by ring closure of a lysine side chain [27]. Presumably, the terminal amino group of the lysine side chain attacks the reactive N=Cα acylimine bond and, concomitantly, the protein backbone is cleaved, leaving the N-terminal fragment with a terminal carboxamide (Fig. 2). The smaller red shift compared with the acylimine of DsRed has been rationalized by the less effective charge delocalization by the cyclic imine [27]. In AsRed, the mutant Ala143Ser of asFP595 from *Anemonia sulcata* [28], the X-ray structure analysis revealed that chromophore maturation is accompanied by a break in the polypeptide chain [29–31], and a carbonyl group, generated by hydrolysis of the intermediately formed acylimine group, extends the conjugated π-electron system of the *p*-HBI chromophore (Fig. 2).

In addition to these spontaneously occurring chemical alterations of the green chromophore, photoinduced modifications can occur, as will be discussed in detail in Sect. 3.5.2.

2.4 Quaternary Structure

GFP has only a weak tendency to form dimers; the dissociation coefficient has been estimated as 100 μM [32]. The homodimeric interface includes a small patch of hydrophobic residues (Ala206, Leu221, Phe223) as well as a large number of

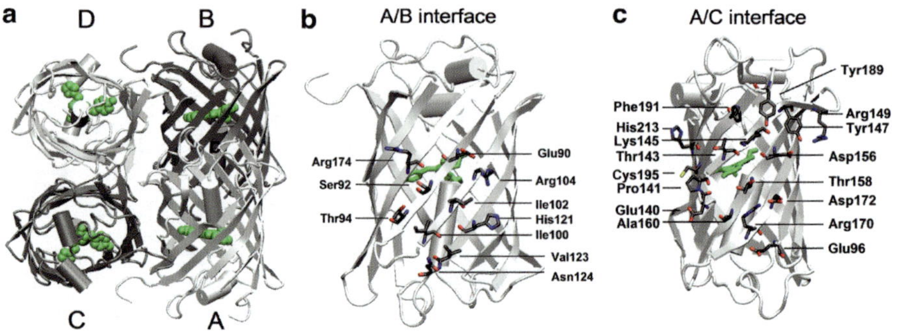

Fig. 3 Quaternary structure of tetrameric EosFP (cartoon representation). (**a**) Tetramer. (**b**) A/B interface. (**c**) A/C interface. The chromophore is highlighted in *green*

hydrophilic contacts (Tyr39, Glu142, Asn144, Ser147, Asn149, Tyr151, Arg168, Asn170, Glu172, Tyr200, Ser202, Gln204, Ser208).

Natural anthozoan FPs are frequently found to form tightly binding tetramers that can only be broken up by harsh procedures, causing irreversible denaturation of the polypeptide chains [33–35]. The four protomers A–D are arranged as dimers of dimers [36], so that two different interfaces between dimers can be distinguished (Fig. 3a). The A/C (B/D) interface (Fig. 3c) is more extended than the A/B (C/D) interface (Fig. 3b). It is hydrophilic and stabilized by salt bridges and hydrogen bonds between polar residues and structural waters [37, 38]. It is reinforced by the C-terminal end of one chain intertwined with the other. The A/B interface features a cluster of hydrophobic amino acids encircled by polar amino acids.

3 Protein Engineering

3.1 Maturation and Thermotolerance

Wild-type GFP folds fairly efficiently when expressed at or below room temperature, but its folding efficiency declines steeply at higher temperatures. However, once it has matured, i.e., turned into its functional form at low temperature, GFP is thermodynamically stable and fluoresces at temperatures up to at least 65°C. For applications that involve mammalian cell cultures, it is necessary that the FP expresses and functions well at 37°C. Codon optimization of GFP has led to an increase in the levels of protein expression by fourfold at 37°C [22, 39]. By extensive mutational studies, a number of mostly bulky amino acids have been identified in GFP that apparently contribute to the lack of expression at higher temperature [5]. Some of these residues are located close to the chromophore (Phe64Leu and Ser72Ala), whereas others are more remote, either buried inside the protein (Val163Ala) or in contact with the solvent (Met153Thr, Ser175Gly,

Phe99Ser). Mutations close to the chromophore may enhance the efficiency of chromophore formation, whereas the more distant mutations Val163Ala and Ser175Gly are likely to prevent protein misfolding at elevated temperatures; mutation Met153Thr may improve solubility by decreasing the surface hydrophobicity [40]. Additional protein engineering yielded the "superfolder GFP", which displays improved tolerance toward circular permutation, greater resistance to chemical denaturants, and enhanced folding kinetics [41].

For other FPs, sometimes only a few amino acid replacements have been observed to enhance protein expression at elevated temperatures. For example, the Ile57Val and Phe102Leu mutations endowed eqFP611 with reliable expression at 37°C, as was revealed by random mutagenesis with subsequent screening of the resulting clones [42]. The monomeric version of EosFP (mEosFP) could even be rendered thermotolerant by introducing a single point mutation, Ala69Val [43].

3.2 Cellular Toxicity

Aggregation of FPs is a general problem that may cause cytotoxic effects similar to the toxicity associated with aggregate formation in neurodegenerative disease [44]. For example, Link et al. [45] have enhanced aggregation of GFP by adding a specific 16-peptide sequence to the C-terminus, which caused paralysis upon expression in body wall muscle cells of *C. elegans*. Interestingly, optimized codon usage significantly reduced the cytotoxicity of EosFP when expressed in murine stem cells, indicating that also rather unspecific effects of overexpression might account for the cytotoxicity of some FPs [46].

Under illumination, FPs produce toxic reactive oxygen species (ROS) [47] including singlet oxygen, $^1O_2$. It attacks preferentially aromatic and sulfur-containing amino acids of proteins, but other cellular components might be affected as well. For many applications, the phototoxicity of typical FPs is low enough to be ignored, which is not surprising considering that FPs evolved in organisms exposed to O_2 and plenty of sunlight.

KillerRed, however, is an engineered FP with an increased production of ROS. It has been created by mutagenesis of a nonfluorescent chromoprotein [48]. KillerRed is orders of magnitude more phototoxic than GFP. Ever under mild illumination conditions, this genetically encoded photosensitizer is capable of killing most cells within minutes. KillerRed can also inactivate an enzyme to which it is joined in a fusion construct. This property is exploited in a technique termed "chromophore-assisted light inactivation" (CALI) [49].

3.3 Monomerization

Essentially all naturally occurring FPs discovered to date tend to form oligomers. For a range of applications, the oligomeric nature of an FP is entirely irrelevant, including tissue imaging, monitoring of gene activity, and highlighting of cells and cellular compartments. However, oligomerization can be detrimental in studies involving FP fusions with other proteins because it may interfere with the function of the fusion partner [46]. Once expressed, formation of dimers or higher-order oligomers induced by the FP part of the fusion construct may produce atypical localization, or alter normal function. The basic strategy for overcoming oligomerization is to modify the FP amino acid sequence to include residues that disrupt the binding interfaces between the protomers, a procedure that may greatly vary in complexity depending on the nature and origin of the FP. For GFP-based variants, dimerization can be either significantly reduced or even completely eliminated by replacing hydrophobic amino acid side chains in the dimer interface with positively charged residues at several key sequence positions. The three most successful mutations, in increasing order of effectiveness, are Phe223Arg, Leu221Lys, and Ala206Lys, where the nonpolar amino acids phenylalanine, leucine, and alanine are replaced by hydrophilic alternatives (arginine or lysine) [50]. In GFP, a single mutation (Ala206Lys) was sufficient for monomerization [51].

Creating FP monomers from tetrameric reef coral and sea anemone proteins has proven far more difficult. Only in a few favorable cases, tetrameric anthozoan FPs have been turned into fairly bright monomers by single amino acid exchanges in each of the two interfaces between β-barrels [52]. In less favorable cases, however, disruption of the tetramer resulted in drastic, sometimes even complete loss of fluorescence [53, 54]. To recover the emission, the dysfunctional protein was subsequently subjected to multiple turns of random mutagenesis followed by screening for brighter clones.

The red FP (RFP) DsRed, for example, is a tightly associated tetramer even at exceedingly low concentrations, and cannot be dissociated without irreversible denaturation of the individual polypeptides [33]. To remove the interactions of each DsRed protomer with its two neighbors, one through a hydrophobic interface and the other through a hydrophilic interface, 33 mutations were required [54]. Other Anthozoa proteins have weaker subunit interactions that, at first sight, may prove easier to break apart into functional monomers. For eqFP611, an RFP from the sea anemone *Entacmaea quadricolor*, functional dimers, denoted by d1eqFP611 and d2eqFP611 were simply obtained by the single mutations Val124Thr or Thr122Arg in the A/B interface, respectively [53]. However, a functional monomer was only obtained after a total of seven rounds of random mutagenesis and four rounds of multiple site-directed mutagenesis. Compared with wild-type eqFP611, the bright red-fluorescent monomer, denoted as mRuby, contains altogether 28 amino acid replacements and is shorter by four amino acids [55].

Another technique for generating "pseudo" monomers from dimeric FPs involves linking two copies of the FP cDNA with a short intervening DNA sequence encoding simple neutral or hydrophilic amino acids (glycine, alanine, and serine) to form so-called tandem dimers. Upon expression in live cells, the fused FPs preferentially interact with each other to form intramolecular dimeric units that perform essentially like monomers, although they have twice the molecular mass. This method was successfully applied with HcRed by fusing two copies of the DNA sequence, separated by a short linker of four amino acids, to several subcellular localization proteins [56]. Tandem dimer constructs have also been developed with DsRed [54] and EosFP [57].

Other procedures for reducing FP oligomerization and aggregation include removing several basic residues from the N-terminus and simultaneous co-expression of FP-tagged proteins with an excess of a nonfluorescent mutant of the marker protein to generate heterodimers or heterotetramers that contain only a single target polypeptide and can thus be considered pseudo-monomeric [58].

3.4 Photostability

Since practically all FP applications rely on fluorescence measurements, photostability is a key parameter for the applicability of any FP. The photostability of a chromophore is quantified by its quantum yield of photobleaching, Φ_b, which, for an ensemble, is the ratio between the number of photobleached molecules and the total number of photons absorbed within a certain time interval. Typically, the fluorescent chromophore will emit 10^4–10^5 photons until it falls victim to permanent photodestruction.

Although insufficient photostability frequently limits the performance of a particular FP, this particular property has often only been an afterthought in FP optimization. At present, the relation between structure and photostability is only poorly understood. Therefore, to improve photostability, multiple rounds of (directed or random) mutagenesis are typically performed, with screening for photostability after each round. Tsien and coworkers noticed the importance of residue 163 in influencing the photostability of mRFP1 variants [59]. For mTFP1, the Asn63Thr mutation resulted in a particularly large increase in photostability [60]. For red FPs, photobleaching probabilities of $\sim 10^{-6}$ are currently being achieved [42].

3.5 Color Tuning

3.5.1 Modifying the Chromophore Environment

The *p*-HBI chromophore can exist as an anionic phenolate species or a neutral hydroxyphenyl form, depending on pH and/or the local environment in its binding pocket. Under physiological conditions, the neutral form with an absorption peak at 395 nm predominates in GFP; the absorption maximum of the anionic chromophore is at 475 nm.

By amino acid modifications in the chromophore microenvironment, the equilibrium between the neutral and anionic chromophores can be markedly changed. Shifting the equilibrium toward the anionic form leads to bright, enhanced GFP (EGFP) variants. Interestingly, in a particular GFP variant (photoactivatable GFP, paGFP) [10], the protonation equilibrium of p-HBI can be shifted irreversibly from the neutral to the anionic species by intense light irradiation, causing decarboxylation of Glu212 [61]. Stabilization of the neutral form, by contrast, is a strategy to generate blue-fluorescent variants (in the absence of ESPT) [62].

Blue-Shifted Chromophore Bands

Polar interactions of the *p*-HBI chromophore with residues in its vicinity may reduce the extent of charge delocalization in the chromophore, resulting in hypsochromic band shifts of the anionic chromophore bands to the blue and cyan regions of the spectrum. This effect explains the blue shift of the absorption and emission of Dendra2 with respect to the very similar EosFP protein [63]. A comparison of the X-ray structures revealed a single structural difference close to the chromophore. In EosFP, the Nη1 atom of Arg66 forms a hydrogen bond to the carbonyl oxygen of the imidazolinone. In contrast, the Arg66 side chain in Dendra2 is pulled away from the imidazolinone ring, held in place by, among other interactions, a weak hydrogen bond between the Arg66 Nε and the hydroxyl side chain of Thr69, resulting in reduced negative charge stabilization on the imidazolinone ring. Electron density is known to shift from the phenyl to the imidazolinone upon electronic excitation of the *p*-HBI chromophore, so that the missing charge stabilization on the imidazolinone by Arg66 in Dendra raises the energy of the excited state with respect to the ground state, leading to a blue shift of the transition. In complete agreement with this explanation, the Thr69Ala mutation turns Dendra2 into a more EosFP-like protein, whereas the Ala69Thr mutation of EosFP elicits the opposite effect [63]. A threonine residue at position 69 has been found in a number of blue or cyan FPs, suggesting a general role of this blue-shifting mechanism (Table 3 in [63]).

In another coral FP, anobCFP, Glu167 is involved in the blue-shifted fluorescence, as its replacement by glycine shifts the emission maximum by 10 nm to the red [35]. Because this mutation also leads to the disappearance of the 425-nm excitation peak, Glu167 most likely stabilizes the neutral form of the chromophore. mKate is a far-red FP emitting at 635 nm [64]. By replacing Met167 in close vicinity to the chromophore by either Glu or Asp, a hydrogen bond to the hydroxyphenyl is established. Because the pK_a values of Glu and Asp are expected to be lower than that of the chromophore, these mutations stabilize the neutral chromophore, which has an absorption peak at ~460 nm [65]. Upon excitation, the chromophore pK_a drops, resulting in ESPT from the hydroxyphenyl to residue 167, and red emission occurs. The blue-shift in the red emission maximum with respect to mKate reflects the stabilization of the negative charge on the hydroxyphenyl ring.

Red-Shifted Chromophore Bands

The chromophore environment may also induce bathochromic shifts of the fluorescence to the yellow or red region of the spectrum. For example, in the YFP variant of *A. victoria* GFP, Thr203 is replaced by tyrosine and, moreover, residue 65 is Gly or Thr instead of Ser to promote ionization of the chromophore [66]. The Tyr203 hydroxyphenyl side chain is stacked on top of the phenol ring of the chromophore, thereby effectively extending the delocalized π-electron system by π-stacking interactions. In fact, any aromatic residue at position 203 (His, Trp, Phe, and Tyr) will increase the excitation and emission wavelengths by up to 20 nm, with the extent of shift increasing in the stated order [5]. These mutants were rationally designed, based on the crystal structure of S65T GFP, with the expectation that the additional polarizability around the chromophore and π–π interactions would reduce the excited state energy and, thereby, increase both the excitation and emission wavelengths. The same design of a red-shifting chromophore environment has been found in nature in the yellow hydromedusan FP phiYFP [67].

The tetrameric eqFP611 displays the most red-shifted fluorescence emission of any unmodified FP studied so far, with an emission peak at 611 nm (Fig. 4a) that is Stokes-shifted by 52 nm from its excitation peak at 559 nm [23]. The 2-iminomethyl-5-(4-hydroxybenzylidene)imidazolinone chromophore formed from the tripeptide Met63-Tyr64-Gly65 assumes a coplanar *trans* configuration (Fig. 4b), stabilized by hydrogen-bonding interactions to Ser158 and Asn143 [68]. As yet, eqFP611 is the only naturally occurring FP known to have a highly fluorescent *trans* chromophore. For the dimeric variant d1eqFP611, we had previously noticed that a red-shifted species can be enhanced by irradiation with pulsed 532-nm light, and concomitant changes in the Raman spectrum suggested a *trans–cis* isomerization of the chromophore [69]. Because the *cis* isomer is energetically preferred by an isolated chromophore, we rationalized that destabilization of the *trans* conformation should result in a shift of the equilibrium conformation toward *cis*. Consequently, we introduced the Asn143Ser substitution into d2eqFP611, so that one of the two hydrogen bonds holding the phenolate moiety in the *trans* form is removed. This highly fluorescent variant, denoted as d2RFP630, has its absorption (emission) maximum shifted to the red by 24 (19) nm to 583 (630) nm (Fig. 4a). The optical spectra of d2RFP630 are significantly broader than those of eqFP611, suggesting that the protein ensemble contains a mixture of spectrally distinct species. This hypothesis was subsequently confirmed by X-ray crystallography (Fig. 4c) [70]. Encouraged by these findings, we anticipated a further destabilization of the *trans* form and possibly a further red shift of the emission wavelength by weakening the remaining hydrogen bond. Indeed, the mutation Ser158Cys caused an additional red shift of the excitation (emission) to 588 (639) nm (Fig. 4a), and the X-ray structure confirmed that the chromophores were all in the *cis* conformation (Fig. 4d), stabilized by the hydrogen bond to Ser143 [70].

For chromophores for which the dipole moment increases significantly upon optical excitation, the fluorescence emission often depends strongly on solvent polarity and shifts to longer wavelengths during the excited-state lifetime (dynamic Stokes shift). Boxer and coworkers have attributed the very large Stokes shift in the far-red emitting DsRed variant mPlum to a picosecond solvation response [25]. This time-dependent shift in emission was not observed in its parental proteins, implying that mPlum has a peculiar chromophore environment that allows such

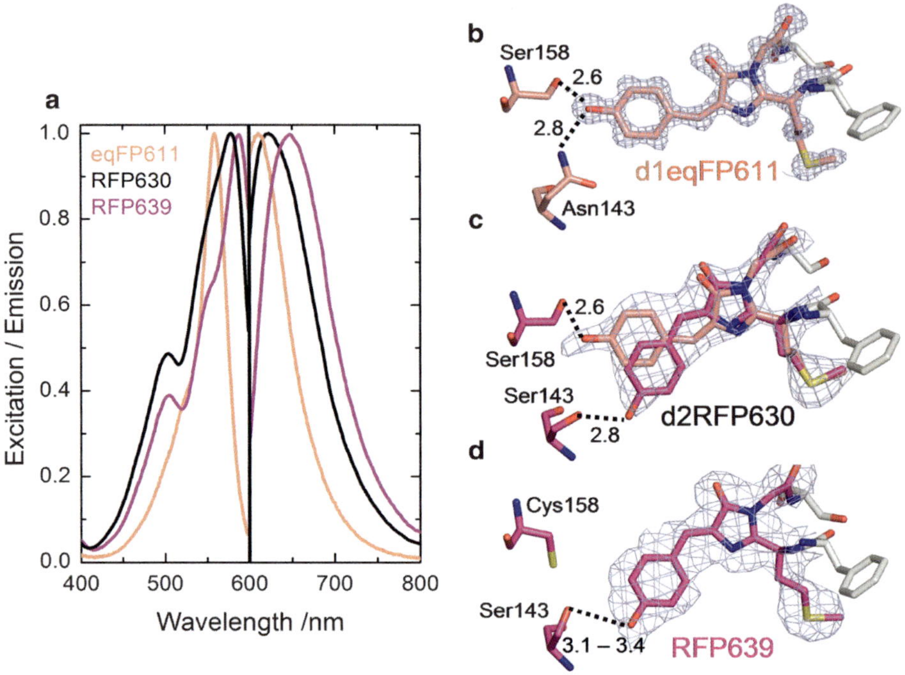

Fig. 4 Red-shifting the absorption and emission of eqFP611 by rational engineering. (**a**) Excitation and emission spectra of d1eqFP611, d2RFP630, and RFP639. (**b**) Stick representation of the d1eqFP611 chromophore (PDB code 3E5T). (**c**) Stick representation of the d2RFP630 chromophore (PDB code 3E5V). (**d**) Stick representation of the RFP639 chromophore (PDB code 3E5W)

a reorganization. Mutational studies have identified a hydrogen bond between Glu16 and the chromophore as the major cause for the red shift [25]. The temporal shift of the fluorescence emission in mPlum was explained by a time-dependent interaction between Glu16 and the excited state of the chromophore, with the glutamic acid side chain performing a rotation in response to the modified charge distribution in the excited state.

Fluorescent Timers

Fluorescent timers, initially introduced by Terskikh et al. in 2000 [71], are FPs which change their emission wavelengths as a function of time. Subach et al. used mCherry as a template to generate monomeric fluorescent timers [72]. Maturation of their chromophores proceeds via a blue intermediate with a single Cα–Cβ bond of the chromophore-forming Tyr. Oxidation of this particular bond is the last step in the maturation of the red chromophore. The timer property is based on a delayed oxidation of this bond. Site-directed mutagenesis revealed that, in mCherry, the delay was achieved by replacing Lys70 and Leu83 by Arg and Trp, respectively [72].

3.5.2 Covalent Modifications of the Chromophore Structure

An alternative approach to color tuning involves modifying the structure of the chromophore itself. For some species, additional chromophore modifications occur spontaneously (autocatalytically) during folding and maturation of the protein (see Sect. 2.3). Others require irradiation with light of suitable wavelengths and thus are photochemical in nature (see Sect. 3.5.2). In addition, color tuning can be achieved by varying the tripeptide composition using site-directed mutagenesis.

Site-Directed Mutagenesis

Substitution of Trp for Tyr at position 66 in the chromophore tripeptide produces a chromophore with an indole instead of a phenol or phenolate moiety [73], with excitation and emission wavelengths intermediate between those of the neutral phenol and the anionic phenolate chromophores. However, the increased bulk of the indole requires many additional mutations to regain a reasonable brightness of these proteins [74], called cyan-fluorescent proteins, or CFPs, because of their blue–green or cyan emission (Fig. 5). Substitution of His for Tyr shifts the emission maximum to even shorter wavelengths than Trp66, generating the blue-fluorescent proteins, BFPs (Fig. 5) [73]. Introduction of the nonnatural amino acids *O*-methyltyrosine and *p*-aminophenylalanine at position 66 resulted in blue-shifted, weak fluorescence with respect to wild-type GFP [75]. Incorporation of the nonnatural residues 3-amino-L-tyrosine and 3-fluoro-L-tyrosine into the chromophore of the DsRed monomer caused a 12 nm red shift and a 12 nm blue shift in fluorescence emission, respectively [76].

Verkhusha and coworkers generated a BFP with a neutral Tyr at position 66. Assuming that the red chromophore of FPs such as TagRFP or mCherry is formed from a blue intermediate, they used a rational design strategy to suppress formation of the acylimine double bond specific to the red chromophore. It involved replacing the first amino acid of the chromophore tripeptide Met65 by an aliphatic residue such as Leu to reduce the polarization of the residue 65αC–H bond in the chromophore [77]. The crystal structure of mTagBFP revealed that the chromophore is a 4-(4-hydroxybenzyl)-1H-imidazole-5-ol [78]. The Cα–Cβ bond of Tyr66 is fully reduced, and the hydroxyphenyl ring is rotated out of the chromophore plane.

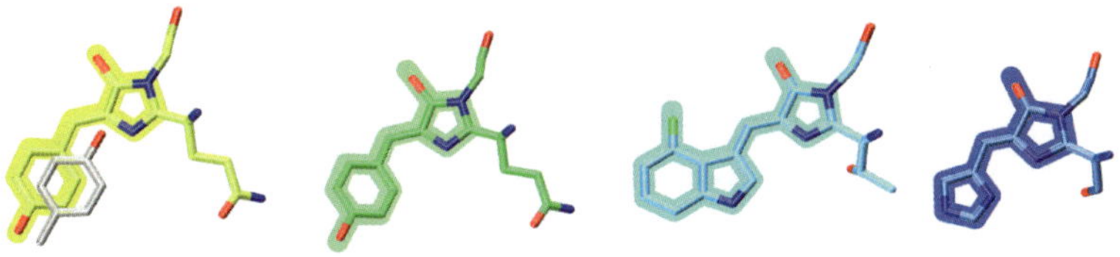

Fig. 5 Variations of and next to the chromophore-forming residue 66 in GFP: *yellow*-fluorescent protein (YFP), GFP, *cyan*-fluorescent protein (CFP), *blue*-fluorescent protein (BFP)

Photoactivated Modifications of the Chromophore

In addition to the spontaneous chemical alterations of the green chromophore described above, strictly photoinduced modifications can occur in a variety of FPs. The first photoactivatable (PA) FP – PA-GFP – was derived from wild-type GFP by a single point mutation, Thr203His [10]. This variant is devoid of fluorescence until activated by irradiation with intense violet light. X-ray structures of PA-GFP determined before and after irradiation showed that photoactivation is the result of a light-induced decarboxylation of the Glu222 side chain that shifts the chromophore equilibrium from the neutral to the anionic form [79], which shows bright green fluorescence. Because the mutant is not visible before light irradiation, it is difficult to identify suitable regions to be photoactivated in a biological sample, which limits the applicability of this protein.

Irreversible Photoconversion from Green to Red Emission

In another class of photoconvertible FPs that includes Kaede [80], EosFP [52, 81], and a few other proteins [67, 82–84], irradiation into the absorption band of the neutral green chromophore (~400 nm) shifts the emission maximum from the green to the red. This process is irreversible and associated with a backbone cleavage between the Nα and Cα atoms of the first amino acid in the chromophore-forming tripeptide, which is always a histidine in this class (Fig. 6). Concomitantly, a double bond forms between its Cα and Cβ atoms, so that the *p*-HBI π-conjugation is extended via an *all-trans* ethenylene moiety into the histidine imidazole ring. Similar to DsRed and AsRed, the protein emits orange-colored fluorescence at ~580 nm after photoconversion, suggesting that the increased delocalization is mainly due to the double bond extension by the ethenylene.

The exact mechanism governing photoconversion is still unclear, although several hypotheses have been proposed. Mizuno et al. suggested initially that the imidazole of His62 could become biprotonated and thus facilitate backbone cleavage via β-elimination, accompanied by an extension of the conjugated

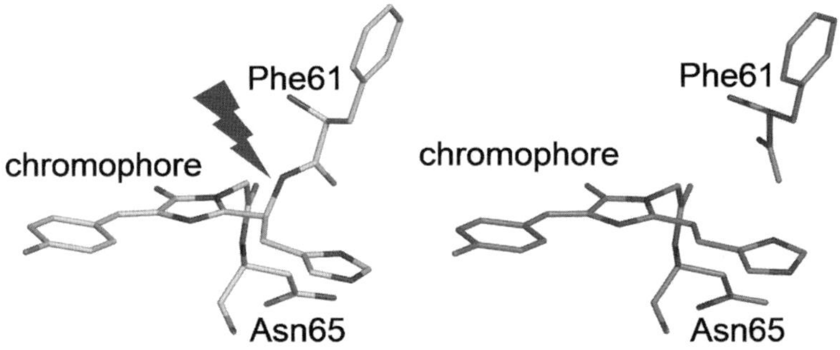

Fig. 6 Backbone cleavage upon green-to-red photoconversion of EosFP. *Left*: green EosFP. *Right*: red EosFP

π-electron system. This reaction creates the red 2-[(1E)-2-(5-imidazolyl)ethenyl]-4-(*p*-hydroxybenzylidene)-5-imidazolinone chromophore [85]. Nienhaus et al. proposed an ESPT from the hydroxyl group of Tyr63 to the Nε of His62 to generate the doubly protonated His62, followed by the β-elimination step in which Glu212 acts as a proton acceptor [81]. Hayashi et al. proposed a water-assisted mechanism to explain the loss of a water molecule (W1) in the red form of Kaede [86], which is also observed in EosFP [81]. They all agreed that His62 is an essential group in the conversion mechanism, which is underscored by the observation that EosFP [81] and also Kaede [85] lose their photoconversion abilities after replacement of His62 by other amino acids [52]. Moreover, there was a general agreement that the neutral hydroxyphenyl is the reactive species, which loses its proton upon excitation. Recently, a theoretical study has suggested a mechanism in which the hydroxyphenyl moiety of the chromophore remains protonated, and there is an ESPT from His62 to Phe61 that promotes peptide bond cleavage [87]. It is impressive to see how this intricate covalent chemistry takes place in the interior of the EosFP β-can without causing any further changes in the protein structure.

Photoconvertible proteins of this class have found widespread use as “pulse-chase” labels in live-cell experiments [57, 82, 88]. Activated molecules, turned red by photoactivation, yield bright red-fluorescent signals that are spectrally well separated from the signals by FPs that have not been photoconverted. Especially, these measurements are not adversely affected by freshly synthesized FP molecules because those will emit in the green range of the spectrum. Targeted photoconversion thus enables spatial and temporal marking of specific structures and tracking of their signals in time in living cells.

Reversible Photoswitching Between Bright and Dark States

In 2000, Lukyanov et al. found a nonfluorescent homolog of GFP in the sea anemone *A. sulcata* [28] that, upon illumination with green light, emitted increasing intensities of red fluorescence at 595 nm. Therefore, it was termed asFP595. The mutation Ala148Gly improved the “kindling” properties [89]; this variant was called KFP-1 [89]. An entirely new generation of improved optical highlighters with reversible on–off switching capabilities was introduced with the photoswitchable FP Dronpa [11]. By excitation with light of two different wavelengths, the protein can be toggled between a bright fluorescent and a dark state. The structural basis of this reversible photoswitching has been elucidated by crystallographic studies of several FPs including Dronpa [90, 91], mTFP0.7 [92], and asFP595 [29, 31, 93]. Under equilibrium conditions, the chromophores of Dronpa and mTFP0.7 assume the *cis* conformation, with the anionic state of the chromophore being the fluorescent species. Photoactivation of the anionic chromophore triggers a *cis–trans* photoisomerization accompanied by a change of the protonation state. As a result, the chromophore becomes essentially nonfluorescent. In KFP1 and asFP595, the nonfluorescent *trans* state is thermodynamically more stable than the *cis* state, and light irradiation induces a *trans–cis* isomerization to the fluorescent *cis* state. When kept in the dark, all FPs relax within minutes to several hours to the thermodynamically more stable isomeric state. This relaxation process can be markedly enhanced with light.

Combining Reversible and Irreversible Photoactivation in IrisFP

The Phe173Ser variant of tetrameric EosFP, denoted as IrisFP, incorporates both reversible and irreversible phototransformations (Fig. 7). In its green fluorescent state, IrisFP displays reversible photoswitching between a bright and a dark state, based on a *cis–trans* isomerization of the chromophore [12]. Like its parent protein EosFP, IrisFP also photoconverts irreversibly to a red-emitting state under violet light. The red form of IrisFP again exhibits reversible photoswitching, which also involves a *cis–trans* isomerization of the chromophore.

With this combination of two modes of photoactivation within one FP, photoconversion from a green to a red emitting form, and photoswitching between a fluorescent and a nonfluorescent state in both forms, it has become possible to perform pulse-chase experiments with super-resolution using photoactivation localization microscopy (PALM, Fig. 8). PALM is an imaging technique based on the

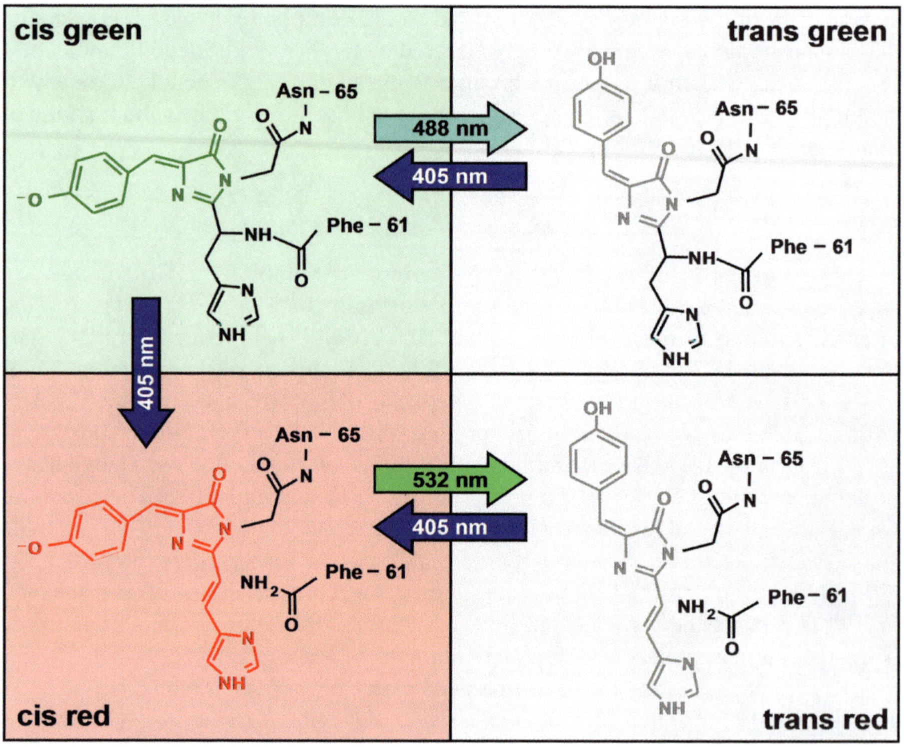

Fig. 7 Chromophore species in IrisFP, generated by illumination with light of specific wavelengths (as indicated by the *arrows*)

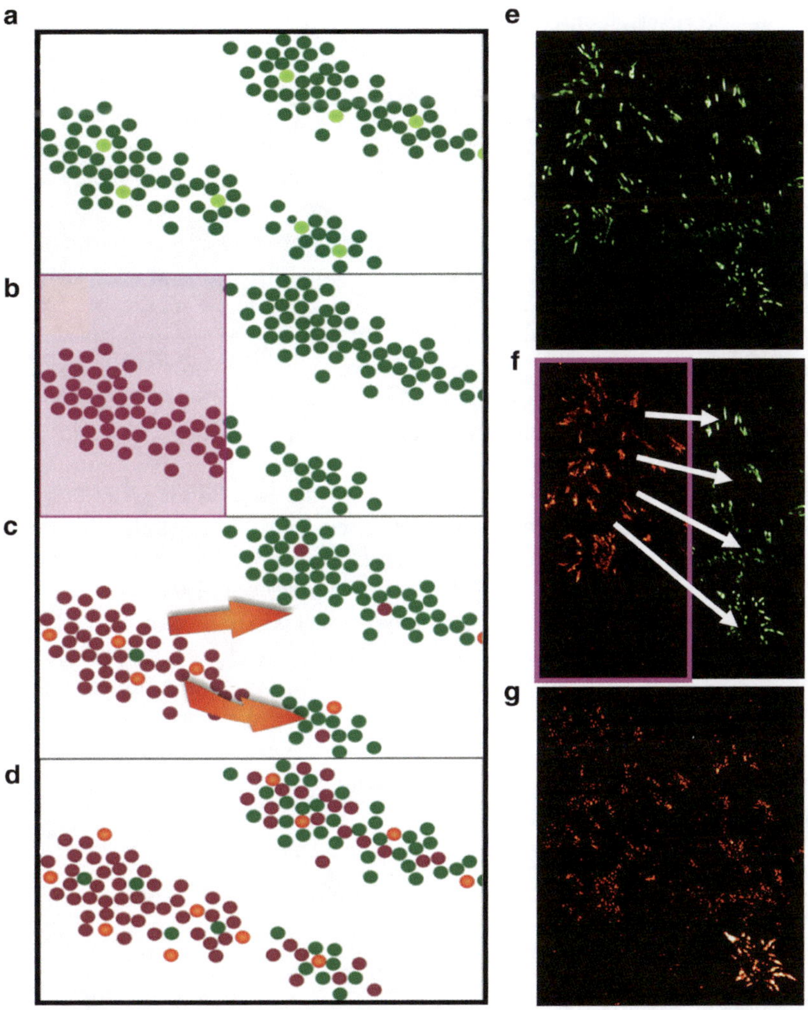

Fig. 8 Pulse-chase imaging with super-resolution PALM using fusion proteins with mIrisFP. (**a**)–(**d**) Schematic depiction of the experiment. (**a**) Only a few protein molecules fluoresce in the green at any given point in time (*bright green dots*); the majority is in the dark state (*dark-green dots*). By sequential activation and localization, a PALM image can be taken. (**b**) Molecules in the area marked by the *violet rectangle* are photoconverted to the red form by 400-nm light irradiation. (**c**) Again, only a small subfraction is in the fluorescent state at any point in time (*bright red dots*), the majority is in the dark state (*dark-red dots*). (**d**) *Red molecules* migrate to other regions of the cell that were not irradiated with 400-nm light, where they are subsequently localized. (**e**) PALM image using *green IrisFP*. (**f**) Targeted photoconversion of a mIrisFP subpopulation to the *red state*. (**g**) PALM image using *red mIrisFP*, showing that *red-converted molecules* migrate to other regions of the cell

precise localization of individual fluorophores that yields an image resolution of typically 20–50 nm.

Because the tetrameric nature of IrisFP may create problems in fusion protein applications, we have developed an advanced, monomeric variant denoted as mIrisFP, which displays excellent properties as a genetically encoded fluorescence marker [43]. Compared with mEosFP, mIrisFP contains four additional mutations: Ala69Val, Phe173Ser, Lys145Ile, and Tyr189Ala.

In a pulse-chase PALM experiment, green mIrisFP is fused to a protein of interest and expressed by a living cell. Subsequently, many (typically a few thousand) CCD camera frames of ~50 ms exposure time are collected. These images are taken with 473-nm light for both excitation and off-switching of the green form of mIrisFP. In each individual frame, only s small number (~100) of mIrisFP molecules are detected that have thermally reverted to their fluorescent state (Fig. 8a). Consequently, their point-spread functions do not overlap, and they can be localized to within 20–50 nm. The super-resolution PALM image is finally reconstructed from all the molecular locations in all CCD frames (Fig. 8e). Later, a sub-ensemble of mIrisFP molecules is photoconverted to the red form by targeted irradiation of a selected cell region (marked by the violet rectangles) with a pulse of 405-nm light (Fig. 8b, f). Migration of the tagged proteins (Fig. 8c) out of the conversion region into other regions of the cell can be observed by super-resolution PALM imaging, now exploiting the photoswitching capability of the red species (Fig. 8d, g). Individual frames of the red form are collected with 561-nm light for excitation and off-switching, and weak 473-nm light is also applied to enhance on-switching of a few molecules to the red-fluorescent state within each individual camera frame. Again, all individual frames are combined to yield the super-resolution PALM image (Fig. 8g).

4 Conclusions

In this chapter, we have discussed structure–dynamics–function relationships in FP marker proteins. This knowledge forms the basis for the continuing rational development of these important tools for the life sciences. Further extension of the color palette toward the near-infrared region would be highly desirable for many biomedical applications, especially for deep tissue and whole organism imaging. Additional bathochromic shifts by extension of the conjugated π-electron system are limited by the size of the β-can. Additional red shifts may, however, still be possible by modification of the chromophore environment, for example, by π-stacking of aromatic amino acids. Advances will also be welcome in the area of photoactivatable proteins. The applicability of green-to-red photoconvertible FPs is limited by the nonreversible nature of photoconversion, and reversibly photoswitching FPs are not visible in their off state. New highlighters such as IrisFP, which combine the two modes of photoactivation, may alleviate these drawbacks. The capability of performing multiple phototransformations will undoubtedly lead to new applications, including sequential photoactivation schemes that can be advantageous for unraveling complex protein–protein interactions. Optimization of conversion efficiencies, thermal rates, and dynamic ranges are still important tasks to be performed on these proteins. While this research is strongly motivated by applications in cell biology, it will also yield new basic insights into the physical chemistry of chromophore–solvent interactions and protein dynamics.

Acknowledgments G.U.N. was supported by the Deutsche Forschungsgemeinschaft (DFG) and the State of Baden-Württemberg through the Center for Functional Nanostructures (CFN), by DFG grant Ni 291/9 and by the Baden-Württemberg Stiftung. J.W. acknowledges funding by DFG grant Wi 1990/2, the Network Fluorescence Applications in Biotechnology and Life Sciences, FABLS, Australia, the Landesstiftung Baden-Württemberg (Elite Postdoc Program); the Natural Environment Research Council, UK (NE/G009643/1) and the University of Southampton.

References

1. Nienhaus GU (2008) The green fluorescent protein: a key tool to study chemical processes in living cells. Angew Chem Int Ed Engl 47:8992–8994
2. Shimomura O, Johnson FH, Saiga Y (1962) Extraction, purification and properties of aequorin, a bioluminescent protein from the luminous hydromedusan, Aequorea. J Cell Comp Physiol 59:223–239
3. Chalfie M, Tu Y, Euskirchen G et al (1994) Green fluorescent protein as a marker for gene expression. Science 263:802–805
4. Prasher DC, Eckenrode VK, Ward WW et al (1992) Primary structure of the Aequorea victoria green-fluorescent protein. Gene 111:229–233
5. Tsien RY (1998) The green fluorescent protein. Annu Rev Biochem 67:509–544
6. Matz MV, Fradkov AF, Labas YA et al (1999) Fluorescent proteins from nonbioluminescent Anthozoa species. Nat Biotechnol 17:969–973
7. Wiedenmann J, Elke C, Spindler KD et al (2000) Cracks in the beta -can: fluorescent proteins from Anemonia sulcata (Anthozoa, Actinaria). Proc Natl Acad Sci USA 97:14091–14096
8. Wiedenmann J (1997) Die Anwendung eines fluoreszierenden Proteins und weiterer farbiger Proteine und der zugehörigen Gene aus der Artengruppe Anemonia sp. (sulcata) Pennant, (Cnidaria, Anthozoa, Actinaria) in Gentechnologie und Molekularbiologie. Offenlegungsschrift DE 197 18 640 A1, Deutsches Patent- und Markenamt, pp 1–18. In, Offenlegungsschrift DE 197 18 640 A1: Deutsches Patent- und Markenamt; 1997:1–18
9. Wiedenmann J, Nienhaus GU (2006) Live-cell imaging with EosFP and other photoactivatable marker proteins of the GFP family. Expert Rev Proteomics 3:361–374
10. Patterson GH, Lippincott-Schwartz J (2002) A photoactivatable GFP for selective photolabeling of proteins and cells. Science 297:1873–1877
11. Habuchi S, Ando R, Dedecker P et al (2005) Reversible single-molecule photoswitching in the GFP-like fluorescent protein Dronpa. Proc Natl Acad Sci U S A 102:9511–9516
12. Adam V, Lelimousin M, Boehme S et al (2008) Structural characterization of IrisFP, an optical highlighter undergoing multiple photo-induced transformations. Proc Natl Acad Sci U S A 105:18343–18348
13. Ormö M, Cubitt AB, Kallio K et al (1996) Crystal structure of the Aequorea victoria green fluorescent protein. Science 273:1392–1395
14. Yang F, Moss LG, Phillips GN Jr (1996) The molecular structure of green fluorescent protein. Nat Biotechnol 14:1246–1251

15. Kummer AD, Kompa C, Niwa H et al (2002) Viscosity-dependent fluorescence decay of the GFP chromophore in solution due to fast internal conversion. J Phys Chem B 106:7554–7559
16. Voityuk AA, Michel-Beyerle ME, Rösch N (1998) Quantum chemical modeling of structure and absorption spectra of the chromophore in green fluorescent proteins. Chem Phys Lett 296:269–276
17. Pouwels LJ, Zhang L, Chan NH et al (2008) Kinetic isotope effect studies on the de novo rate of chromophore formation in fast- and slow-maturing GFP variants. Biochemistry 47: 10111–10122
18. Katranidis A, Atta D, Schlesinger R et al (2009) Fast biosynthesis of GFP molecules: a single-molecule fluorescence study. Angew Chem Int ed 48:1758–1761
19. Youvan DC, Michel-Beyerle ME (1996) Structure and fluorescence mechanism of GFP. Nat Biotechnol 14:1219–1220
20. Nienhaus GU (2010) The "wiggling and jiggling of atoms" leading to excited-state proton transfer in green fluorescent protein. ChemPhysChem 11:971–974
21. Lossau H, Kummer A, Heinecke R et al (1996) Time-resolved spectroscopy of wild-type and mutant green fluorescent proteins reveals excited state deprotonation consistent with fluorophore-protein interactions. Chem Phys 213:1–16
22. Gross LA, Baird GS, Hoffman RC et al (2000) The structure of the chromophore within DsRed, a red fluorescent protein from coral. Proc Natl Acad Sci U S A 97:11990–11995
23. Wiedenmann J, Schenk A, Röcker C et al (2002) A far-red fluorescent protein with fast maturation and reduced oligomerization tendency from Entacmaea quadricolor (Anthozoa, Actinaria). Proc Natl Acad Sci USA 99:11646–11651
24. Nienhaus GU, Wiedenmann J (2009) Structure, dynamics and optical properties of fluorescent proteins: perspectives for marker development. ChemPhysChem 10:1369–1379
25. Abbyad P, Childs W, Shi X et al (2007) Dynamic Stokes shift in green fluorescent protein variants. Proc Natl Acad Sci U S A 104:20189–20194
26. Shu X, Shaner NC, Yarbrough CA et al (2006) Novel chromophores and buried charges control color in mFruits. Biochemistry 45:9639–9646
27. Remington SJ, Wachter RM, Yarbrough DK et al (2005) zFP538, a yellow-fluorescent protein from Zoanthus, contains a novel three-ring chromophore. Biochemistry 44:202–212
28. Lukyanov KA, Fradkov AF, Gurskaya NG et al (2000) Natural animal coloration can be determined by a nonfluorescent green fluorescent protein homolog. J Biol Chem 275:25879–25882
29. Quillin ML, Anstrom DM, Shu X et al (2005) Kindling fluorescent protein from Anemonia sulcata: dark-state structure at 1.38 A resolution. Biochemistry 44:5774–5787
30. Wilmann PG, Petersen J, Devenish RJ et al (2005) Variations on the GFP chromophore: a polypeptide fragmentation within the chromophore revealed in the 2.1-A crystal structure of a nonfluorescent chromoprotein from Anemonia sulcata. J Biol Chem 280:2401–2404
31. Andresen M, Wahl MC, Stiel AC et al (2005) Structure and mechanism of the reversible photoswitch of a fluorescent protein. Proc Natl Acad Sci U S A 102:13070–13074
32. Phillips GN Jr (1997) Structure and dynamics of green fluorescent protein. Curr Opin Struct Biol 7:821–827
33. Baird GS, Zacharias DA, Tsien RY (2000) Biochemistry, mutagenesis, and oligomerization of DsRed, a red fluorescent protein from coral. Proc Natl Acad Sci U S A 97:11984–11989
34. Vrzheshch PV, Akovbian NA, Varfolomeyev SD et al (2000) Denaturation and partial renaturation of a tightly tetramerized DsRed protein under mildly acidic conditions. FEBS Lett 487:203–208
35. Alieva NO, Konzen KA, Field SF et al (2008) Diversity and evolution of coral fluorescent proteins. PLoS One 3:e2680
36. Gurskaya NG, Fradkov AF, Terskikh A et al (2001) GFP-like chromoproteins as a source of far-red fluorescent proteins. FEBS Lett 507:16–20
37. Verkhusha VV, Lukyanov KA (2004) The molecular properties and applications of Anthozoa fluorescent proteins and chromoproteins. Nat Biotechnol 22:289–296

38. Yarbrough D, Wachter RM, Kallio K et al (2001) Refined crystal structure of DsRed, a red fluorescent protein from coral, at 2.0-Å resolution. Proc Natl Acad Sci U S A 98:462–467
39. Yang TT, Sinai P, Green G et al (1998) Improved fluorescence and dual color detection with enhanced blue and green variants of the green fluorescent protein. J Biol Chem 273: 8212–8216
40. Cubitt AB, Heim R, Adams SR et al (1995) Understanding, improving and using green fluorescent proteins. Trends Biochem Sci 20:448–455
41. Pedelacq JD, Cabantous S, Tran T et al (2006) Engineering and characterization of a superfolder green fluorescent protein. Nat Biotechnol 24:79–88
42. Kredel S, Nienhaus K, Wolff M et al (2008) Optimized and far-red emitting variants of fluorescent protein eqFP611. Chem Biol 15:224–233
43. Fuchs J, Böhme S, Oswald F et al (2010) Imaging protein movements in live cells with superresolution using mIrisFP. Nat Methods 7:627–630
44. Stefani M, Dobson CM (2003) Protein aggregation and aggregate toxicity: new insights into protein folding, misfolding diseases and biological evolution. J Mol Med (Berlin, Germany) 81:678–699
45. Link CD, Fonte V, Hiester B et al (2006) Conversion of green fluorescent protein into a toxic, aggregation-prone protein by C-terminal addition of a short peptide. J Biol Chem 281: 1808–1816
46. Wiedenmann J, Oswald F, Nienhaus GU (2009) Fluorescent proteins for live cell imaging: opportunities, limitations, and challenges. IUBMB life 61:1029–1042
47. Remington SJ (2006) Fluorescent proteins: maturation, photochemistry and photophysics. Curr Opin Struct Biol 16:714–721
48. Bulina ME, Chudakov DM, Britanova OV et al (2006) A genetically encoded photosensitizer. Nat Biotechnol 24:95–99
49. Tour O, Meijer RM, Zacharias DA et al (2003) Genetically targeted chromophore-assisted light inactivation. Nat Biotechnol 21:1505–1508
50. Day RN, Davidson MW (2009) The fluorescent protein palette: tools for cellular imaging. Chem Soc Rev 38:2887–2921
51. Zacharias DA, Violin JD, Newton AC et al (2002) Partitioning of lipid-modified monomeric GFPs into membrane microdomains of live cells. Science 296:913–916
52. Wiedenmann J, Ivanchenko S, Oswald F et al (2004) EosFP, a fluorescent marker protein with UV-inducible green-to-red fluorescence conversion. Proc Natl Acad Sci USA 101: 15905–15910
53. Wiedenmann J, Vallone B, Renzi F et al (2005) The red fluorescent protein eqFP611 and its genetically engineered dimeric variants. J Biomed Optics 10:014003 (014007 pages)
54. Campbell RE, Tour O, Palmer AE et al (2002) A monomeric red fluorescent protein. Proc Natl Acad Sci USA 99:7877–7882
55. Kredel S, Oswald F, Nienhaus K et al (2009) mRuby, a bright monomeric red fluorescent protein for labeling of subcellular structures. PLoS One 4:e4391
56. Fradkov AF, Verkhusha VV, Staroverov DB et al (2002) Far-red fluorescent tag for protein labelling. Biochem J 368:17–21
57. Nienhaus GU, Nienhaus K, Hölzle A et al (2006) Photoconvertible fluorescent protein EosFP-biophysical properties and cell biology applications. Photochem Photobiol 82:351–358
58. Bulina ME, Verkhusha VV, Staroverov DB et al (2003) Hetero-oligomeric tagging diminishes non-specific aggregation of target proteins fused with Anthozoa fluorescent proteins. Biochem J 371:109–114
59. Shaner NC, Lin MZ, McKeown MR et al (2008) Improving the photostability of bright monomeric orange and red fluorescent proteins. Nat Methods 5:545–551
60. Ai HW, Henderson JN, Remington SJ et al (2006) Directed evolution of a monomeric, bright and photostable version of Clavularia cyan fluorescent protein: structural characterization and applications in fluorescence imaging. Biochem J 400:531–540

61. van Thor JJ, Gensch T, Hellingwerf KJ et al (2002) Phototransformation of green fluorescent protein with UV and visible light leads to decarboxylation of glutamate 222. Nat Struct Biol 9:37–41
62. Ai HW, Shaner NC, Cheng Z et al (2007) Exploration of new chromophore structures leads to the identification of improved blue fluorescent proteins. Biochemistry 46:5904–5910
63. Adam V, Nienhaus K, Bourgeois D et al (2009) Structural basis of enhanced photoconversion yield in green fluorescent protein-like protein Dendra2. Biochemistry 48:4905–4915
64. Shcherbo D, Merzlyak EM, Chepurnykh TV et al (2007) Bright far-red fluorescent protein for whole-body imaging. Nat Methods 4:741–746
65. Piatkevich KD, Hulit J, Subach OM et al (2010) Monomeric red fluorescent proteins with a large Stokes shift. Proc Natl Acad Sci U S A 107:5369–5374
66. Wachter RM, Elsliger MA, Kallio K et al (1998) Structural basis of spectral shifts in the yellow-emission variants of green fluorescent protein. Structure 6:1267–1277
67. Shagin DA, Barsova EV, Yanushevich YG et al (2004) GFP-like proteins as ubiquitous metazoan superfamily: evolution of functional features and structural complexity. Mol Biol Evol 21:841–850
68. Petersen J, Wilmann PG, Beddoe T et al (2003) The 2.0-Å crystal structure of eqFP611, a far red fluorescent protein from the sea anemone Entacmaea quadricolor. J Biol Chem 278:44626–44631
69. Loos DC, Habuchi S, Flors C et al (2006) Photoconversion in the red fluorescent protein from the sea anemone Entacmaea quadricolor: is cis-trans isomerization involved? J Am Chem Soc 128:6270–6271
70. Nienhaus K, Nar H, Heilker R et al (2008) Trans-cis isomerization is responsible for the red-shifted fluorescence in variants of the red fluorescent protein eqFP611. J Am Chem Soc 130:12578–12579
71. Terskikh A, Fradkov A, Ermakova G et al (2000) "Fluorescent timer": protein that changes color with time. Science 290:1585–1588
72. Subach FV, Subach OM, Gundorov IS et al (2009) Monomeric fluorescent timers that change color from blue to red report on cellular trafficking. Nat Chem Biol 5:118–126
73. Heim R, Prasher DC, Tsien RY (1994) Wavelength mutations and posttranslational autoxidation of green fluorescent protein. Proc Natl Acad Sci U S A 91:12501–12504
74. Heim R, Tsien RY (1996) Engineering green fluorescent protein for improved brightness, longer wavelengths and fluorescence resonance energy transfer. Curr Biol 6:178–182
75. Kajihara D, Hohsaka T, Sisido M (2005) Synthesis and sequence optimization of GFP mutants containing aromatic non-natural amino acids at the Tyr66 position. Protein Eng Des Sel 18:273–278
76. Goulding A, Shrestha S, Dria K et al (2008) Red fluorescent protein variants with incorporated non-natural amino acid analogues. Protein Eng Des Sel 21:101–106
77. Subach OM, Gundorov IS, Yoshimura M et al (2008) Conversion of red fluorescent protein into a bright blue probe. Chem Biol 15:1116–1124
78. Subach OM, Malashkevich VN, Zencheck WD et al (2010) Structural characterization of acylimine-containing blue and red chromophores in mTagBFP and TagRFP fluorescent proteins. Chem Biol 17:333–341
79. Henderson JN, Gepshtein R, Heenan JR et al (2009) Structure and mechanism of the photoactivatable green fluorescent protein. J Am Chem Soc 131:4176–4177
80. Ando R, Hama H, Yamamoto-Hino M et al (2002) An optical marker based on the UV-induced green-to-red photoconversion of a fluorescent protein. Proc Natl Acad Sci USA 99:12651–12656
81. Nienhaus K, Nienhaus GU, Wiedenmann J et al (2005) Structural basis for photo-induced protein cleavage and green-to-red conversion of fluorescent protein EosFP. Proc Natl Acad Sci U S A 102:9156–9159

82. Gurskaya NG, Verkhusha VV, Shcheglov AS et al (2006) Engineering of a monomeric green-to-red photoactivatable fluorescent protein induced by blue light. Nat Biotechnol 24: 461–465
83. Labas YA, Gurskaya NG, Yanushevich YG et al (2002) Diversity and evolution of the green fluorescent protein family. Proc Natl Acad Sci U S A 99:4256–4261
84. Oswald F, Schmitt F, Leutenegger A et al (2007) Contributions of host and symbiont pigments to the coloration of reef corals. FEBS J 274:1102–1109
85. Mizuno H, Mal TK, Tong KI et al (2003) Photo-induced peptide cleavage in the green-to-red conversion of a fluorescent protein. Mol Cell 12:1051–1058
86. Hayashi I, Mizuno H, Tong KI et al (2007) Crystallographic evidence for water-assisted photo-induced peptide cleavage in the stony coral fluorescent protein Kaede. J Mol Biol 372:918–926
87. Lelimousin M, Adam V, Nienhaus GU et al (2009) Photoconversion of the fluorescent protein EosFP: a hybrid potential simulation study reveals intersystem crossings. J Am Chem Soc 131:16814–16823
88. Lukyanov KA, Chudakov DM, Lukyanov S et al (2005) Innovation: photoactivatable fluorescent proteins. Nat Rev Mol Cell Biol 6:885–891
89. Chudakov DM, Belousov VV, Zaraisky AG et al (2003) Kindling fluorescent proteins for precise in vivo photolabeling. Nat Biotechnol 21:191–194
90. Wilmann PG, Turcic K, Battad JM et al (2006) The 1.7 A crystal structure of Dronpa: a photoswitchable green fluorescent protein. J Mol Biol 364:213–224
91. Andresen M, Stiel AC, Trowitzsch S et al (2007) Structural basis for reversible photoswitching in Dronpa. Proc Natl Acad Sci U S A 104:13005–13009
92. Henderson JN, Ai HW, Campbell RE et al (2007) Structural basis for reversible photobleaching of a green fluorescent protein homologue. Proc Natl Acad Sci U S A 104:6672–6677
93. Schäfer LV, Groenhof G, Klingen AR et al (2007) Photoswitching of the fluorescent protein asFP595: mechanism, proton pathways, and absorption spectra. Angew Chem Int Ed 46: 530–536

Index

G. Jung (ed.), *Fluorescent Proteins I*, Springer Ser Fluoresc (2012) 11: 265–268,
DOI 10.1007/4243_2011, © Springer-Verlag Berlin Heidelberg 2011

M

N

O

P

Q

R

S

T